Active Plasmonics and Tuneable Plasmonic Metamaterials

Active Plasmonics and Tuneable Plasmonic Metamaterials

Edited by

Anatoly V. Zayats
Stefan A. Maier

Co-Published by John Wiley & Sons, Inc., and ScienceWise Publishing Hoboken, New Jersey.
Published simultaneously in Canada.

For general information on our other products and services or for technical support, please contact our Customer Care Department within the United States at (800) 762-2974, outside the United States at (317) 572-3993 or fax (317) 572-4002.

Wiley also publishes its books in a variety of electronic formats. Some content that appears in print may not be available in electronic formats. For more information about Wiley products, visit our web site at www.wiley.com.

Library of Congress Cataloging-in-Publication Data:

Active plasmonics and tuneable plasmonic metamaterials / edited by Anatoly V. Zayats, Stefan A. Maier.
pages cm
Includes bibliographical references and index.
ISBN 978-1-118-09208-8 (hardback)
1. Plasmons (Physics) 2. Metamaterials. I. Zayats, A. V. (Anatoly V.), editor of compilation.
II. Maier, Stefan A., editor of compilation.
QC176.8.P55A32 2013
530.4′4–dc23

2012047943

Printed in the United States of America

10 9 8 7 6 5 4 3 2 1

Contents

Preface

Plasmonics provides unique advantages in all areas of science and technology where the manipulation of light at the nanoscale is a prominent ingredient. It relies on light-driven coherent oscillations of electrons near a metal surface, called surface plasmons, which can trap electromagnetic waves at subwavelength and nanoscales. Thus, light can be efficiently manipulated using these plasmonic excitations. In recent years, nanostructured metals have allowed unprecedented control over optical properties and linear as well as nonlinear optical processes. As a consequence, nanoplasmonics has become a major research area and is making important advances in several main fields of applications, such as information technologies, energy, high-density data storage, life sciences, and security. In parallel, in the field of metamaterials, plasmonic composites are indispensable for achieving new optical properties in the visible and infrared spectral range, such as negative refractive index, super-resolution, and cloaking. Various types of metamaterials based on split-ring resonators, fishnet-type structures, nanorods, and others have been proposed in the whole spectral range, from microwaves to visible wavelengths. A variety of passive plasmonic elements such as mirrors, lenses, waveguides, and resonators have been demonstrated. After these initial successes, the development of active plasmonic elements capable of controlling light on the nanoscale dimensions with external electronic or optical stimuli is now on the agenda. For example, the availability of tuneable metamaterials with optical properties controlled by the external signals will immensely broaden their possible field of applications.

The number of research publications in both passive and active plasmonics and metamaterials is growing very fast. We felt that it is important to provide for readers interested in these fields a "one-stop shop" with information on the fundamentals and the state of the art in this field. Active plasmonics and metamaterials are fastmoving fields of research and we could not possibly provide exhaustive coverage of all topics in this field. Instead, this book contains chapters from world-leading authorities in the field, covering active plasmonics from basic principles to the most recent application breakthroughs. The former covers quantum and nonlinear plasmonics, amplification of plasmonic signals and spasers, transformation optics for design of plasmonic nanostructures, active and nonlinear plasmonic metamaterials, and light control via designed phase discontinuities. The latter includes integrated plasmonic detectors, subwavelength imaging with anisotropic metamaterials, tuneable plasmonic lenses, as well as terahertz plasmonic surfaces for sensing.

The book begins with a review of gain in nanoplasmonics. This includes topics such as the spaser and plasmonic gain, amplification, and loss compensation. Both fundamental theoretical concepts and experimental developments have been reviewed. In Chapter 2, physical mechanisms responsible for nonlinearities in plasmonic nanostructures are discussed. Nonlinear surface plasmon polaritons, plasmon solitons, and nonlinear waveguide devices are presented, together with a survey of nonlinearities enhanced by localized surface plasmons and nanoantennas. Plasmonic nanorod metamaterials are introduced in Chapter 3: their optical properties as well as active and nonlinear properties allow tuneable optical responses. Transformation optics provides an extremely powerful tool for designing plasmonic nanostructures with a desired optical response, such as broadband light harvesting and nanofocusing structures, as well as cavities tailored for fluorescence enhancement in the vicinity of complex nanostructures. This new toolkit, as well as the impact of nonlocality on the optical properties of complex plasmonic nanostructures, is presented in Chapter 4.

The next chapter, building on the theoretical paradigms discussed earlier, reviews work conducted on SPP loss compensation and amplification in various types of SPP waveguides. Chapter 6 introduces the principles of light control via phase response engineering in plasmonic antennas. Generalized laws of reflection and refraction in the presence of linear interfacial phase distributions, demonstrations of giant and tuneable birefringence, and generation of optical vortices that carry orbital angular momentum are discussed. The integration of various types of nanoscale photodetectors with plasmonic waveguides and nanostructures is overviewed in Chapter 7. The adaptation of plasmonic concepts from the optical region of the spectrum to the THz band is briefly surveyed from the point of view of sensing applications in Chapter 8. The plasmonic guiding modalities in spectral range are realized either with semiconductors or via patterned metal surfaces. This chapter concludes with a description of possibilities offered by the exciting new material, graphene. A detailed review of subwavelength imaging using extremely anisotropic media in the canalization regime is presented in Chapter 9. Finally, Chapter 10 presents an overview of planar plasmonic diffractive focusing devices. The imaging properties of such plasmonic lenses can be externally tuned by controlling refractive index of the dielectric components of the nanostructure.

There are a variety of new and exciting developments outside the scope of this book, ranging from control over single-photon emission with quantum dots coupled to metallic nanostructures and the exploration of the quantum regime of SPPs, to more applied topics such as integrated nanobiosensors with optical or electrical readout. In general, we anticipate that the combination of plasmonic structures, be it waveguides or nanoparticle cavities, with active media, will greatly accelerate the transition of plasmonics into devices. The chapters in this book provide a snapshot of the exciting possibilities that lie ahead.

ANATOLY V. ZAYATS
King's College London

STEFAN A. MAIER
Imperial College London

Contributors

Francesco Aieta, School of Engineering and Applied Sciences, Harvard University, Cambridge, Massachusetts, USA

Guillaume Aoust, School of Engineering and Applied Sciences, Harvard University, Cambridge, Massachusetts, USA

Alexandre Aubry, Institut Langevin, ESPCI ParisTech, Paris, France

Pavel A. Belov, Queen Mary University of London, London, UK; National Research University ITMO, St. Petersburg, Russia

Pierre Berini, School of Electrical Engineering and Computer Science, University of Ottawa, Ottawa, Canada

Romain Blanchard, School of Engineering and Applied Sciences, Harvard University, Cambridge, Massachusetts, USA

Federico Capasso, School of Engineering and Applied Sciences, Harvard University, Cambridge, Massachusetts, USA

Wayne Dickson, Department of Physics, King's College London, Strand, London, UK

Vladimir P. Drachev, Birck Nanotechnology Center and School of Electrical and Computer Engineering, Purdue University, West Lafayette, Indiana, USA

Zeno Gaburro, School of Engineering and Applied Sciences, Harvard University, Cambridge, Massachusetts, USA

Patrice Genevet, School of Engineering and Applied Sciences, Harvard University, Cambridge, Massachusetts, USA

Pavel Ginzburg, Department of Physics, King's College London, Strand, London, UK

Stephen M. Hanham, Department of Materials, Imperial College London, London, UK

Satoshi Ishii, Birck Nanotechnology Center and School of Electrical and Computer Engineering, Purdue University, West Lafayette, Indiana, USA

Mikhail A. Kats, School of Engineering and Applied Sciences, Harvard University, Cambridge, Massachusetts, USA

Alexander V. Kildishev, Birck Nanotechnology Center and School of Electrical and Computer Engineering, Purdue University, West Lafayette, Indiana, USA

Stefan A. Maier, Experimental Solid State Group, Department of Physics, Imperial College London, London, UK

Antony Murphy, Centre for Nanostructured Media, Queen's University of Belfast, Belfast, UK

Pieter Neutens, Department ESAT—IMEC, Heverlee, Leuven, Belgium

Xingjie Ni, Birck Nanotechnology Center and School of Electrical and Computer Engineering, Purdue University, West Lafayette, Indiana, USA

Meir Orenstein, Department of Electrical Engineering, Technion – IIT, Haifa, Israel

John B. Pendry, Blackett Laboratory, Department of Physics, Imperial College London, London, UK

Robert J. Pollard, Centre for Nanostructured Media, Queen's University of Belfast, Belfast, UK

Vladimir M. Shalaev, Birck Nanotechnology Center and School of Electrical and Computer Engineering, Purdue University, West Lafayette, Indiana, USA

Mark I. Stockman, Department of Physics and Astronomy, Georgia State University, Atlanta, Georgia, USA

Mark D. Thoreson, Birck Nanotechnology Center and School of Electrical and Computer Engineering, Purdue University, West Lafayette, Indiana, USA

Paul Van Dorpe, Department ESAT—IMEC, Heverlee, Leuven, Belgium

Gregory A. Wurtz, Department of Physics, King's College London, Strand, London, UK

Nanfang Yu, School of Engineering and Applied Sciences, Harvard University, Cambridge, Massachusetts, USA

Anatoly V. Zayats, Department of Physics, King's College London, Strand, London, UK

1

Spaser, Plasmonic Amplification, and Loss Compensation

MARK I. STOCKMAN
Department of Physics and Astronomy, Georgia State University, Atlanta, Georgia, USA

1.1 INTRODUCTION TO SPASERS AND SPASING

Not just a promise anymore [1], nanoplasmonics has delivered a number of important applications: ultrasensing [2], scanning near-field optical microscopy [3, 4], surface plasmon (SP)-enhanced photodetectors [5], thermally assisted magnetic recording [6], generation of extreme UV (EUV) [7], biomedical tests [2, 8], SP-assisted thermal cancer treatment [9], plasmonic-enhanced generation of EUV pulses [7] and extreme ultraviolet to soft x-ray (XUV) pulses [10], and many others—see also Reference 11 and 12.

To continue its vigorous development, nanoplasmonics needs an active device—near-field generator and amplifier of nanolocalized optical fields, which has until recently been absent. A nanoscale amplifier in microelectronics is the metal-oxide-semiconductor field effect transistor (MOSFET) [13, 14], which has enabled all contemporary digital electronics, including computers and communications, and the present-day technology as we know it. However, the MOSFET is limited by frequency and bandwidth to $\lesssim$ 100 GHz, which is already a limiting factor in further technological development. Another limitation of the MOSFET is its high sensitivity to temperature, electric fields, and ionizing radiation, which limits its use in extreme environmental conditions and nuclear technology and warfare.

An active element of nanoplasmonics is the spaser (Surface Plasmon Amplification by Stimulated Emission of Radiation), which was proposed [15, 16] as a nanoscale quantum generator of nanolocalized coherent and intense optical fields. The idea of spaser has been further developed theoretically [17–26]. Spaser effect has recently

Active Plasmonics and Tuneable Plasmonic Metamaterials, First Edition. Edited by Anatoly V. Zayats and Stefan A. Maier.

been observed experimentally [27]. Also a number of surface plasmon polariton (SPP) spasers (also called nanolasers) have been experimentally observed [28–33], see also References 34–37. Closely related to the spaser are nanolasers built on deep subwavelength metal nanocavities [38, 39].

1.2 SPASER FUNDAMENTALS

Spaser is a nanoplasmonic counterpart of laser [15, 17]: It is a quantum generator and nanoamplifier where photons as the participating quanta are replaced by SPs. Spaser consists of a metal nanoparticle, which plays the role of a laser cavity (resonator), and the gain medium. Figure 1.1 schematically illustrates the geometry of a spaser as introduced in the original article [15], which contains a V-shaped metal nanoparticle surrounded by a layer of semiconductor nanocrystal quantum dots (QDs).

The laser has two principal elements: resonator (or cavity) that supports photonic mode(s) and the gain (or active) medium that is population-inverted and supplies energy to the lasing mode(s). An inherent limitation of the laser is that the size of the laser cavity in the propagation direction is at least half the wavelength and practically more than that even for the smallest lasers developed [28, 29, 40].

In a true spaser [15, 18], this limitation is overcome. The spasing modes are SPs whose localization length is on the nanoscale [41] and is only limited by the minimum inhomogeneity scale of the plasmonic metal and the nonlocality radius [42] $l_{nl} \sim 1$ nm. This nonlocality length is the distance that an electron with the Fermi velocity v_F moves in space during a characteristic period of the optical field:

$$l_{nl} \sim v_F/\omega \sim 1 \text{ nm}, \tag{1.1}$$

where ω is the optical frequency, and the estimate is shown for the optical spectral region. So, the spaser is truly nanoscopic—its minimum total size can be just a few nanometers.

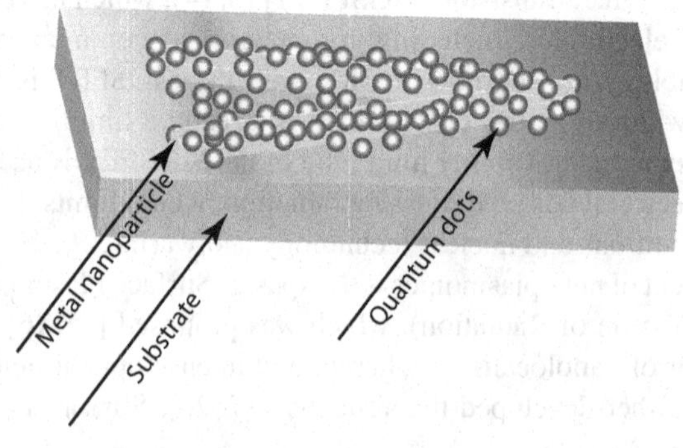

FIGURE 1.1 Schematic of the spaser as originally proposed in Reference 15. The resonator of the spaser is a metal nanoparticle shown as a gold V-shape. It is covered by the gain medium depicted as nanocrystal quantum dots. This active medium is supported by a neutral substrate.

The resonator of a spaser can be any plasmonic metal nanoparticle whose total size R is much less than the wavelength λ and whose metal thickness is between l_{nl} and l_s, which supports an SP mode with required frequency ω_n. Here l_s is the skin depth:

$$l_s = \bar{\lambda}\left[\text{Re}\left(\frac{-\varepsilon_m^2}{\varepsilon_m + \varepsilon_d}\right)^{1/2}\right]^{-1}, \tag{1.2}$$

where $\bar{\lambda} = \lambda/(2\pi) = \omega/c$ is the reduced vacuum wavelength, ε_m is the dielectric function (or, permittivity) of the metal, and ε_d is that of the embedding dielectric. For single-valence plasmonic metals (silver, gold, copper, alkaline metals) $l_s \approx 25$ nm in the entire optical region.

This metal nanoparticle should be surrounded by the gain medium that overlaps with the spasing SP eigenmode spatially and whose emission line overlaps with this eigenmode spectrally [15]. As an example, we consider in more detail a model of a nanoshell spaser [17, 18, 43], which is illustrated in Figure 1.2. Panel (a) shows a silver nanoshell carrying a single SP (plasmon population number $N_n = 1$) in the dipole eigenmode. It is characterized by a uniform field inside the core and hot spots at the poles outside the shell with the maximum field reaching $\sim 10^6$ V/cm. Similarly, Figure 1.2b shows the quadrupole mode in the same nanoshell. In this case, the mode electric field is nonuniform, exhibiting hot spots of $\sim 1.5 \times 10^6$ V/cm of the modal electric field at the poles. These high values of the modal fields, which are related to the small modal volume, are the underlying physical reason for a very strong feedback in the spaser. Under our conditions, the electromagnetic retardation within the spaser volume can be safely neglected. Also, the radiation of such a spaser is a weak effect: The decay rate of plasmonic eigenmodes is dominated by the internal loss in the metal. Therefore, it is sufficient to consider only quasistatic eigenmodes [41, 44] and not their full electrodynamic counterparts [45].

For the sake of numerical illustrations of our theory, we will use the dipole eigenmode (Fig. 1.2a). There are two basic ways to place the gain medium: (i) outside the nanoshell, as shown in panel (c), and (ii) in the core, as in panel (d), which was originally proposed in Reference 43. As we have verified, these two designs lead to comparable characteristics of the spaser. However, the placement of the gain medium inside the core illustrated in Figure 1.2d has a significant advantage because the hot spots of the local field are not covered by the gain medium and are sterically available for applications.

Note that any l-multipole mode of a spherical particle is, indeed, $2l + 1$-times degenerate. This may make the spasing mode to be polarization unstable, like in lasers without polarizing elements. In reality, the polarization may be clamped and become stable due to deviations from the perfect spherical symmetry, which exist naturally or can be introduced deliberately. More practical shape for a spaser may be a nanorod [24], which has a mode with the stable polarization along the major axis. However, a nanorod is a more complicated geometry for theoretical treatment.

The level diagram of the spaser gain medium and the plasmonic metal nanoparticle is displayed in Figure 1.2e along with a schematic of the relevant energy transitions in

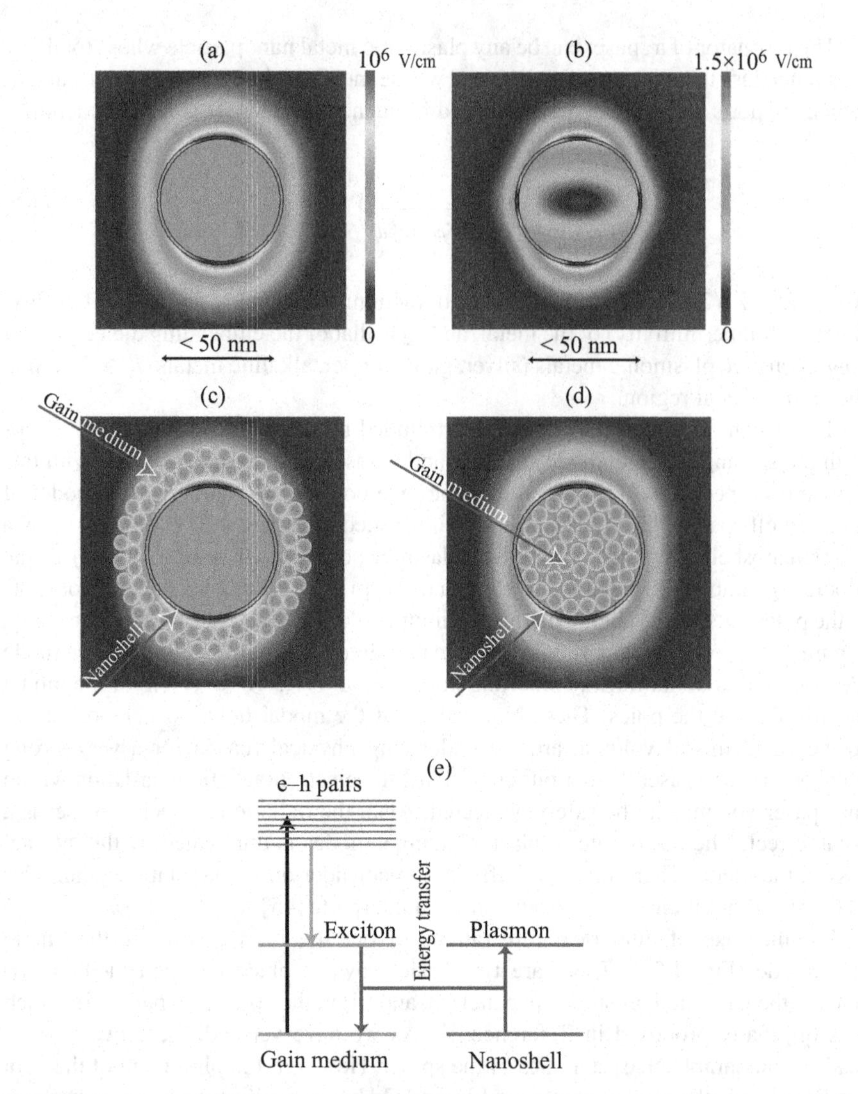

FIGURE 1.2 Schematic of spaser geometry, local fields, and fundamental processes leading to spasing. (a) Nanoshell geometry and the local optical field distribution for one SP in an axially symmetric dipole mode. The nanoshell has aspect ratio $\eta = 0.95$. The local-field magnitude is color-coded by the scale bar in the right-hand side of the panel. (b) The same as (a) but for a quadrupole mode. (c) Schematic of a nanoshell spaser where the gain medium is outside of the shell, on the background of the dipole-mode field. (d) The same as (c) but for the gain medium inside the shell. (e) Schematic of the spasing process. The gain medium is excited and population-inverted by an external source, as depicted by the black arrow, which produces electron–hole pairs in it. These pairs relax, as shown by the green arrow, to form the excitons. The excitons undergo decay to the ground state emitting SPs into the nanoshell. The plasmonic oscillations of the nanoshell stimulate this emission, supplying the feedback for the spaser action. Adapted from Reference 18.

the system. The gain medium chromophores may be semiconductor nanocrystal QDs [15,46], dye molecules [47,48], rare-earth ions [43], or electron–hole excitations of an unstructured semiconductor [28,40]. For certainty, we will use a semiconductor-science language of electrons and holes in QDs.

The pump excites electron–hole pairs in the chromophores (Fig. 1.2e), as indicated by the vertical black arrow, which relax to form excitons. The excitons constitute the two-level systems that are the donors of energy for the SP emission into the spasing mode. In vacuum, the excitons would recombine emitting photons. However, in the spaser geometry, the photoemission is strongly quenched due to the resonance energy transfer to the SP modes, as indicated by the red arrows in the panel. The probability of the radiativeless energy transfer to the SPs relative to that of the radiative decay (photon emission) is given by the so-called Purcell factor

$$\sim \frac{\bar{\lambda}^3 Q}{R^3} \gg 1, \tag{1.3}$$

where R is a characteristic size of the spaser metal core and Q is the plasmonic quality factor [12], and $Q \sim 100$ for a good plasmonic metal such as silver. Thus, this radiativeless energy transfer to the spaser mode is the dominant process whose probability is by orders of magnitude greater than that of the free-space (far-field) emission.

The plasmons already in the spaser mode create the high local fields that excite the gain medium and stimulate more emission to this mode, which is the feedback mechanism. If this feedback is strong enough, and the lifetime of the spaser SP mode is long enough, then an instability develops leading to the avalanche of the SP emission in the spasing mode and spontaneous symmetry breaking, establishing the phase coherence of the spasing state. Thus the establishment of spasing is a nonequilibrium phase transition, as in the physics of lasers.

1.2.1 Brief Overview of the Latest Progress in Spasers

After the original theoretical proposal and prediction of the spaser [15], there has been an active development in this field, both theoretical [17–26] and experimental [27–33]; see also [11,12]. There has also been a US patent issued on spaser [16].

Among theoretical developments, a nanolens spaser has been proposed [49], which possesses a nanofocus ("the hottest spot") of the local fields. In References 15 and 49, the necessary condition of spasing has been established on the basis of the perturbation theory.

There have been theories published describing the SPP spasers (or "nanolasers" as sometimes they are called) phenomenologically, on the basis of classic linear electrodynamics by considering the gain medium as a dielectric with a negative imaginary part of the permittivity (e.g., [43]). Very close fundamentally and technically are works on the loss compensation in metamaterials [50–53]. Such linear-response approaches do not take into account the nature of the spasing as a nonequilibrium phase transition, at the foundation of which is spontaneous symmetry breaking: establishing coherence with an arbitrary but sustained phase of the SP quanta in the system [18]. Spaser is necessarily a deeply nonlinear (nonperturbative) phenomenon where the coherent SP field always saturates the gain medium, which eventually brings about establishment of the stationary [or continuous wave (CW)] regime of the spasing [18]. This leads to principal differences of the linear-response results from the microscopic quantum-mechanical theory in the region of spasing, as we discuss below in conjunction with Figure 1.4.

There has also been a theoretical publication on a bow tie spaser (nanolaser) with electrical pumping [54]. It is based on balance equations and only the CW spasing generation intensity is described. Yet another theoretical development has been a proposal of the lasing spaser [55], which is made of a plane array of spasers.

There has also been a theoretical proposal of a spaser ("nanolaser") consisting of a metal nanoparticle coupled to a single chromophore [56]. In this paper, a dipole–dipole interaction is illegitimately used at very small distances r where it has a singularity (diverging for $r \to 0$), leading to a dramatically overestimated coupling with the SP mode. As a result, a completely unphysical prediction of CW spasing due to single chromophore has been obtained [56]. In contrast, our theory [18] is based on the full (exact) field of the spasing SP mode without the dipole (or any multipole) approximation. As our results of Section 1.3.4 below show, hundreds of chromophores per metal nanoparticle are realistically required for the spasing even under the most favorable conditions.

There has been a vigorous experimental investigation of the spaser and the concepts of spaser. Stimulated emission of SPPs has been observed in a proof-of-principle experiment using pumped dye molecules as an active (gain) medium [47]. There have also been later experiments that demonstrated strong stimulated emission compensating a significant part of the SPP loss [48, 57–61]. As a step toward the lasing spaser, the first experimental demonstration has been reported of a partial compensation of the Joule losses in a metallic photonic metamaterial using optically pumped PbS semiconductor QDs [46]. There have also been experimental investigations reporting the stimulated emission effects of SPs in plasmonic metal nanoparticles surrounded by gain media with dye molecules [62, 63]. The full loss compensation and amplification of the long-range SPPs at $\lambda = 882$ nm in a gold nanostrip waveguide with a dye solution as a gain medium has been observed [64].

At the present time, there have been a considerable number of successful experimental observations of spasers and SPP spasers (also called nanolasers). An electrically pumped nanolaser with semiconductor gain medium has been demonstrated [28] where the lasing modes are SPPs with a one-dimensional (1d) confinement to a ~ 50 nm size. A nanolaser with an optically pumped semiconductor gain medium and a hybrid semiconductor/metal (CdS/Ag) SPP waveguide has been demonstrated with an extremely tight transverse (2d) mode confinement to ~ 10 nm size [29]. This has been followed by the development of a CdS/Ag nanolaser generating a visible single mode at room temperature with a tight 1d confinement (~ 20 nm) and a 2d confinement in the plane of the structure to an area $\sim 1\ \mu m^2$ [30]. A highly efficient SPP spaser in the communication range ($\lambda = 1.46\ \mu m$) with an optical pumping based on a gold film and an InGaAs semiconductor quantum-well gain medium has recently been reported [31]. Another nanolaser (spaser) has been reported based on gold as a plasmonic metal and InGaN/GaN nanorods as gain medium [32]. This spaser generates in the green optical range. Also a promising type of spasers has been introduced [33] based on distributed feedback (DFB). The nanolaser demonstrated in Reference 33 generates at room temperature and has lower threshold than other spasers—see also the corresponding discussion in Section 1.4.6.

There has been an observation published of a nanoparticle spaser [27]. This spaser is a chemically synthesized gold nanosphere of radius 7 nm surrounded by a dielectric shell of 21 nm outer radius containing immobilized dye molecules. Under nanosecond optical pumping in the absorption band of the dye, this spaser develops a relatively narrow-spectrum and intense visible emission that exhibits a pronounced threshold in pumping intensity. The observed characteristics of this spaser are in an excellent qualitative agreement and can be fully understood on the basis of the corresponding theoretical results described in Section 1.3.4.

1.3 QUANTUM THEORY OF SPASER

1.3.1 Surface Plasmon Eigenmodes and Their Quantization

Here we will follow References 41, 65, and 66 to introduce SPs as eigenmodes and Reference 15 to quantize them. Assuming that a nanoplasmonic system is small enough, $R \ll \bar{\lambda}$, $R \lesssim l_s$, we employ the so-called quasistatic approximation where the Maxwell equations reduce to the continuity equation for the electrostatic potential $\varphi(\mathbf{r})$:

$$\frac{\partial}{\partial \mathbf{r}}\varepsilon(\mathbf{r})\frac{\partial}{\partial \mathbf{r}}\varphi(\mathbf{r}) = 0. \tag{1.4}$$

The systems permittivity (dielectric function) varying in space and frequency-dependent is expressed as

$$\varepsilon(\mathbf{r}, \omega) = \varepsilon_m(\omega)\Theta(\mathbf{r}) + \varepsilon_d[1 - \Theta(\mathbf{r})] = \varepsilon_d\left[1 - \frac{\Theta(\mathbf{r})}{s(\omega)}\right]. \tag{1.5}$$

Here $\Theta(\mathbf{r})$ is the so-called characteristic function of the nanosystem, which is equal to 1 when $\mathbf{r}$ belongs to the metal and 0 otherwise. We have also introduced Bergman's spectral parameter [44]:

$$s(\omega) = \frac{\varepsilon_d}{\varepsilon_d - \varepsilon_m(\omega)}. \tag{1.6}$$

A classical-field SP eigenmode $\varphi_n(\mathbf{r})$ is defined by the following generalized eigenproblem, which is obtained from Equation (1.4) by substituting Equations (1.5) and (1.6):

$$\frac{\partial}{\partial \mathbf{r}}\Theta(\mathbf{r})\frac{\partial}{\partial \mathbf{r}}\varphi_n(\mathbf{r}) - s_n\frac{\partial^2}{\partial \mathbf{r}^2}\varphi_n(\mathbf{r}) = 0, \tag{1.7}$$

where ω_n is the corresponding eigenfrequency and $s_n = s(\omega_n)$ is the corresponding eigenvalue.

To be able to carry out the quantization procedure, we must neglect losses, that is, consider a purely Hamiltonian system. This requires that we neglect $\mathrm{Im}\,\varepsilon_m$, which we do only in this subsection. Then the eigenvalues s_n and the corresponding SP wave functions φ_n, as defined by Equation (1.7), are all real. Note that for good metals in the plasmonic region, $\mathrm{Im}\,\varepsilon_m \ll |\mathrm{Re}\,\varepsilon_m|$, cf. Reference 12, so this procedure is meaningful.

The eigenfunctions $\varphi_n(\mathbf{r})$ satisfy the homogeneous Dirichlet–Neumann boundary conditions on a surface S surrounding the system. These we set as

$$\varphi_1(\mathbf{r})|_{\mathbf{r}\in S} = 0, \text{ or } \mathbf{n}(\mathbf{r})\frac{\partial}{\partial \mathbf{r}}\varphi_1(\mathbf{r})\bigg|_{\mathbf{r}\in S} = 0, \tag{1.8}$$

with $\mathbf{n}(\mathbf{r})$ denoting a normal to the surface S at a point of $\mathbf{r}$.

From Equations (1.4), (1.5), (1.6), (1.7), and (1.8) it is straightforward to obtain that

$$\int_V \varepsilon(\mathbf{r},\omega)\,|\nabla\varphi_n(\mathbf{r})|^2\,dV = \varepsilon_d\left[1 - \frac{s_n}{s(\omega)}\right], \tag{1.9}$$

where V is the volume of the system.

To quantize the SPs, we write the operator of the electric field of an SP eigenmode as a sum over the eigenmodes:

$$\hat{\mathbf{E}}(\mathbf{r}) = -\sum_n A_n \nabla\varphi_n(\mathbf{r})(\hat{a}_n + \hat{a}_n^\dagger), \tag{1.10}$$

where $\hat{a}_n^\dagger$ and $\hat{a}_n$ are the SP creation and annihilation operators, $-\nabla\varphi_n(\mathbf{r}) = \mathbf{E}_n(\mathbf{r})$ is the modal field of an nth mode, and A_n is an unknown normalization constant. Note that $\hat{a}_n^\dagger$ and $\hat{a}_n$ satisfy the Bose–Einstein canonical commutation relations,

$$\left[\hat{a}_n, \hat{a}_m^\dagger\right] = \delta_{mn}, \tag{1.11}$$

where δ_{mn} is the Kronecker symbol.

To find normalization constant A_n, we invoke Brillouin's expression [67] for the average energy $\langle \hat{H}_{SP}\rangle$ of SPs as a frequency-dispersive system:

$$\left\langle \hat{H}_{SP}\right\rangle = \frac{1}{8\pi}\int_V \frac{\partial}{\partial\omega}\left[\omega\varepsilon(\mathbf{r},\omega)\right]\sum_n \left\langle \hat{\mathbf{E}}_n^\dagger(\mathbf{r})\hat{\mathbf{E}}_n(\mathbf{r})\right\rangle dV, \tag{1.12}$$

where

$$\hat{H}_{SP} = \sum_n \hbar\omega_n\left(\hat{a}_n^\dagger\hat{a}_n + \frac{1}{2}\right) \tag{1.13}$$

is the SP Hamiltonian in the second quantization.

Finally, we substitute the field expansion (1.10) into Equation (1.12) and take into account Equation (1.9) to carry out the integration. Comparing the result with Equation (1.13), we immediately obtain an expression for the quantization constant:

$$A_n = \left(\frac{4\pi\hbar s_n}{\varepsilon_d s'_n}\right)^{1/2}, \; s'_n \equiv \mathrm{Re}\frac{ds(\omega)}{d\omega}\bigg|_{\omega=\omega_n} \tag{1.14}$$

Note that we have corrected a misprint in Reference 15 by replacing the coefficient 2π by 4π.

1.3.2 Quantum Density Matrix Equations (Optical Bloch Equations) for Spaser

Here we follow Reference 18. The spaser Hamiltonian has the form

$$\hat{H} = \hat{H}_g + \hat{H}_{SP} - \sum_p \hat{\mathbf{E}}(\mathbf{r}_p)\hat{\mathbf{d}}^{(p)}, \tag{1.15}$$

where $\hat{H}_g$ is the Hamiltonian of the gain medium, p is a number (label) of a gain medium chromophore, $\mathbf{r}_p$ is its coordinate vector, and $\hat{\mathbf{d}}^{(p)}$ is its dipole-moment operator. In this theory, we treat the gain medium quantum mechanically but the SPs quasi-classically, considering $\hat{a}_n$ as a classical quantity (c-number) a_n with time dependence as $a_n = a_{0n}\exp(-i\omega t)$, where a_{0n} is a slowly varying amplitude. The number of coherent SPs per spasing mode is then given by $N_p = |a_{0n}|^2$. This approximation neglects quantum fluctuations of the SP amplitudes. However, when necessary, we will take into account these quantum fluctuations, in particular, to describe the spectrum of the spaser.

Introducing $\rho^{(p)}$ as the density matrix of a pth chromophore, we can find its equation of motion in a conventional way by commutating it with the Hamiltonian (1.15) as

$$i\hbar\dot{\rho}^{(p)} = [\rho^{(p)}, \hat{H}], \tag{1.16}$$

where the dot denotes temporal derivative. We use the standard rotating wave approximation (RWA), which only takes into account the resonant interaction between the optical field and chromophores. We denote $|1\rangle$ and $|2\rangle$ as the ground and excited states of a chromophore, with the transition $|2\rangle \rightleftharpoons |1\rangle$ resonant to the spasing plasmon mode n. In this approximation, the time dependence of the nondiagonal elements of the density matrix is $(\rho^{(p)})_{12} = \bar{\rho}_{12}^{(p)}\exp(i\omega t)$ and $(\rho^{(p)})_{21} = \bar{\rho}_{12}^{(p)*}\exp(-i\omega t)$, where $\bar{\rho}_{12}^{(p)}$ is an amplitude slowly varying in time, which defines the coherence (polarization) for the $|2\rangle \rightleftharpoons |1\rangle$ spasing transition in a pth chromophore of the gain medium.

Introducing a rate constant Γ_{12} to describe the polarization relaxation and a difference $n_{21}^{(p)} = \rho_{22}^{(p)} - \rho_{11}^{(p)}$ as the population inversion for this spasing transition, we derive an equation of motion for the nondiagonal element of the density matrix as

$$\dot{\bar{\rho}}_{12}^{(p)} = -\left[i\left(\omega - \omega_{12}\right) + \Gamma_{12}\right]\bar{\rho}_{12}^{(p)} + ia_{0n}n_{21}^{(p)}\tilde{\Omega}_{12}^{(p)*}, \tag{1.17}$$

where

$$\tilde{\Omega}_{12}^{(p)} = -A_n \mathbf{d}_{12}^{(p)} \nabla\varphi_n(\mathbf{r}_p)/\hbar \qquad (1.18)$$

is the one-plasmon Rabi frequency for the spasing transition in a pth chromophore and $\mathbf{d}_{12}^{(p)}$ is the corresponding transitional dipole element. Note that always $\mathbf{d}_{12}^{(p)}$ is either real or can be made real by a proper choice of the quantum state phases, making the Rabi frequency $\tilde{\Omega}_{12}^{(p)}$ also a real quantity.

An equation of motion for n_{21}^{p} can be found in a standard way by commutating it with $\hat{H}$ and adding the corresponding decay and excitation rates. To provide conditions for the population inversion ($n_{21}^{p} > 0$), we imply existence of a third level. For simplicity, we assume that it very rapidly decays into the excited state $|2\rangle$ of the chromophore, so its own population is negligible. It is pumped by an external source from the ground state (optically or electrically) with some rate that we will denote g. In this way, we obtain the following equation of motion:

$$\dot{n}_{21}^{(p)} = -4\mathrm{Im}\left[a_{0n}\bar{\rho}_{12}^{(p)}\tilde{\Omega}_{21}^{(p)}\right] - \gamma_2\left(1 + n_{21}^{(p)}\right) + g\left(1 - n_{21}^{(p)}\right), \qquad (1.19)$$

where γ_2 is the decay rate $|2\rangle \rightarrow |1\rangle$.

The stimulated emission of the SPs is described as their excitation by the local field created by the coherent polarization of the gain medium. The corresponding equation of motion can be obtained using Hamiltonian (1.15) and adding the SP relaxation with a rate of γ_n as

$$\dot{a}_{0n} = \left[i\left(\omega - \omega_n\right) - \gamma_n\right] a_{0n} + i a_{0n} \sum_p \rho_{12}^{(p)*}\tilde{\Omega}_{12}^{(p)}. \qquad (1.20)$$

As an important general remark, the system of Equations (1.17), (1.19), and (1.20) is highly nonlinear: Each of these equations contains a quadratic nonlinearity: a product of the plasmon-field amplitude a_{0n} by the density matrix element ρ_{12} or population inversion n_{21}. Altogether, this is a six-order nonlinearity. This nonlinearity is a fundamental property of the spaser equations, which makes the spaser generation always a fundamentally nonlinear process. This process involves a nonequilibrium phase transition and a spontaneous symmetry breaking: establishment of an arbitrary but sustained phase of the coherent SP oscillations.

A relevant process is spontaneous emission of SPs by a chromophore into a spasing SP mode. The corresponding rate $\gamma_2^{(p)}$ for a chromophore at a point $\mathbf{r}_p$ can be found in a standard way using the quantized field (1.10) as

$$\gamma_2^{(p)} = 2\frac{A_n^2}{\hbar\gamma_n}\left|\mathbf{d}_{12}\nabla\varphi_n(\mathbf{r}_p)\right|^2 \frac{\left(\Gamma_{12} + \gamma_n\right)^2}{\left(\omega_{12} - \omega_n\right)^2 + \left(\Gamma_{12} + \gamma_n\right)^2}. \qquad (1.21)$$

As in Schawlow-Townes theory of laser-line width [68], this spontaneous emission of SPs leads to the diffusion of the phase of the spasing state. This defines width γ_s of the spasing line as

$$\gamma_s = \frac{\sum_p \left(1 + n_{21}^{(p)}\right) \gamma_2^{(p)}}{2(2N_p + 1)}. \tag{1.22}$$

This width is small for a case of developed spasing when $N_p \gg 1$. However, for $N_p \sim 1$, the predicted width may be too high because the spectral diffusion theory assumes that $\gamma_s \lesssim \gamma_n$. To take into account this limitation in a simplified way, we will interpolate to find the resulting spectral width Γ_s of the spasing line as $\Gamma_s = (\gamma_n^{-2} + \gamma_s^{-2})^{-1/2}$.

We will also examine the spaser as a bistable (logical) amplifier. One of the ways to set the spaser in such a mode is to add a saturable absorber. This is described by the same Equations (1.17), (1.18), (1.19), and (1.20) where the chromophores belonging to the absorber are not pumped by the external source directly, that is, for them in Equation (1.19) one has to set $g = 0$.

Numerical examples are given for a silver nanoshell where the core and the external dielectric have the same permittivity of $\varepsilon_d = 2$; the permittivity of silver is adopted from Reference 69. The following realistic parameters of the gain medium are used (unless indicated otherwise): $d_{12} = 1.5 \times 10^{-17}$ esu, $\hbar\Gamma_{12} = 10$ meV, $\gamma_2 = 4 \times 10^{12}\,\mathrm{s}^{-1}$ (this value takes into account the spontaneous decay into SPs), and density of the gain medium chromophores is $n_c = 2.4 \times 10^{20}\,\mathrm{cm}^{-3}$, which is realistic for dye molecules but may be somewhat high for semiconductor QDs that were proposed as the chromophores [15] and used in experiments [46]. We will assume a dipole SP mode and chromophores situated in the core of the nanoshell as shown in Figure 1.2d. This configuration is of advantage both functionally (because the region of the high local fields outside the shell is accessible for various applications) and computationally (the uniformity of the modal fields makes the summation of the chromophores trivial, thus greatly facilitating numerical procedures).

1.3.3 Equations for CW Regime

Physically, the spaser action is a result of spontaneous symmetry breaking when the phase of the coherent SP field is established from the spontaneous noise. Mathematically, the spaser is described by homogeneous differential Equations (1.17), (1.18), (1.19), and (1.20). These equations become homogeneous algebraic equations for the CW case. They always have a trivial, zero solution. However, they may also possess a nontrivial solution describing spasing. An existence condition of such a nontrivial solution is

$$(\omega_s - \omega_n + i\gamma_n)^{-1} \times \\ (\omega_s - \omega_{21} + i\Gamma_{12})^{-1} \sum_p \left|\tilde{\Omega}_{12}^{(p)}\right|^2 n_{21}^{(p)} = -1, \tag{1.23}$$

where ω_s is the generation (spasing) frequency. Here, the population inversion of a pth chromophore $n_{21}^{(p)}$ is explicitly expressed as

$$n_{21}^{(p)} = (g - \gamma_2) \times \left\{ g + \gamma_2 + 4N_n \left|\tilde{\Omega}_{12}^{(p)}\right|^2 / \left[(\omega_s - \omega_{21})^2 + \Gamma_{12}^2\right] \right\}^{-1}. \tag{1.24}$$

From the imaginary part of Equation (1.24) we immediately find the spasing frequency ω_s,

$$\omega_s = (\gamma_n \omega_{21} + \Gamma_{12}\omega_n) / (\gamma_n + \Gamma_{12}), \tag{1.25}$$

which generally does not coincide with either the gain transition frequency ω_{21} or the SP frequency ω_n, but is between them. Note that this is a frequency walk-off phenomenon similar to that well known in laser physics. Substituting Equation (1.25) back into Equations (1.24) and (1.25), we obtain a system of equations:

$$\frac{(\gamma_n + \Gamma_{12})^2}{\gamma_n \Gamma_{12}\left[(\omega_{21} - \omega_n)^2 + (\Gamma_{12} + \gamma_n)^2\right]} \times \sum_p \left|\tilde{\Omega}_{12}^{(p)}\right|^2 n_{21}^{(p)} = 1, \tag{1.26}$$

$$n_{21}^{(p)} = (g - \gamma_2) \times \left[g + \gamma_2 + \frac{4N_n \left|\tilde{\Omega}_{12}^{(p)}\right|^2 (\Gamma_{12} + \gamma_n)}{(\omega_{12} - \omega_n)^2 + (\Gamma_{12} + \gamma_n)^2} \right]^{-1}. \tag{1.27}$$

This system defines the stationary (for the CW generation) number of SPs per spasing mode, N_n.

Since $n_{21}^{(p)} \leq 1$, from Equations (1.26) and (1.27) we immediately obtain a necessary condition of the existence of spasing:

$$\frac{(\gamma_n + \Gamma_{12})^2}{\gamma_n \Gamma_{12}\left[(\omega_{21} - \omega_n)^2 + (\Gamma_{12} + \gamma_n)^2\right]} \sum_p \left|\tilde{\Omega}_{12}^{(p)}\right|^2 \geq 1. \tag{1.28}$$

This expression is fully consistent with Reference 15. The following order of magnitude estimate of this spasing condition has a transparent physical meaning and is of heuristic value:

$$\frac{d_{12}^2 Q N_c}{\hbar \Gamma_{12} V_n} \gtrsim 1, \tag{1.29}$$

where $Q = \omega/\gamma_n$ is the quality factor of SPs, V_n is the volume of the spasing SP mode, and N_c is the number of gain medium chromophores within this volume. Deriving this estimate, we have neglected the detuning, that is, set $\omega_{21} - \omega_n = 0$. We also used the definitions of A_n of Equation (1.10) and $\tilde{\Omega}_{12}^{(p)}$ given by Equation (1.18) and the estimate $|\nabla\varphi_n(\mathbf{r})|^2 \sim 1/V$ following from the normalization of the SP eigenmodes $\int |\nabla\varphi_n(\mathbf{r})|^2 d^3r = 1$ of Reference 41. The result of Equation (1.29) is, indeed, in agreement with Reference 15 where it was obtained in different notations.

It follows from Equation (1.29) that for the existence of spasing it is beneficial to have a high quality factor Q, a high density of the chromophores, and a large transition dipole (oscillator strength) of the chromophore transition. The small modal volume V_n (at a given number of the chromophores N_c) is beneficial for this spasing condition: Physically, it implies strong feedback in the spaser. Note that for the given density of the chromophores $n_c = N_c/V_n$, this spasing condition does not explicitly depend on the spaser size, which opens up a possibility of spasers of a very small size limited from the bottom by only the nonlocality radius $l_{nl} \sim 1$ nm. Another important property of Equation (1.29) is that it implies the quantum-mechanical nature of spasing and spaser amplification: This condition fundamentally contains the Planck constant $\hbar$ and, thus, does not have a classical counterpart. Note that in contrast to lasers, the spaser theory and Equations (1.28) and (1.29) in particular do not contain speed of light, that is, they are quasistatic.

Now we will examine the spasing condition and reduce it to a requirement for the gain medium. First, we substitute into Equation (1.28) all the definitions and assume perfect resonance between the generating SP mode and the gain medium, that is, $\omega_n = \omega_{21}$. As a result, we obtain from Equation (1.28),

$$\frac{4\pi}{3}\frac{s_n\,|\mathbf{d}_{12}|^2}{\hbar\gamma_n\Gamma_{12}\varepsilon_d s_n'}\int_V [1-\Theta(\mathbf{r})]\,|\mathbf{E}_n(\mathbf{r})|^2\,d^3r \geq 1, \tag{1.30}$$

where the integral is extended over the volume V of the system, and the Θ-function takes into account a simplifying realistic assumption that the gain medium occupies the entire space free from the core's metal. We also assume that the orientations of the transition dipoles $\mathbf{d}_{12}^{(p)}$ are random and average over them, which results in the factor of 3 in the denominator in Equation (1.30).

From Equations (1.7) or (1.9) it can be obtained that

$$\int_V [1-\Theta(\mathbf{r})]\,|\mathbf{E}_n(\mathbf{r})|^2\,d^3r = 1 - s_n. \tag{1.31}$$

Next, we give approximate expressions for the spectral parameter (1.6), which are very accurate for the realistic case of $Q \gg 1$:

$$\operatorname{Im} s(\omega) = \frac{s_n^2}{\varepsilon_d}\operatorname{Im}\varepsilon_m(\omega) = \frac{1}{Q}s_n\,(1-s_n)\,. \tag{1.32}$$

Taking into account Equations (1.31) and (1.32), we obtain from Equation (1.30) a necessary condition of spasing at a frequency ω as

$$\frac{4\pi}{3}\frac{|\mathbf{d}_{12}|^2 n_c [1 - \mathrm{Re}\, s(\omega)]}{\hbar\Gamma_{12}\mathrm{Re}\, s(\omega)\mathrm{Im}\, \varepsilon_m(\omega)} \geq 1. \tag{1.33}$$

This condition can also be given an alternative form conventional in laser physics in the following way. For the sake of comparison, consider a continuous gain medium comprised of the same chromophores as the gain shell of the spaser. Its gain g (it is the linear gain whose dimensionality is cm^{-1}) is given by a standard expression

$$g = \frac{4\pi}{3}\frac{\omega}{c}\frac{\sqrt{\varepsilon_d}\,|\mathbf{d}_{12}|^2 n_c}{\hbar\Gamma_{12}}. \tag{1.34}$$

Taking this into account, from Equation (1.33), we obtain the spasing criterion in terms of the gain as

$$g \geq g_{th},\, g_{th} = \frac{\omega}{c\sqrt{\varepsilon_d}}\frac{\mathrm{Re}\, s(\omega)}{1 - \mathrm{Re}\, s(\omega)}\mathrm{Im}\, \varepsilon_m(\omega) \tag{1.35}$$

where g_{th} has a meaning of the threshold gain needed for spasing. Importantly, this gain depends only on the dielectric properties of the system and spasing frequency but not on the geometry of the system or the distribution of the local fields of the spasing mode (hot spots, etc.) explicitly. However, note that the system's geometry (along with the permittivities) does define the spasing frequency.

In Figures 1.3a and 1.3b, we illustrate the analytical expression (1.35) for gold and silver, correspondingly, embedded in a dielectric with $\varepsilon_d = 2$ (simulating a

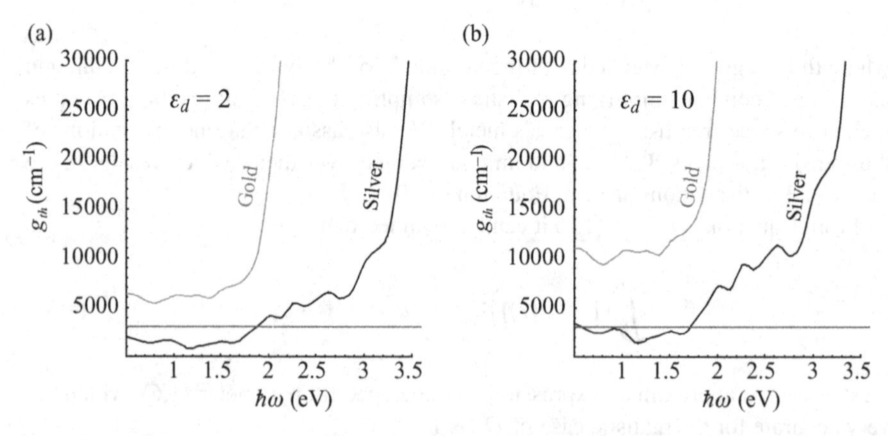

FIGURE 1.3 Threshold gain for spasing g_{th} for silver and gold, as indicated in the graphs, as a function of the spasing frequency ω. The red line separates the area $g_{th} < 3 \times 10^3\ \mathrm{cm}^{-1}$, which can relatively easily be achieved with direct band-gap semiconductors (DBGSs). The real part of the gain medium permittivity is denoted in the corresponding panels as ε_d.

light glass) and $\varepsilon_d = 10$ (simulating a semiconductor), correspondingly. These are computed from Equation (1.35) assuming that the metal core is embedded into the gain medium with the real part of the dielectric function equal to ε_d. As we see from Figure 1.3, the spasing is possible for silver in the near-IR communication range and the adjacent red portion of the visible spectrum for a gain $g < 3000\,\mathrm{cm}^{-1}$ (regions below the red line in Figure 1.3), which is realistically achievable with direct band-gap semiconductors (DBGSs).

1.3.4 Spaser operation in CW Mode

The "spasing curve" (a counterpart of the light–light curve, or L–L curve, for lasers), which is the dependence of the coherent SP population N_n on the excitation rate g, obtained by solving Equations (1.26) and (1.27), is shown in Figure 1.4a for four types of the silver nanoshells with the frequencies of the spasing dipole modes as indicated, which are in the range from near-IR ($\hbar\omega_s = 1.2$ eV) to mid-visible ($\hbar\omega_s = 2.2$ eV). In all cases, there is a pronounced threshold of the spasing at an excitation rate $g_{th} \sim 10^{12}\,\mathrm{s}^{-1}$. Soon above the threshold, the dependence $N_n(g)$ becomes linear, which means that every quantum of excitation added to the active medium with a high probability is stimulated to be emitted as an SP, adding to the coherent SP population, or is dissipated to the heat due to the metal loss with a constant branching ratio between these two processes.

While this is similar to conventional lasers, there is a dramatic difference for the spaser. In lasers, a similar relative rate of the stimulated emission is achieved at a photon population of $\sim 10^{18} - 10^{20}$, while in the spaser the SP population is $N_n \lesssim 100$. This is due to the much stronger feedback in spasers because of the much smaller modal volume V_n—see discussion of Equation (1.29). The shape of the spasing curves of Figure 1.4a (the well-pronounced threshold with the linear dependence almost immediately above the threshold) is in a qualitative agreement with the experiment [27].

The population inversion number n_{21} as a function of the excitation rate g is displayed in Figure 1.4b for the same set of frequencies (and with the same color coding) as in panel (a). Before the spasing threshold, n_{21} increases with g to become positive with the onset of the population inversion just before the spasing threshold. For higher g, after the spasing threshold is exceeded, the inversion n_{21} becomes constant (the inversion clamping). The clamped levels of the inversion are very low, $n_{21} \sim 0.01$, which again is due to the very strong feedback in the spaser.

The spectral width Γ_s of the spaser generation is due to the phase diffusion of the quantum SP state caused by the noise of the spontaneous emission of the SPs into the spasing mode, as described by Equation (1.22). This width is displayed in Figure 1.4c as a function of the pumping rate g. At the threshold, Γ_s is that of the SP line γ_n but for stronger pumping, as the SPs accumulate in the spasing mode, it decreases $\propto N_n^{-1}$, as given by Equation (1.22). This decrease of Γ_s reflects the higher coherence of the spasing state with the increased number of SP quanta and, correspondingly, lower quantum fluctuations. As we have already mentioned, this is similar to the lasers as described by the Schawlow–Townes theory [68].

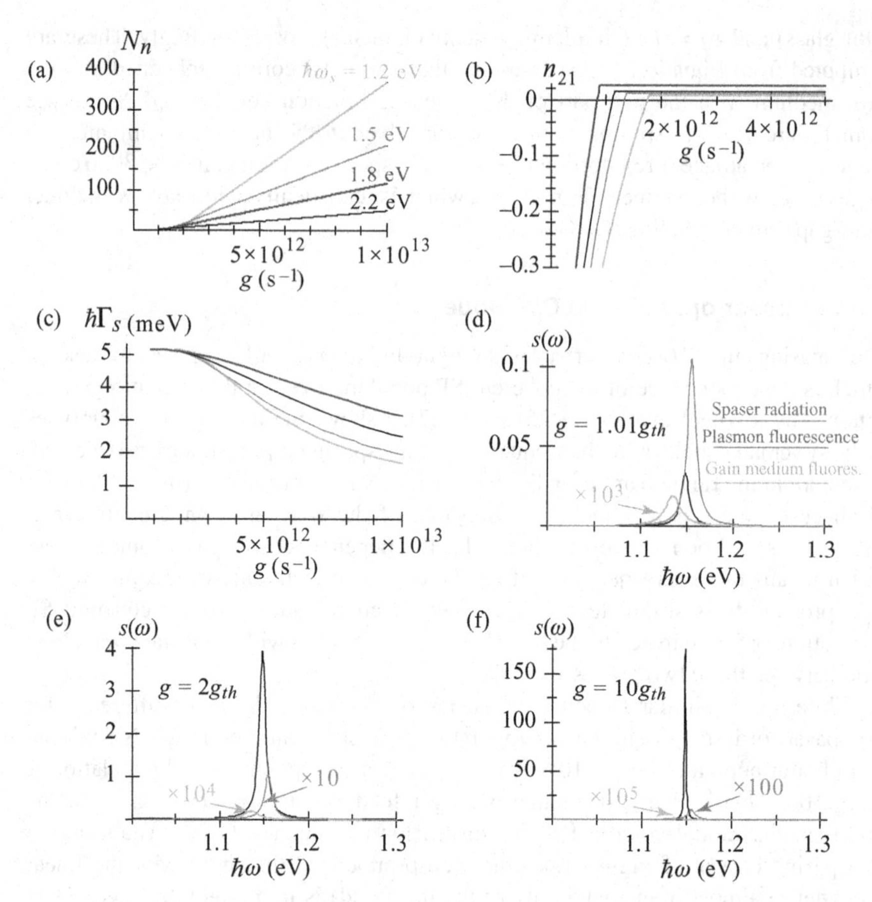

FIGURE 1.4 Spaser SP population and spectral characteristics in the stationary state. The computations are done for a silver nanoshell with the external radius $R_2 = 12$ nm; the detuning of the gain medium from the spasing SP mode is $\hbar(\omega_{21} - \omega_n) = -0.02$ eV. The other parameters are indicated in the text in Section 1.3.2. (a) Number N_n of plasmons per spasing mode as a function of the excitation rate g (per one chromophore of the gain medium). Computations are done for the dipole eigenmode with the spasing frequencies ω_s as indicated, which were chosen by the corresponding adjustment of the nanoshell aspect ratio. (b) Population inversion n_{12} as a function of the pumping rate g. The color coding of the lines is the same as in panel (a). (c) The spectral width Γ_s of the spasing line (expressed as $\hbar\Gamma_s$ in meV) as a function of the pumping rate g. The color coding of the lines is the same as in panel (a). (d)–(f) Spectra of the spaser for the pumping rates g expressed in the units of the threshold rate g_{th}, as indicated in the panels. The curves are color-coded and scaled as indicated.

The developed spasing in a dipole SP mode will show itself in the far field as an anomalously narrow and intense radiation line. The shape and intensity of this line in relation to the lines of the spontaneous fluorescence of the isolated gain medium and its SP-enhanced fluorescence line in the spaser is illustrated in Figures 1.4d–1.4f. Note that for the system under consideration, there is a 20 meV red shift of the gain medium fluorescence with respect to the SP line center. It is chosen so as to illustrate the spectral walk-off of the spaser line. For 1% in the excitation rate above the threshold of the spasing [panel (d)], a broad spasing line (red color) appears comparable in intensity to the SP-enhanced spontaneous fluorescence line (blue color). The width of this spasing line is approximately the same as of the fluorescence, but its position is shifted appreciably (spectral walk-off) toward the isolated gain medium line (green color). For the pumping twice more intense [panel (e)], the spaser-line radiation dominates, but its width is still close to that of the SP line due to significant quantum fluctuations of the spasing state phase. Only when the pumping rate is an order of magnitude above the threshold, the spaser line strongly narrows [panel (f)], and it also completely dominates the spectrum of the radiation. This is a regime of small quantum fluctuations, which is desired in applications.

These results in the spasing region are different in the most dramatic way from previous phenomenological models [43, 51]. For instance, in a "toy model" [51], the width of the resonance line tends to zero at the threshold of spasing and then broadens up again. This distinction of the present theory is due to the nature of the spasing as a spontaneous symmetry breaking (nonequilibrium phase transition with a randomly established but sustained phase) leading to the establishment of a coherent SP state. This nonequilibrium phase transition to spasing and the spasing itself are contained in the present theory due to the fact that the fundamental equations of the spasing (1.17), (1.19), and (1.20) are nonlinear, as we have already discussed above in conjunction with these equations—see the text after Equation (1.20). The previous publications on gain compensation by loss [43,51,53] based on linear electrodynamic equations do not contain spasing. Therefore, they are not applicable in the region of the complete loss compensation and spasing.

1.3.5 Spaser as Ultrafast Quantum Nanoamplifier

As we have already mentioned in Section 1.1, a fundamental and formidable problem is that, in contrast to the conventional lasers and amplifiers in quantum electronics, the spaser has an inherent feedback, which is due to the localization of SP modes, which fundamentally cannot be removed. Such a spaser will develop generation and accumulation of the macroscopic number of coherent SPs in the spasing mode. This leads to the population inversion clamping in the CW regime at a very low level—cf. Figure 1.4b. This CW regime corresponds to the net amplification equals zero, which means that the gain exactly compensates the loss, whose condition is expressed by Equation (1.26). This is a consequence of the nonlinear gain saturation. This holds for any stable CW generator in the CW regime (including any spaser or laser) and precludes using them as amplifiers.

There are several ways to set a spaser as a quantum amplifier. One of them is to reduce the feedback, that is, to allow some or most of the SP energy in the spaser to escape from the active region, so the spaser will not generate in the region of amplification. Such a root has successfully been employed to build an SPP plasmonic amplifier on the long-range plasmon polaritons [64]. A similar root for the SP spasers would be to allow some optical energy to escape either by a near-field coupling or by a radiative coupling to far-field radiation. The near-field coupling approach is promising for building integrated active circuits of the spasers.

Following Reference 18, we consider here two distinct approaches for setting the spasers as quantum nanoamplifiers. The first is a transient regime based on the fact that the establishment of the CW regime and the consequent inversion clamping and the total gain vanishing require some time that is determined mainly by the rate of the quantum feedback and depends also on the relaxation rates of the SPs and the gain medium. After the population inversion is created by the onset of pumping and before the spasing spontaneously develops, as we show below in this section, there is a time interval of approximately 250 fs, during which the spaser provides usable (and as predicted, quite high) amplification—see Section 1.3.6 below.

The second approach to set the spaser as a logical quantum nanoamplifier is a bistable regime that is achieved by introducing a saturable absorber into the active region, which prevents the spontaneous spasing. Then injection of a certain above-threshold number of SP quanta will saturate the absorber and initiate the spasing. Such a bistable quantum amplifier will be considered below in Section 1.3.6.1.

The temporal behavior of the spaser has been found by direct numerical solution of Equations (1.17), (1.18), (1.19), and (1.20). This solution is facilitated by the fact that in the model under consideration all the chromophores experience the same local field inside the nanoshell, and there are only two types of such chromophores: belonging to the gain medium and the saturable absorber, if it is present.

1.3.6 Monostable Spaser as a Nanoamplifier in Transient Regime

Here we consider a monostable spaser in a transient regime. This implies that no saturable absorber is present. We will consider two pumping regimes: stationary and pulse.

Starting with the stationary regime, we assume that the pumping at a rate (per one chromophore) of $g = 5 \times 10^{12}\,\mathrm{s}^{-1}$ starts at a moment of time $t = 0$ and stays constant after that. Immediately at $t = 0$, a certain number of SPs are injected into the spaser. We are interested in its temporal dynamics from this moment on.

The dynamical behavior of the spaser under this pumping regime is illustrated in Figures 1.5a and 1.5b. As we see, the spaser, which starts from an arbitrary initial population N_n, rather rapidly, within a few hundred femtoseconds approaches the same stationary ("logical") level. At this level, an SP population of $N_n = 67$ is established, while the inversion is clamped at a low level of $n_{21} = 0.02$. On the way to this stationary state, the spaser experiences relaxation oscillations in both the SP numbers and inversion, which have a trend to oscillate out of phase [compare panels (a) and (b)]. This temporal dynamics of the spaser is quite complicated and highly nonlinear (unharmonic). It is controlled not by a single relaxation time but by a set of relaxation rates. Clearly, among these are the energy transfer rate from the gain medium to the SPs and the relaxation rates of the SPs and the chromophores.

In this mode, the main effect of the initial injection of the SPs (described theoretically as different initial values of N_n) is in the interval of time required for the spaser to reach the final (CW) state. For very small N_n, which in practice can be supplied by

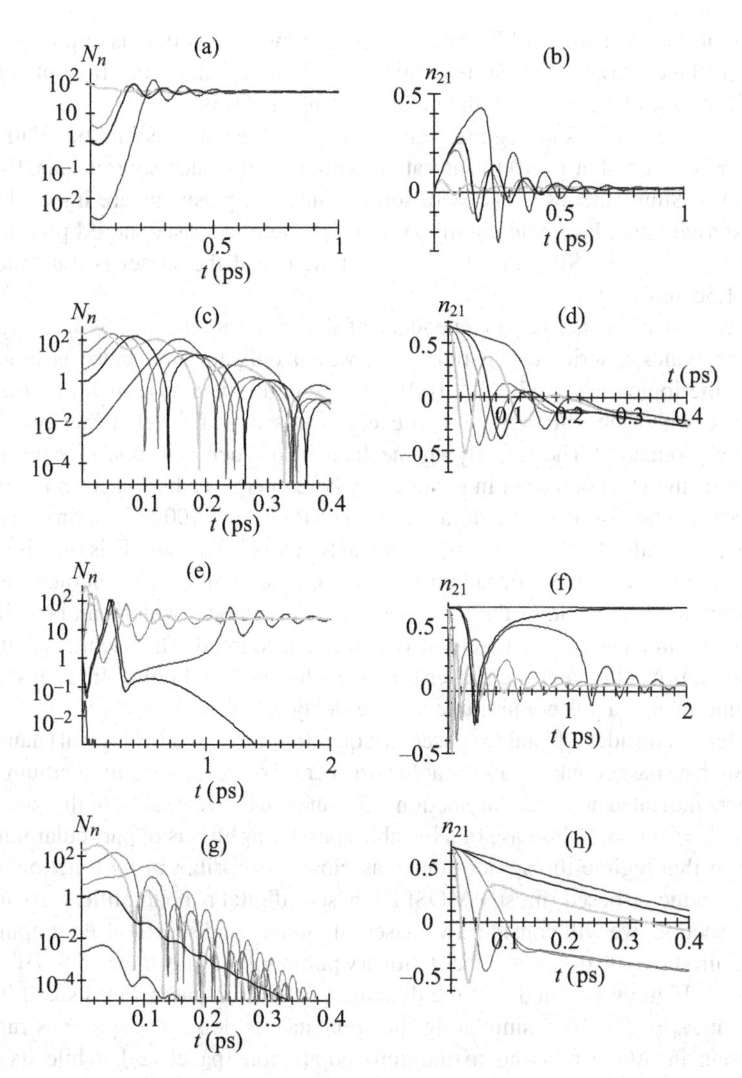

FIGURE 1.5 Ultrafast dynamics of spaser. (a) For monostable spaser (without a saturable absorber), dependence of SP population in the spasing mode N_n on time t. The spaser is stationary pumped at a rate of $g = 5 \times 10^{12}\ \mathrm{s}^{-1}$. The color-coded curves correspond to the initial conditions with the different initial SP populations, as shown in the graph. (b) The same as (a) but for the temporal behavior of the population inversion n_{21}. (c) Dynamics of a monostable spaser (no saturable absorber) with the pulse pumping described as the initial inversion $n_{21} = 0.65$. Coherent SP population N_n is displayed as a function of time t. Different initial populations are indicated by color-coded curves. (d) The same as (c) but for the corresponding population inversion n_{21}. (e) The same as (a) but for bistable spaser with the saturable absorber in concentration $n_a = 0.66n_c$. (f) The same as (b) but for the bistable spaser. (g) The same as (e) but for the pulse pumping with the initial inversion $n_{21} = 0.65$. (h) The same as (g) but for the corresponding population inversion n_{21}.

the noise of the spontaneous SP emission into the mode, this time is approximately 250 fs (cf. the corresponding SP relaxation time is less than 50 fs). In contrast, for the initial values of $N_n = 1 - 5$, this time shortens to 150 fs.

Now consider the second regime: pulse pumping. The gain medium population of the spaser is inverted at $t = 0$ to saturation with a short (much shorter than 100 fs) pump pulse. Simultaneously, at $t = 0$, some number of plasmons are injected (say, by an external nanoplasmonic circuitry). In response, the spaser should produce an amplified pulse of the SP excitation. Such a function of the spaser is illustrated in Figures 1.5c and 1.5d.

As we see from panel (c), independent of the initial number of SPs, the spaser always generates a series of SP pulses, of which only the first pulse is large (at or above the logical level of $N_n \sim 100$). (An exception is a case of little practical importance when the initial $N_n = 120$ exceeds this logical level, when two large pulses are produced.) The underlying mechanism of such a response is the rapid depletion of the inversion seen in panel (d), where energy is dissipated in the metal of the spaser. The characteristic duration of the SP pulse ~ 100 fs is defined by this depletion, controlled by the energy transfer and SP relaxation rates. This time is much shorter than the spontaneous decay time of the gain medium. This acceleration is due to the stimulated emission of the SPs into the spasing mode (which can be called a "stimulated Purcell effect"). There is also a pronounced trend: The lower is the initial SP population N_n, the later the spaser produces the amplified pulse. In a sense, this spaser functions as a pulse-amplitude to time-delay converter.

Now let us consider a bistable spaser as a quantum threshold (or logical) nanoamplifier. Such a spaser contains a saturable absorber mixed with the gain medium with parameters indicated at the end of Section 1.3.2 and the concentration of the saturable absorber $n_a = 0.66n_c$. This case of a bistable spaser amplifier is of particular interest because in this regime the spaser comes as close as possible in its functioning to the semiconductor-based (mostly MOSFET-based) digital nanoamplifiers. As in the previous section, we will consider two cases: stationary and short-pulse pumping.

We again start with the case of the stationary pumping at a rate of $g = 5 \times 10^{12}\,\mathrm{s}^{-1}$. We show in Figures 1.5e and 1.5f the dynamics of such a spaser. For a small initial population $N_n = 5 \times 10^{-3}$ simulating the spontaneous noise, the spaser is rapidly (faster than in 50 fs) relaxing to the zero population [panel (e)], while its gain medium population is equally rapidly approaching a high level [panel (f)] $n_{21} = 0.65$ that is defined by the competition of the pumping and the enhanced decay into the SP mode (the purple curves). This level is so high because the spasing SP mode population vanishes and the stimulated emission is absent. After reaching this stable state (which one can call, say, "logical zero"), the spaser stays in it indefinitely long despite the continuing pumping.

In contrast, for initial values N_n of the SP population large enough (for instance, for $N_n = 5$, as shown by the blue curves in Figs. 1.5e and 1.5f), the spaser tends to the "logical one" state where the stationary SP population reaches the value of $N_n \approx 60$. Due to the relaxation oscillations, it actually exceeds this level within a short time of $\lesssim 100$ fs after the seeding with the initial SPs. As the SP population N_n

reaches its stationary (CW) level, the gain medium inversion n_{21} is clamped down at a low level of a few percent, as typical for the CW regime of the spaser. This "logical one" state also persists indefinitely, as long as the inversion is supported by the pumping.

There is a critical curve (separatrix) that divides the two stable dynamics types (leading to the logical levels of zero and one). For the present set of parameters this separatrix starts with the initial population of $N_n \approx 1$. For a value of the initial N_n slightly below 1, the SP population N_n experiences a slow (hundreds of femtoseconds in time) relaxation oscillation but eventually relaxes to zero (Fig. 1.5e, black curve), while the corresponding chromophore population inversion n_{21} relaxes to the high value $n_{21} = 0.65$ [panel (f), black curve]. In contrast, for a value of N_n slightly higher than 1 [light blue curves in panels (e) and (f)], the dynamics is initially close to the separatrix but eventually the initial slow dynamics tends to the high SP population and low chromophore inversion through a series of the relaxation oscillations. The dynamics close to the separatrix is characterized by a wide range of oscillation times due to its highly nonlinear character. The initial dynamics is slowest (the "decision stage" of the bistable spaser that lasts $\gtrsim 1$ ps). The "decision time" is diverging infinitesimally close to the separatrix, as is characteristic of any threshold (logical) amplifier.

The gain (amplification coefficient) of the spaser as a logical amplifier is the ratio of the high CW level to the threshold level of the SP population N_n. For this specific spaser with the chosen set of parameters, this gain is ≈ 60, which is more than sufficient for the digital information processing. Thus this spaser can make a high gain, ~ 10 THz bandwidth logical amplifier or dynamical memory cell with excellent prospects of applications.

The last but not the least regime to consider is that of the pulse pumping in the bistable spaser. In this case, the population inversion ($n_{21} = 0.65$) is created by a short pulse at $t = 0$ and simultaneously initial SP population N_n is created. Both are simulated as the initial conditions in Equations (1.17), (1.18), (1.19), and (1.20). The corresponding results are displayed in Figures 1.5g and 1.5h.

When the initial SP population exceeds the critical one of $N_n = 1$ (the blue, green, and red curves), the spaser responds with generating a short (duration less than 100 fs) pulse of the SP population (and the corresponding local fields) within a time $\lesssim 100$ fs [panel (g)]. Simultaneously, the inversion is rapidly (within ~ 100 fs) exhausted [panel (h)].

In contrast, when the initial SP population N_n is less than the critical one (i.e., $N_n < 1$ in this specific case), the spaser rapidly (within a time $\lesssim 100$ fs) relaxes as $N_n \to 0$ through a series of relaxation oscillations—see the black and magenta curves in Figure 1.5g. The corresponding inversion decays in this case almost exponentially with a characteristic time ~ 1 ps determined by the enhanced energy transfer to the SP mode in the metal—see the corresponding curves in panel (h). Note that the SP population decays faster when the spaser is above the generation threshold due to the stimulated SP emission leading to the higher local fields and enhanced relaxation.

1.4 COMPENSATION OF LOSS BY GAIN AND SPASING

1.4.1 Introduction to Loss Compensation by Gain

Here, we will mostly follow References 19 and 20. A problem for many applications of plasmonics and metamaterials is posed by losses inherent in the interaction of light with metals. There are several ways to bypass, mitigate, or overcome the detrimental effects of these losses, which we briefly discuss below:

(i) The most common approach consists in employing effects where the losses are not fundamentally damaging such as SPP propagation used in sensing [11], ultramicroscopy [70, 71], and solar energy conversion [72]. For realistic losses, there are other effects and applications that are not prohibitively suppressed by the losses and are useful, in particular, sensing based on SP resonances and surface-enhanced Raman scattering (SERS) [2, 11, 73–75].

(ii) Another promising idea is to use superconducting plasmonics to dramatically reduce losses [76–80]. However, this is only applicable for frequencies below the superconducting gaps, that is, in the terahertz region.

(iii) Yet another proposed direction is using highly doped semiconductors where the ohmic losses can be significantly lower due to much lower free carrier concentrations [81]. However, a problem with this approach may lie in the fact that the usefulness of plasmonic modes depends not on the loss *per se* but on the quality factor Q, which for doped semiconductors may not be higher than for the plasmonic metals.

(iv) One of the alternative approaches to low-loss plasmonic metamaterials is based on our idea of the spaser: It is using a gain to compensate the dielectric (ohmic) losses [82, 83]. In this case the gain medium is included into the metamaterials. It surrounds the metal plasmonic component in the same manner as in the spasers. The idea is that the gain will provide quantum amplification compensating the loss in the metamaterials, which is quite analogous to the spaser.

We will consider the theory of the loss compensation in the plasmonic metamaterials containing gain [19, 20]. Below we show analytically that the full compensation or overcompensation of the optical loss in a dense resonant gain metamaterial leads to an instability that is resolved by its spasing (i.e., by becoming a generating spaser). We further show analytically that the conditions of the complete loss compensation by gain and the threshold condition of spasing—see Equations (1.33) and (1.35)—are identical. Thus the full compensation (overcompensation) of the loss by gain in such a metamaterial will cause spasing. This spasing limits (clamps) the gain—see Section 1.3.4—and, consequently, inhibits the complete loss compensation (overcompensation) at any frequency.

1.4.2 Permittivity of Nanoplasmonic Metamaterial

We consider, for certainty, an isotropic and uniform metamaterial that, by definition, in a range of frequencies ω can be described by the effective permittivity $\bar{\varepsilon}(\omega)$ and permeability $\bar{\mu}(\omega)$. We will concentrate below on the loss compensation for the optical electric responses; Similar consideration for the optical magnetic responses is straightforward. Our theory is applicable for the true 3d metamaterials whose size is much greater than the wavelength λ (ideally, an infinite metamaterial).

Consider a small slab of such a metamaterial with sizes much greater that the unit cell but much smaller than λ. Such a piece is a metamaterial itself. Let us subject this metamaterial to a uniform electric field $\mathbf{E}(\omega) = -\nabla\phi(\mathbf{r}, \omega)$ oscillating with frequency ω. Note that $\mathbf{E}(\omega)$ is the amplitude of the macroscopic electric field inside the metamaterial. A true periodic metamaterial is a crystal where the eigenmodes are Bloch states [84]. In such a state, the field magnitude is periodic on the lattice in accord with the Bloch theorem. Consequently, the influx and outflow of energy balance each other. Thus, selecting a metamaterial slab instead of an infinite crystal will not affect the loss and its compensation.

We will denote the local field at a point $\mathbf{r}$ inside this metamaterial as $\mathbf{e}(\mathbf{r}, \omega) = -\nabla\varphi(\mathbf{r}, \omega)$. We assume standard boundary conditions

$$\varphi(\mathbf{r}, \omega) = \phi(\mathbf{r}, \omega), \tag{1.36}$$

for $\mathbf{r}$ belonging to the surface S of the slab under consideration.

To present our results in a closed form, we first derive a homogenization formula used in Reference 85 (see also references cited therein). By definition, the electric displacement in the volume V of the metamaterial is given by a formula

$$\mathbf{D}(\mathbf{r}, \omega) = \frac{1}{V}\int_V \varepsilon(\mathbf{r}, \omega)\mathbf{e}(\mathbf{r}, \omega)dV, \tag{1.37}$$

where $\varepsilon(\mathbf{r}, \omega)$ is a position-dependent permittivity. This can be identically expressed (by multiplying and dividing by the conjugate of the macroscopic field E^*) and, using the Gauss theorem, transformed to a surface integral as

$$\begin{aligned} D = & \frac{1}{VE^*(\omega)}\int_V \mathbf{E}^*(\omega)\varepsilon(\mathbf{r}, \omega)\mathbf{e}(\mathbf{r}, \omega)dV = \\ & \frac{1}{VE^*(\omega)}\int_S \phi^*(\mathbf{r}, \omega)\varepsilon(\mathbf{r}, \omega)\mathbf{e}(\mathbf{r}, \omega)d\mathbf{S}, \end{aligned} \tag{1.38}$$

where we took into account the Maxwell continuity equation $\nabla[\varepsilon(\mathbf{r}, \omega)\mathbf{e}(\mathbf{r}, \omega)] = 0$. Now, using the boundary conditions of Equation (1.36), we can transform it back to the volume integral as

$$\begin{aligned} D = & \frac{1}{VE^*(\omega)}\int_S \varphi^*(\mathbf{r})\varepsilon(\mathbf{r}, \omega)\mathbf{e}(\mathbf{r}, \omega)d\mathbf{S} = \\ & \frac{1}{VE^*(\omega)}\int_V \varepsilon(\mathbf{r}, \omega)\left|\mathbf{e}(\mathbf{r}, \omega)\right|^2 dV. \end{aligned} \tag{1.39}$$

From the last equality, we obtain the required homogenization formula as an expression for the effective permittivity of the metamaterial:

$$\bar{\varepsilon}(\omega) = \frac{1}{V\,|E(\omega)|^2} \int_V \varepsilon(\mathbf{r}, \omega)\,|\mathbf{e}(\mathbf{r}, \omega)|^2\, dV. \tag{1.40}$$

1.4.3 Plasmonic Eigenmodes and Effective Resonant Permittivity of Metamaterials

This piece of the metamaterial with the total size $R \ll \lambda$ can be treated in the quasistatic approximation. The local field inside the nanostructured volume V of the metamaterial is given by the eigenmode expansion [41, 66, 86]

$$\mathbf{e}(\mathbf{r}, \omega) = \mathbf{E}(\omega) - \sum_n \frac{a_n}{s(\omega) - s_n} \mathbf{E}_n(\mathbf{r}), \tag{1.41}$$

$$a_n = \mathbf{E}(\omega) \int_V \theta(\mathbf{r}) \mathbf{E}_n(\mathbf{r}) dV,$$

where we remind that $\mathbf{E}(\omega)$ is the macroscopic field. In the resonance, $\omega = \omega_n$, only one term at the pole in Equation (1.41) dominates, and it becomes

$$\mathbf{e}(\mathbf{r}, \omega) = \mathbf{E}(\omega) + i \frac{a_n}{\mathrm{Im}\, s(\omega_n)} \mathbf{E}_n(\mathbf{r}). \tag{1.42}$$

The first term in this equation corresponds to the mean (macroscopic) field and the second one describes the deviations of the local field from the mean field containing contributions of the hot spots [87]. The mean root square ratio of the second term (local field) to the first (mean field) is estimated as

$$\sim \frac{f}{\mathrm{Im}\, s(\omega_n)} = \frac{fQ}{s_n(1 - s_n)}, \tag{1.43}$$

where we took into account an estimate $E_n \sim V^{-1/2}$, which follows from the eigenmode field normalization $\int_V |\mathbf{E}_n|^2 dV = 1$, and

$$f = \frac{1}{V} \int_V \theta(\mathrm{r}) dV, \tag{1.44}$$

where f is the metal fill factor of the system and Q is the plasmonic quality factor. Deriving expression (1.43), we have also taken into account an equality $\mathrm{Im}\, s(\omega_n) = s_n(1 - s_n)/Q$, which is valid in the assumed limit of the high quality factor, $Q \gg 1$ (see the next paragraph).

For a good plasmonic metal $Q \gg 1$. For most metal-containing metamaterials, the metal fill factor is not small, typically $f \gtrsim 0.5$. The eigenvalues s_n are limited [12,41], $1 > s_n > 0$. Thus, it is very realistic to assume the following condition:

$$\frac{fQ}{s_n(1-s_n)} \gg 1. \tag{1.45}$$

If so, the second (local) term of the field (1.42) dominates and, with a good precision, the local field is approximately the eigenmode's field:

$$\mathbf{e}(\mathbf{r},\omega) = i\frac{a_n}{\operatorname{Im} s(\omega_n)}\mathbf{E}_n(\mathbf{r}). \tag{1.46}$$

Substituting this into Equation (1.40), we obtain a homogenization formula

$$\bar{\varepsilon}(\omega) = b_n \int_V \varepsilon(\mathbf{r},\omega)\,[\mathbf{E}_n(\mathbf{r})]^2\,dV, \tag{1.47}$$

where $b_n > 0$ is a real positive coefficient whose specific value is

$$b_n = \frac{1}{3V}\left(\frac{Q\int_V \theta(\mathbf{r})\mathbf{E}_n(\mathbf{r})dV}{s_n\,(1-s_n)}\right)^2. \tag{1.48}$$

Using Equation (1.47), it is straightforward to show that the effective permittivity (1.47) simplifies exactly to

$$\bar{\varepsilon}(\omega) = b_n\left[s_n\varepsilon_m(\omega) + (1-s_n)\varepsilon_h(\omega)\right]. \tag{1.49}$$

1.4.4 Conditions of Loss Compensation by Gain and Spasing

In the case of full inversion (maximum gain) and in exact resonance, the host medium permittivity acquires the imaginary part describing the stimulated emission as given by the standard expression

$$\varepsilon_h(\omega) = \varepsilon_d - i\frac{4\pi}{3}\frac{|\mathbf{d}_{12}|^2 n_c}{\hbar\Gamma_{12}}, \tag{1.50}$$

where $\varepsilon_d = \operatorname{Re}\varepsilon_h$, $\mathbf{d}_{12}$ is a dipole matrix element of the gain transition in a chromophore center of the gain medium, Γ_{12} is a spectral width of this transition, and n_c is the concentration of these centers (these notations are consistent with those used above in Sections 1.3.2–1.3.6.1). Note that if the inversion is not maximum, then this and subsequent equations are still applicable if one sets as the chromophore concentration n_c the inversion density: $n_c = n_2 - n_1$, where n_2 and n_1 are the concentrations of the chromophore centers of the gain medium in the upper and lower states of the gain transition, respectively.

The condition for the full electric loss compensation in the metamaterial and amplification (overcompensation) at the resonant frequency $\omega = \omega_n$ is

$$\operatorname{Im} \bar{\varepsilon}(\omega) \leq 0. \tag{1.51}$$

Taking Equation (1.49) into account, this reduces to

$$s_n \operatorname{Im} \varepsilon_m(\omega) - \frac{4\pi}{3} \frac{|\mathbf{d}_{12}|^2 n_c (1 - s_n)}{\hbar \Gamma_{12}} \leq 0. \tag{1.52}$$

Finally, taking into account that $\operatorname{Im} \varepsilon_m(\omega) > 0$, we obtain from Equation (1.52) the condition of the loss (over)compensation as

$$\frac{4\pi}{3} \frac{|\mathbf{d}_{12}|^2 n_c \left[1 - \operatorname{Re} s(\omega)\right]}{\hbar \Gamma_{12} \operatorname{Re} s(\omega) \operatorname{Im} \varepsilon_m(\omega)} \geq 1, \tag{1.53}$$

where the strict inequality corresponds to the overcompensation and net amplification. In Equation (1.50) we have assumed nonpolarized gain transitions. If these transitions are all polarized along the excitation electric field, the concentration n_c should be multiplied by a factor of 3.

Equation (1.53) is a fundamental condition, which is precise [assuming that the requirement (1.45) is satisfied, which is very realistic for metamaterials] and general. Moreover, it is fully analytical and, actually, very simple. Remarkably, it depends only on the material characteristics and does not contain any geometric properties of the metamaterial system or the local fields. (Note that the system's geometry does affect the eigenmode frequencies and thus enters the problem implicitly.) In particular, the hot spots, which are prominent in the local fields of nanostructures [41, 87], are completely averaged out due to the integrations in Equations (1.40) and (1.47).

The condition (1.53) is completely nonrelativistic (quasistatic)—it does not contain speed of light c, which is characteristic also of the spaser. It is useful to express this condition also in terms of the total stimulated emission cross section $\sigma_e(\omega)$ (where ω is the central resonance frequency) of a chromophore of the gain medium as

$$\frac{c \sigma_e(\omega) \sqrt{\varepsilon_d} n_c \left[1 - \operatorname{Re} s(\omega)\right]}{\omega \operatorname{Re} s(\omega) \operatorname{Im} \varepsilon_m(\omega)} \geq 1. \tag{1.54}$$

We see that Equation (1.53) *exactly* coincides with a spasing condition expressed by Equation (1.33). This brings us to an important conclusion: The full compensation (overcompensation) of the optical losses in a metamaterial [which is resonant and dense enough to satisfy condition (1.45)] and the spasing occur under precisely the same conditions.

We have considered above in Section 1.3.3 the conditions of spasing, which are equivalent to Equation (1.54). These are given by one of equivalent conditions of Equations (1.33), (1.35), and (1.53). It is also illustrated in Figure 1.3. We stress that exactly the same conditions are for the full loss compensation (overcompensation) of a dense resonant plasmonic metamaterial with gain.

We would also like to point out that the criterion given by the equivalent conditions of Equations (1.33), (1.35), (1.53), and (1.54) is derived for localized SPs, which are describable in the quasistatic approximation, and is not directly applicable to the propagating SPP modes. However, we expect that very short wavelength SPPs, whose wave vector $k \lesssim l_s$, can be described by these conditions because they are, basically, quasistatic. For instance, the SPPs on a thin metal wire of radius $R \lesssim l_s$ are described by a dispersion relation [88]

$$k \approx \frac{1}{R}\left[-\frac{\varepsilon_m}{2\varepsilon_d}\left(\ln\sqrt{-\frac{4\varepsilon_m}{\varepsilon_d}} - \gamma\right)\right]^{-1/2}, \tag{1.55}$$

where $\gamma \approx 0.57721$ is the Euler constant. This relation is obviously quasistatic because it does not contain speed of light c.

1.4.5 Discussion of Spasing and Loss Compensation by Gain

This fact of the equivalence of the full loss compensation and spasing is intimately related to the general criteria of the thermodynamic stability with respect to small fluctuations of electric and magnetic fields—see Chapter IX of Reference 67:

$$\mathrm{Im}\,\bar{\varepsilon}(\omega) > 0, \, \mathrm{Im}\,\bar{\mu}(\omega) > 0, \tag{1.56}$$

which must be *strict* inequalities for all frequencies for electromagnetically stable systems. For systems in thermodynamic equilibrium, these conditions are automatically satisfied.

However, for the systems with gain, the conditions (1.56) can be violated, which means that such systems can be electromagnetically unstable. The first of conditions (1.56) is opposite to Equations (1.51) and (1.53). This has a transparent meaning: The electrical instability of the system is resolved by its spasing.

The significance of these stability conditions for gain systems can be elucidated by the following *gedanken* experiment. Take a small isolated piece of such a metamaterial (which is a metamaterial itself). Consider that it is excited at an optical frequency ω by a weak external optical field $\mathbf{E}$ or acquires such a field due to fluctuations (thermal or quantum). The energy density $\mathcal{E}$ of such a system is given by the Brillouin formula [67]

$$\mathcal{E} = \frac{1}{16\pi}\frac{\partial \omega \mathrm{Re}\,\bar{\varepsilon}}{\partial \omega}|\mathbf{E}|^2 . \tag{1.57}$$

Note that for the energy of the system to be definite, it is necessary to assume that the loss is not too large, $|\mathrm{Re}\,\bar{\varepsilon}| \gg \mathrm{Im}\,\bar{\varepsilon}$. This condition is realistic for many metamaterials, including all potentially useful ones.

The internal optical energy-density loss per unit time Q (i.e., the rate of the heat-density production in the system) is [67]

$$Q = \frac{\omega}{8\pi} \operatorname{Im} \bar{\varepsilon} \, |\mathbf{E}|^2 . \tag{1.58}$$

Assume that the internal (ohmic) loss dominates over other loss mechanisms such as the radiative loss, which is also a realistic assumption since the ohmic loss is very large for the experimentally studied systems and the system itself is very small (the radiative loss rate is proportional to the volume of the system). In such a case of the dominating ohmic losses, we have $d\mathcal{E}/dt = Q$. Then Equations (1.57) and (1.58) can be resolved together yielding the energy $\mathcal{E}$ and electric field $|\mathbf{E}|$ of this system to evolve with time t exponentially as

$$|\mathbf{E}| \propto \sqrt{\mathcal{E}} \propto e^{-\Gamma t}, \Gamma = \omega \operatorname{Im} \bar{\varepsilon} / \frac{\partial(\omega \operatorname{Re} \bar{\varepsilon})}{\partial \omega} . \tag{1.59}$$

We are interested in a resonant case when the metamaterial possesses a resonance at some eigenfrequency $\omega_n \approx \omega$. For this to be true, the system's behavior must be plasmonic, that is, $\operatorname{Re} \bar{\varepsilon}(\omega) < 0$. Then the dominating contribution to $\bar{\varepsilon}$ comes from a resonant SP eigenmode n with a frequency $\omega_n \approx \omega$. In such a case, the dielectric function [41] $\bar{\varepsilon}(\omega)$ has a simple pole at $\omega = \omega_n$. As a result, $\partial(\omega \operatorname{Re} \bar{\varepsilon})/\partial\omega \approx \omega \partial \operatorname{Re} \bar{\varepsilon}/\partial\omega$ and, consequently, $\Gamma = \gamma_n$, where γ_n is the SP decay rate [12, 15]:

$$\gamma_n = \left. \frac{\operatorname{Im} \varepsilon_m(\omega)}{\frac{\partial \operatorname{Re} \varepsilon_m(\omega)}{\partial \omega}} \right|_{\omega = \omega_n} , \tag{1.60}$$

and the metal dielectric function ε_m is replaced by the effective permittivity $\bar{\varepsilon}$ of the metamaterial. Thus, Equation (1.59) is fully consistent with the spectral theory of SPs [12, 15].

If the losses are not very large so that the energy of the system is meaningful, the Kramers–Kronig causality requires [67] that $\partial(\omega \operatorname{Re} \bar{\varepsilon})/\partial\omega > 0$. Thus, $\operatorname{Im} \bar{\varepsilon} < 0$ in Equation (1.59) would lead to a negative decrement,

$$\Gamma < 0, \tag{1.61}$$

implying that the initial small fluctuation starts to grow exponentially in time in its field and energy, which is an instability. Such an instability is indeed not impossible: It will result in spasing that will eventually stabilize $|\mathbf{E}|$ and $\mathcal{E}$ at finite stationary (CW) levels of the spaser generation.

Note that the spasing limits (clamps) the gain and population inversion making *the net gain to be precisely zero* [18] in the stationary (CW) regime; see Section 1.3.5 and Figure 1.4b. Above the threshold of the spasing, the population inversion of the gain medium is clamped at a rather low level $n_{21} \sim 1\%$. The corresponding net amplification in the CW spasing regime is exactly zero, which is a condition for the

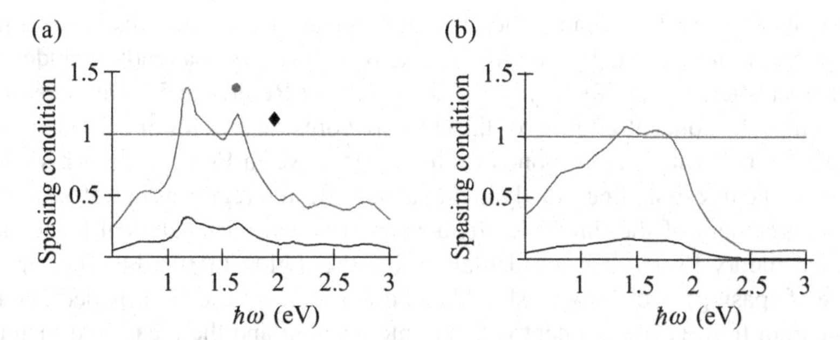

FIGURE 1.6 Spasing criterion as a function of optical frequency ω. The straight line (red line) represents the threshold for the spasing and full loss compensation, which take place for the curve segments above it. (a) Computations for silver. The chromophore concentration is $n_c = 6 \times 10^{18}$ cm^{-3} for the lower curve (black) and $n_c = 2.9 \times 10^{19}$ cm^{-3} for the upper curve (blue line). The black diamond shows the value of the spasing criterion for the conditions of Reference 48—see the text. (b) Computations for gold. The chromophore concentration is $n_c = 3 \times 10^{19}$ cm^{-3} for the lower curve (black) and $n_c = 2 \times 10^{20}$ cm^{-3} for the upper curve (blue line).

CW regime. This makes the complete loss compensation and its overcompensation impossible in a dense resonant metamaterial where the feedback is created by the internal inhomogeneities (including its periodic structure) and the facets of the system.

Because the loss (over)compensation condition (1.53), which is also the spasing condition, is geometry-independent, it is useful to illustrate it for commonly used plasmonic metals, gold and silver whose permittivity we adopt from Reference 69. For the gain medium chromophores, we will use a reasonable set of parameters: $\Gamma_{12} = 5 \times 10^{13}$ s^{-1} and $d_{12} = 4.3 \times 10^{-18}$ esu. The results of computations are shown in Figure 1.6. (Note that this figure expresses a condition of spasing equivalent to that of Fig. 1.3.) For silver as a metal and $n_c = 6 \times 10^{18}$ cm^{-3}, the corresponding lower (black) curve in panel (a) does not reach the value of 1, implying that no full loss compensation is achieved. In contrast, for a higher but still very realistic concentration of $n_c = 2.9 \times 10^{19}$ cm^{-3}, the upper curve in Figure 1.6a does cross the threshold line in the near-IR region. Above the threshold area, there will be the instability and the onset of the spasing. As Figure 1.6b demonstrates, for gold the spasing occurs at higher, but still realistic, chromophore concentrations.

1.4.6 Discussion of Published Research on Spasing and Loss Compensations

Now let us discuss the implications of these results for the research published recently on gain metamaterials. To carry out a quantitative comparison with Reference 53, we turn to Figure 1.6a where the lower (black) curve corresponds to the nominal value of $n_c = 6 \times 10^{18}$ cm^{-3} used in Reference 53. There is no full loss compensation and spasing. This is explained by the fact that Reference 53 uses, as a close inspection shows, the gain dipoles parallel to the field (this is equivalent to increasing n_c by a factor of 3) and the local-field enhancement [this is equivalent to increasing n_c by a factor of $(\varepsilon_h + 2)/3$]. Because the absorption cross section of dyes is measured in the appropriate host media (liquid solvents or polymers), it already includes the Lorentz local-field factor. To compare to the results of Reference 53, we increase in our formulas the concentration n_c of the chromophores by a factor of $\varepsilon_h + 2$ to $n_c = 2.9 \times 10^{19}\,\text{cm}^{-3}$, which corresponds to the upper curve in Figure 1.6a. This curve rises above the threshold line exactly in the same (infra)red region as in Reference 53.

This agreement of the threshold frequencies between our analytical theory and numerical theory [53] is not accidental: Inside the region of stability (i.e., in the absence of spasing) both theories should and do give close results, provided that the gain medium transition alignment is taken into account and the local-field factor is incorporated.

However, above the threshold (in the region of the overcompensation), there should be spasing causing the population inversion clamping and zero net gain, and not a loss compensation. To describe this effect, it is necessary to invoke Equation (1.20) for coherent SP amplitude, which is absent in Reference 53. Also fundamentally important, spasing, just like the conventional lasing, is a highly nonlinear phenomenon, which is described by nonlinear equations—see the discussion after Equation (1.20).

The complete loss compensation is stated in a recent experimental paper [89], where the system is actually a nanofilm rather than a 3d metamaterial, to which our theory would have been applicable. For the Rhodamine 800 dye used with extinction cross section [90] $\sigma = 2 \times 10^{-16}\,\text{cm}^2$ at 690 nm in concentration $n_c = 1.2 \times 10^{19}\,\text{cm}^{-3}$, realistically assuming $\varepsilon_d = 2.3$, for frequency $\hbar\omega = 1.7$ eV, we calculate from Equation (1.54) a point shown by the magenta solid circle in Figure 1.6a, which is significantly above the threshold. Because in such a nanostructure the local fields are very nonuniform and confined near the metal similar to the spaser, they likewise cause a feedback. The condition of Equation (1.45) is likely to be well satisfied for Reference 89. Thus, the system may spase, which would cause the clamping of inversion and loss of gain.

In contrast to these theoretical arguments, there is no evidence of spasing indicated in the experiment—see Reference 89, which can be explained by various factors. Among them, the system of Reference 89 is a gain-plasmonic nanofilm and not a true 3d material. This system is not isotropic. Also, the size of the unit cell $a \approx 280$ nm is significantly greater than the reduced wavelength $\bar{\lambda}$, which violates the quasistatic conditions and makes the possibility of homogenization and considering this system as an optical metamaterial problematic. This circumstance may lead to an appreciable spatial dispersion. It may also cause a significant radiative loss and prevent spasing for some modes.

We would also like to point out that the fact that the unit cell of the negative-refracting (or double-negative) metamaterial of Reference 89 is relatively large, $a \approx 280$ nm, is not accidental. As follows from theoretical consideration of Reference 91, optical magnetism and, consequently, negative refraction for metals, is only possible if the minimum scale of the conductor feature (the diameter d of the nanowire) is greater than the skin depth, $d \gtrsim l_s \approx 25$ nm, which allows one to circumvent Landau–Lifshitz's limitation on the existence of optical magnetism [67,91]. Thus, a ring-type resonator structure would have a size $\gtrsim 2l_s$ (two wires forming a loop) and still the same diameter for the hole in the center, which comes to the total of $\gtrsim 4l_s \approx 100$ nm. Leaving the same distance between the neighboring resonator wires, we arrive at an estimate of the size of the unit cell $a \gtrsim 8l_s = 200$ nm, which is, indeed, the case for Reference 89 and other negative-refraction "metamaterials" in the optical region. This makes our theory not directly applicable to them. Nevertheless, if the spasing condition (1.33) [or (1.35) or (1.54)] is satisfied, the system still may spase on the hot-spot defect modes.

In an experimental study of the lasing spaser [46], a nanofilm of PbS QDs was positioned over a 2d metamaterial consisting of an array of negative split ring resonators. When the QDs were optically pumped, the system exhibited an increase of the transmitted light intensity on the background of a strong luminescence of the QDs but apparently did not reach the lasing threshold. The polarization-dependent loss compensation was only $\sim 1\%$. Similarly, for an array of split ring resonators over a resonant quantum well, where the inverted electron–hole population was excited optically [92], the loss compensation did not exceed $\sim 8\%$. The relatively low-loss compensation in these papers may be due to random spasing and/or spontaneous or amplified spontaneous emission enhanced by this plasmonic array, which reduces the population inversion.

A dramatic example of possible random spasing is presented in Reference 48. The system studied was a Kretschmann-geometry SPP setup [93] with an added ~ 1 μm polymer film containing Rhodamine 6G dye in the $n_c = 1.2 \times 10^{19}$ cm^{-3} concentration. When the dye was pumped, there was outcoupling of radiation in a range of angles. This was a threshold phenomenon with the threshold increasing with the Kretschmann angle. At the maximum of the pumping intensity, the widest range of the outcoupling angles was observed, and the frequency spectrum at every angle narrowed to a peak near a single frequency $\hbar\omega \approx 2.1$ eV.

These observations of Reference 48 can be explained by the spasing where the feedback is provided by the roughness of the metal. At high pumping, the localized SPs (hot spots), which possess the highest threshold, start to spase in a narrow frequency range around the maximum of the spasing criterion—the left-hand side of Equation (1.53). Because of the subwavelength size of these hot spots, the Kretschmann phase-matching condition is relaxed, and the radiation is outcoupled into a wide range of angles.

The SPPs of Reference 48 excited by the Kretschmann coupling are short-range SPPs, very close to the antisymmetric SPPs. They are localized at subwavelength distances from the surface, and their wavelength in the plane is much shorter the ω/c. Thus they can be well described by the quasistatic approximation and the present theory is applicable to them. Substituting the above given parameters of the dye and the extinction cross section $\sigma_e = 4 \times 10^{-16}$ cm^2 into Equation (1.54), we obtain a point shown by the black diamond in Figure 1.6, which is clearly above the threshold, supporting our assertion of the spasing. Likewise, the amplified spontaneous emission and possibly spasing appear to have prevented the full loss compensation in an SPP system of Reference 60.

Note that the long-range SPPs of Reference 64 are localized significantly weaker (at distances $\sim \lambda$) than those excited in Kretschmann geometry. Thus the long-range

SPPs experience a much weaker feedback, and the amplification instead of the spasing can be achieved. Generally, the long-range SPPs are fully electromagnetic (non-quasistatic) and are not describable in the present theory.

As we have already discussed in conjunction with Figure 1.3, the spasing is readily achievable with the gain medium containing common DBGSs or dyes. There have been numerous experimental observations of the spaser—see, in particular, References 27–33. Among them is a report of an SP spaser with a 7 nm gold nanosphere as its core and a laser dye in the gain medium [27], observations of the SPP spasers (also known as nanolasers) with silver as a plasmonic-core metal and DBGS as the gain medium with a 1d confinement [28, 31], a tight 2d confinement [29], and a 3d confinement [30]. There has also been a report on observation of an SPP microcylinder spaser [94]. A high-efficiency room-temperature semiconductor spaser with a DBGS InGaAS gain medium operating near 1.5 μm (i.e., in the communication near-IR range) has been reported [31].

The research and development in the area of spasers as quantum nanogenerators is very active and will undoubtedly lead to further rapid advances. The next in line is the spaser as an ultrafast nanoamplifier, which is one of the most important tasks in nanotechnology.

In contrast to this success and rapid development in the field of spasing and spasers, there has so far been a comparatively limited progress in the field of loss compensation by gain in metamaterials, which is based on the same principles of quantum amplification as the spaser. This status exists despite a significant effort in this direction and numerous theoretical publications (e.g., [53,95]). There has been so far a single, not yet confirmed independent, observation of the full loss compensation in a plasmonic metamaterial with gain [89].

In large periodic metamaterials, plasmonic modes are generally propagating waves (SPPs) that satisfy the Bloch theorem [84] and are characterized by quasi-wave vector **k**. These are propagating waves except for the band edges where $\mathbf{ka} = \pm\pi$, where **a** is the lattice vector. At the band edges, the group velocity v_g of these modes is zero, and these modes are localized, that is, they are SPs. Their wave function is periodic with period $2a$, which may be understood as a result of the Bragg reflection from the crystallographic planes. Within this $2a$ period, these band-edge modes can, indeed, be treated quasistatically because $2a \ll l_s, \bar{\lambda}$. If any of the band-edge frequencies is within the range of compensation [where the condition (1.33) [or (1.35)] is satisfied], the system will spase. The same is true for nonpropagating modes whose frequencies are in the band gap. In fact, at the band edge or in the band gap, a metamaterial with gain is similar to a DFB laser [96]—see also the next paragraph.

There has been a recent experimental observation [33] of an electrically pumped DFB nanolaser (spaser) working in the communication frequency range at room temperature. This spaser generates on a stop-band (nonpropagating) plasmonic mode, that is, a mode with frequency within the band gap. The system is a 1d plasmonic-crystal metamaterial with gain, containing a Bragg grating. There is a strong coupling between the unit cells typical for DFB lasers. Because of the suppression of the spontaneous emission and the strong coupling between the unit cells leading to efficient feedback, this spaser has a significantly lower threshold, narrow spectral line, and higher efficiency of the generation than the ones working on SPP reflection from the edges. This observation is in full agreement with our theoretical arguments [97] and in direct contradiction with the contention of Reference 98 that coupling between the unit cells increases the spasing threshold.

Moreover, the SPPs, which are at the band edge or in the band gap, will not just be localized. Due to unavoidable disorder caused by fabrication defects in metamaterials, there will be scattering of the SPPs from these defects. Close to the band edge, the group velocity becomes small, $v_g \to 0$. Because the scattering cross section of any wave is $\propto v_g^{-2}$, the corresponding SPPs experience Anderson localization [99]. Also, there will always be SPs nanolocalized at the defects of the metamaterial, whose local fields are hot spots [87, 100, 101]. Each of such hot spots within the bandwidth of conditions (1.33) or (1.35) will be a generating spaser, which clamps the inversion and precludes the full loss compensation.

Note that for a 2d metamaterial (metasurface), the amplification of the spontaneous emission and spasing may occur in SPP modes propagating *in plane* of the structure, unlike the signal that propagates normally to it as in Reference 89.

ACKNOWLEDGMENTS

This work was supported by Grant No. DEFG02-01ER15213 from the Chemical Sciences, Biosciences and Geosciences Division and by Grant No. DE-FG02-11ER46789 from the Materials Sciences and Engineering Division of the Office of the Basic Energy Sciences, Office of Science, U.S. Department of Energy, and by a grant from the U.S. Israel Binational Science Foundation.

REFERENCES

1. Atwater HA (2007) The promise of plasmonics. *Sci. Am.* 296: 56–63.
2. Anker JN, Hall WP, Lyandres O, Shah NC, Zhao J, Duyne RPV (2008) Biosensing with plasmonic nanosensors. *Nat. Mater.* 7: 442–453.
3. Novotny L, Hecht B (2006) *Principles of Nano-Optics*. Cambridge, NY: Cambridge University Press.
4. Israel A, Mrejen M, Lovsky Y, Polhan M, Maier S, Lewis A (2007) Near-field imaging probes electromagnetic waves. *Laser Focus World* 43: 99–102.
5. Tang L, Kocabas SE, Latif S, Okyay AK, Ly-Gagnon DS, Saraswat KC, Miller DAB (2008) Nanometre-scale germanium photodetector enhanced by a near-infrared dipole antenna. *Nat. Photonics* 2: 226–229.
6. Challener WA, Peng C, Itagi AV, Karns D, Peng W, Peng Y, Yang X, Zhu X, Gokemeijer NJ, Hsia YT, Ju G, Rottmayer RE, Seigler MA, Gage EC (2009) Heat-assisted magnetic recording by a near-field transducer with efficient optical energy transfer. *Nat. Photonics* 3: 220–224.

7. Kim S, Jin JH, Kim YJ, Park IY, Kim Y, Kim SW (2008) High-harmonic generation by resonant plasmon field enhancement. *Nature* 453: 757–760.
8. Nagatani N, Tanaka R, Yuhi T, Endo T, Kerman K, Takamura Y, Tamiya E (2006) Gold nanoparticle-based novel enhancement method for the development of highly sensitive immunochromatographic test strips. *Sci. Technol. Adv. Mater.* 7: 270–275.
9. Hirsch LR, Stafford RJ, Bankson JA, Sershen SR, Rivera B, Price RE, Hazle JD, Halas NJ, West JL (2003) Nanoshell-mediated near-infrared thermal therapy of tumors under magnetic resonance guidance. *Proc. Natl Acad. Sci. USA* 100: 13549–13554.
10. Park I-Y, Kim S, Choi J, Lee D-H, Kim Y-J, Kling MF, Stockman MI, Kim S-W (2011) Plasmonic generation of ultrashort extreme-ultraviolet light pulses. *Nat. Photonics* 5: 677–681.
11. Stockman MI (2011) Nanoplasmonics: the physics behind the applications. *Phys. Today* 64: 39–44.
12. Stockman MI (2011) Nanoplasmonics: past, present, and glimpse into future. *Opt. Express* 19: 22029–22106.
13. Kahng D (1963) Electric field controlled semiconductor device. US Patent 3,102,230.
14. Tsividis Y (1999) *Operation and Modeling of the MOS Transistor*. New York: McGraw-Hill.
15. Bergman DJ, Stockman MI (2003) Surface plasmon amplification by stimulated emission of radiation: quantum generation of coherent surface plasmons in nanosystems. *Phys. Rev. Lett.* 90: 027402.
16. Stockman MI, Bergman DJ (2009) Surface plasmon amplification by stimulated emission of radiation (spaser). US Patent 7,569,188.
17. Stockman MI (2008) Spasers explained. *Nat. Photonics* 2: 327–329.
18. Stockman MI (2010) The spaser as a nanoscale quantum generator and ultrafast amplifier. *J. Opt.* 12: 024004.
19. Stockman MI (2011) Spaser action, loss compensation, and stability in plasmonic systems with gain. *Phys. Rev. Lett.* 106: 156802.
20. Stockman MI (2011) Loss compensation by gain and spasing. *Phil. Trans. R. Soc. A* 369: 3510–3524.
21. Andrianov ES, Pukhov AA, Dorofeenko AV, Vinogradov AP, Lisyansky AA (2011) Dipole response of spaser on an external optical wave. *Opt. Lett.* 36: 4302–4304.
22. Campbell SD, Ziolkowski RW (2011) Impact of strong localization of the incident power density on the nano-amplifier characteristics of active coated nano-particles. *Opt. Commun.* doi: 10.1016/j.optcom.2011.11.006.
23. Lisyansky AA, Nechepurenko IA, Dorofeenko AV, Vinogradov AP, Pukhov AA (2011) Channel spaser: coherent excitation of one-dimensional plasmons from quantum dots located along a linear channel. *Phys. Rev. B* 84: 153409.
24. Liu SY, Li JF, Zhou F, Gan L, Li ZY (2011) Efficient surface plasmon amplification from gain-assisted gold nanorods. *Opt. Lett.* 36: 1296–1298.
25. Fedyanin DY (2012) Toward an electrically pumped spaser. *Opt. Lett.* 37: 404–406.

26. Andrianov ES, Pukhov AA, Dorofeenko AV, Vinogradov AP, Lisyansky AA (2012) Spaser chains. ArXiv: 1202.2925 [cond-mat.mes-hall].

27. Noginov MA, Zhu G, Belgrave AM, Bakker R, Shalaev VM, Narimanov EE, Stout S, Herz E, Suteewong T, Wiesner U (2009) Demonstration of a spaser-based nanolaser. *Nature* 460: 1110–1112.

28. Hill MT, Marell M, Leong ESP, Smalbrugge B, Zhu Y, Sun M, van Veldhoven PJ, Geluk EJ, Karouta F, Oei Y-S, Nötzel R, Ning C-Z, Smit MK (2009) Lasing in metal–insulator–metal sub-wavelength plasmonic waveguides. *Opt. Express* 17: 11107–11112.

29. Oulton RF, Sorger VJ, Zentgraf T, Ma R-M, Gladden C, Dai L, Bartal G, Zhang X, (2009) Plasmon lasers at deep subwavelength scale. *Nature* 461: 629–632.

30. Ma R-M, Oulton RF, Sorger VJ, Bartal G, Zhang X (2010) Room-temperature sub-diffraction-limited plasmon laser by total internal reflection. *Nat. Mater.* 10: 110–113.

31. Flynn RA, Kim CS, Vurgaftman I, Kim M, Meyer JR, Mäkinen AJ, Bussmann K, Cheng L, Choa FS, Long JP (2011) A room-temperature semiconductor spaser operating near 1.5 micron. *Opt. Express* 19: 8954–8961.

32. Wu CY, Kuo CT, Wang CY, He CL, Lin MH, Ahn H, Gwo S (2011) Plasmonic green nanolaser based on a metal-oxide-semiconductor structure. *Nano Lett.* 11: 4256–4260.

33. Marell MJH, Smalbrugge B, Geluk EJ, van Veldhoven PJ, Barcones B, Koopmans B, Nötzel R, Smit MK, Hill MT (2011) Plasmonic distributed feedback lasers at telecommunications wavelengths. *Opt. Express* 19: 15109–15118.

34. Sorger VJ, Zhang X (2011) Spotlight on plasmon lasers. *Science* 333: 709–710.

35. Ma RM, Oulton RF, Sorger VJ, Zhang X (2012) Plasmon lasers: coherent light source at molecular scales. *Laser Photonics Rev.* doi: 10.1002/lpor.201100040.

36. Berini P, Leon ID (2012) Surface plasmon-polariton amplifiers and lasers. *Nat. Photonics* 6: 16–24.

37. Leon ID, Berini P (2011) Measuring gain and noise in active long-range surface plasmon-polariton waveguides. *Rev. Sci. Instrum.* 82: 033107–10.

38. Ding K, Liu ZC, Yin LJ, Hill MT, Marell MJH, van Veldhoven PJ, Noetzel R, Ning CZ (2012) Room-temperature continuous wave lasing in deep-subwavelength metallic cavities under electrical injection. *Phys. Rev. B* 85: 041301.

39. Kwon S-H, Kang J-H, Kim S-K, Park H-G (2011) Surface plasmonic nanodisk/nanopan lasers. *IEEE J. Quantum Elect.* 47: 1346–1353.

40. Hill MT, Oei Y-S, Smalbrugge B, Zhu Y, de Vries T, van Veldhoven PJ, van Otten FWM, Eijkemans TJ, Turkiewicz JP, de Waardt H, Geluk EJ, Kwon S-H, Lee Y-H, Noetzel R, Smit MK (2007) Lasing in metallic-coated nanocavities. *Nat. Photonics* 1: 589–594.

41. Stockman MI, Faleev SV, Bergman DJ (2001) Localization versus delocalization of surface plasmons in nanosystems: can one state have both characteristics? *Phys. Rev. Lett.* 87: 167401.

42. Larkin IA, Stockman MI (2005) Imperfect perfect lens. *Nano Lett.* 5: 339–343.

43. Gordon JA, Ziolkowski RW (2007) The design and simulated performance of a coated nano-particle laser. *Opt. Express* 15: 2622–2653.

44. Bergman DJ, Stroud D (1992) Properties of macroscopically inhomogeneous media. In: Ehrenreich H, Turnbull D, editors. *Solid State Physics.* Vol. 46. Boston: Academic Press. pp 148–270.
45. Bergman DJ, Stroud D (1980) Theory of resonances in the electromagnetic scattering by macroscopic bodies. *Phys. Rev. B* 22: 3527–3539.
46. Plum E, Fedotov VA, Kuo P, Tsai DP, Zheludev NI (2009) Towards the lasing spaser: controlling metamaterial optical response with semiconductor quantum dots. *Opt. Express* 17: 8548–8551.
47. Seidel J, Grafstroem S, Eng L (2005) Stimulated emission of surface plasmons at the interface between a silver film and an optically pumped dye solution. *Phys. Rev. Lett.* 94: 177401.
48. Noginov MA, Zhu G, Mayy M, Ritzo BA, Noginova N, Podolskiy VA (2008) Stimulated emission of surface plasmon polaritons. *Phys. Rev. Lett.* 101: 226806.
49. Li K, Li X, Stockman MI, Bergman DJ (2005) Surface plasmon amplification by stimulated emission in nanolenses. *Phys. Rev. B* 71: 115409.
50. Dong ZG, Liu H, Li T, Zhu ZH, Wang SM, Cao JX, Zhu SN, Zhang X (2008) Resonance amplification of left-handed transmission at optical frequencies by stimulated emission of radiation in active metamaterials. *Opt. Express* 16: 20974–20980.
51. Wegener M, Garcia-Pomar JL, Soukoulis CM, Meinzer N, Ruther M, Linden S (2008) Toy model for plasmonic metamaterial resonances coupled to two-level system gain. *Opt. Express* 16: 19785–19798.
52. Fang A, Koschny T, Wegener M, Soukoulis CM (2009) Self-consistent calculation of metamaterials with gain. *Phys. Rev. B (Rapid Commun.)* 79: 241104(R).
53. Wuestner S, Pusch A, Tsakmakidis KL, Hamm JM, Hess O (2010) Overcoming losses with gain in a negative refractive index metamaterial. *Phys. Rev. Lett.* 105: 127401.
54. Chang SW, Ni CYA, Chuang SL (2008) Theory for bowtie plasmonic nanolasers. *Opt. Express* 16: 10580–10595.
55. Zheludev NI, Prosvirnin SL, Papasimakis N, Fedotov VA (2008) Lasing spaser. *Nat. Photonics* 2: 351–354.
56. Protsenko IE, Uskov AV, Zaimidoroga OA, Samoilov VN, O'Reilly EP (2005) Dipole nanolaser. *Phys. Rev. A* 71: 063812.
57. Ambati M, Nam SH, Ulin-Avila E, Genov DA, Bartal G, Zhang X (2008) Observation of stimulated emission of surface plasmon polaritons. *Nano Lett.* 8: 3998–4001.
58. Zhou ZK, Su XR, Peng XN, Zhou L (2008) Sublinear and superlinear photoluminescence from Nd doped anodic aluminum oxide templates loaded with Ag nanowires. *Opt. Express* 16: 18028–18033.
59. Noginov MA, Podolskiy VA, Zhu G, Mayy M, Bahoura M, Adegoke JA, Ritzo BA, Reynolds K (2008) Compensation of loss in propagating surface plasmon polariton by gain in adjacent dielectric medium. *Opt. Express* 16: 1385–1392.
60. Bolger PM, Dickson W, Krasavin AV, Liebscher L, Hickey SG, Skryabin DV, Zayats AV (2010) Amplified spontaneous emission of surface plasmon polaritons and limitations on the increase of their propagation length. *Opt. Lett.* 35: 1197–1199.

61. Chen Y-H, Li J, Ren M-L, Li Z-Y (2012) Amplified spontaneous emission of surface plasmon polaritons with unusual angle-dependent response. *Small.* doi: 10.1002/smll.201101806.

62. Noginov MA, Zhu G, Bahoura M, Adegoke J, Small C, Ritzo BA, Drachev VP, Shalaev VM (2007) The effect of gain and absorption on surface plasmons in metal nanoparticles. *Appl. Phys. B* 86: 455–460.

63. Noginov MA (2008) Compensation of surface plasmon loss by gain in dielectric medium. *J. Nanophotonics* 2: 021855.

64. Leon ID, Berini P (2010) Amplification of long-range surface plasmons by a dipolar gain medium. *Nat. Photonics* 4: 382–387.

65. Bergman DJ, Stroud D (1992) Properties of macroscopically inhomogeneous media. In: Ehrenreich H, Turnbull D, editors. *Solid State Physics*. Vol. 46. Boston: Academic Press. pp 148–270.

66. Stockman MI, Bergman DJ, Kobayashi T (2004) Coherent control of nanoscale localization of ultrafast optical excitation in nanosystems. *Phys. Rev. B* 69: 054202.

67. Landau LD, Lifshitz EM (1984) *Electrodynamics of Continuous Media.* Oxford: Pergamon.

68. Schawlow AL, Townes CH (1958) Infrared and optical masers. *Phys. Rev.* 112: 1940.

69. Johnson PB, Christy RW (1972) Optical constants of noble metals. *Phys. Rev. B* 6: 4370–4379.

70. De Angelis F, Das G, Candeloro P, Patrini M, Galli M, Bek A, Lazzarino M, Maksymov I, Liberale C, Andreani LC, Di Fabrizio E (2009) Nanoscale chemical mapping using three-dimensional adiabatic compression of surface plasmon polaritons. *Nat. Nanotechnol.* 5: 67–72.

71. Neacsu CC, Berweger S, Olmon RL, Saraf LV, Ropers C, Raschke MB (2010) Near-field localization in plasmonic superfocusing: a nanoemitter on a tip. *Nano Lett.* 10: 592–596.

72. Atwater HA, Polman A (2010) Plasmonics for improved photovoltaic devices. *Nat. Mater.* 9: 205–213.

73. Kneipp K, Moskovits M, Kneipp H, editors (2006) *Surface Enhanced Raman Scattering: Physics and Applications*. Heidelberg: Springer-Verlag.

74. Kneipp J, Kneipp H, Wittig B, Kneipp K (2010) Novel optical nanosensors for probing and imaging live cells. *Nanomed. Nanotechnol. Biol. Med.* 6: 214–226.

75. Stockman MI (2006) Electromagnetic theory of SERS. In: Moskovits M, Kneipp K, Kneipp H, editors. *Surface Enhanced Raman Scattering*. Vol. 103. Heidelberg: Springer-Verlag. pp 47–66.

76. Dunmore FJ, Liu DZ, Drew HD, Dassarma S, Li Q, Fenner DB (1995) Observation of below-gap plasmon excitations in superconducting YBa2Cu3O7 films. *Phys. Rev. B* 52: R731–R734.

77. Schumacher D, Rea C, Heitmann D, Scharnberg K (1998) Surface plasmons and Sommerfeld-Zenneck waves on corrugated surfaces: application to high-T_c superconductors. *Surf. Sci.* 408: 203–211.

78. Fedotov VA, Tsiatmas A, Shi JH, Buckingham R, deGroot P, Chen Y, Wang S, Zheludev NI (2010) Temperature control of Fano resonances and transmission in superconducting metamaterials. *Opt. Express* 18: 9015–9019.

79. Tsiatmas A, Buckingham AR, Fedotov VA, Wang S, Chen Y, deGroot PAJ, Zheludev NI (2010) Superconducting plasmonics and extraordinary transmission. *Appl. Phys. Lett.* 97: 111106.

80. Anlage SM (2011) The physics and applications of superconducting metamaterials. *J. Opt.* 13: 024001.

81. Boltasseva A, Atwater HA (2011) Low-loss plasmonic metamaterials. *Science* 331: 290–291.

82. Shalaev VM (2007) Optical negative-index metamaterials. *Nat. Photonics* 1: 41–48.

83. Zheludev NI (2011) A roadmap for metamaterials. *Opt. Photonics News* 22: 30–35.

84. Bloch F (1929) Über die Quantenmechanik der Elektronen in Kristallgittern. *Z. Phys. A* 52: 555–600.

85. Stockman MI, Kurlayev KB, George TF (1999) Linear and nonlinear optical susceptibilities of Maxwell Garnett composites: dipolar spectral theory. *Phys. Rev. B* 60: 17071–17083.

86. Li X, Stockman MI (2008) Highly efficient spatiotemporal coherent control in nanoplasmonics on a nanometer-femtosecond scale by time reversal. *Phys. Rev. B* 77: 195109.

87. Stockman MI, Pandey LN, George TF (1996) Inhomogeneous localization of polar eigenmodes in fractals. *Phys. Rev. B* 53: 2183–2186.

88. Stockman MI (2004) Nanofocusing of optical energy in tapered plasmonic waveguides. *Phys. Rev. Lett.* 93: 137404.

89. Xiao S, Drachev VP, Kildishev AV, Ni X, Chettiar UK, Yuan H-K, Shalaev VM (2010) Loss-free and active optical negative-index metamaterials. *Nature* 466: 735–738.

90. Gryczynski Z, Abugo OO, Lakowicz JR (1999) Polarization sensing of fluorophores in tissues for drug compliance monitoring. *Anal. Biochem.* 273: 204–211.

91. Merlin R (2009) Metamaterials and the Landau-Lifshitz permeability argument: large permittivity begets high-frequency magnetism. *Proc. Natl Acad. Sci. USA* 106: 1693–1698.

92. Meinzer N, Ruther M, Linden S, Soukoulis CM, Khitrova G, Hendrickson J, Olitzky JD, Gibbs HM, Wegener M (2010) Arrays of Ag split-ring resonators coupled to InGaAs single-quantum-well gain. *Opt. Express* 18: 24140–24151.

93. Kretschmann E, Raether H (1968) Radiative decay of nonradiative surface plasmons excited by light. *Z. Naturforsch. A* 23: 2135–2136.

94. Kitur JK, Podolskiy VA, Noginov MA (2011) Stimulated emission of surface plasmon polaritons in a microcylinder cavity. *Phys. Rev. Lett.* 106: 183903.

95. Hess O, Wuestner S, Pusch A, Tsakmakidis KL, Hamm JM (2011) Gain and plasmon dynamics in active negative-index metamaterials. *Phil. Trans. R. Soc. A* 369: 3525–3550.

96. Ghafouri-Shiraz H (2003) *Distributed Feedback Laser Diodes and Optical Tuneable Filters*. West Sussex, UK: John Wiley & Sons Ltd.

97. Stockman MI (2011) Reply to comment on "Spaser action, loss compensation, and stability in plasmonic systems with gain". *Phys. Rev. Lett.* 107: 259704.

98. Pendry JB, Maier SA (2011) Comment on "Spaser action, loss compensation, and stability in plasmonic systems with gain". *Phys. Rev. Lett.* 107: 259703.

99. Anderson PW (1958) Absence of diffusion in certain random lattices. *Phys. Rev.* 109: 1492–1505.

100. Tsai DP, Kovacs J, Wang Z, Moskovits M, Shalaev VM, Suh JS, Botet R (1994) Photon scanning tunneling microscopy images of optical excitations of fractal metal colloid clusters. *Phys. Rev. Lett.* 72: 4149–4152.

101. Stockman MI, Pandey LN, Muratov LS, George TF (1995) Photon scanning-tunneling-microscopy images of optical-excitations of fractal metal colloid clusters—comment. *Phys. Rev. Lett.* 75: 2450–2450.

97. Stockman MI (2011) [illegible] compensation, and stability in plasmonic systems with gain. Phys Rev Lett [illegible]

98. Stockman MI (2011) [illegible] on "Spaser action, loss compensation, and stability in plasmonic systems with gain". Phys Rev Lett [illegible]

99. Anderson PW (1958) Absence of diffusion in certain random lattices. Phys Rev 109: 1492–1505

100. [illegible] Wang Z, [illegible] Shalaev VM, [illegible] [illegible] of optical [illegible] Phys [illegible]

101. Stockman MI, [illegible] (2001) [illegible] Phys Rev Lett [illegible]

2

Nonlinear Effects in Plasmonic Systems

PAVEL GINZBURG
Department of Physics, King's College London, Strand, London, UK

MEIR ORENSTEIN
Department of Electrical Engineering, Technion – IIT, Haifa, Israel

2.1 INTRODUCTION

Nonlinear photonics celebrated its formal 50th anniversary in 2011, following the first experiment of second harmonic generation by Franken et al. in 1961 [1]. It was always the backbone in the evolution of modern optics, yielding exciting discoveries and understanding of fundamental optical phenomena and also serving as a source for a large variety of applications. In optics, nonlinear interactions are relatively weak, such that in many cases the induced polarization vector is treated perturbatively as a power series in the electrical field and serves as a source term in the Maxwell wave equations. However, non-perturbative nonlinear optics at extremely intense excitation is a field of interest today, for example, for high harmonic generation [2], laser-induced fusion [3], and material processing [4]. The quantum nature of light may give rise to a handful of nonlinear phenomena, where quantized vacuum fluctuations initiate nonlinear interactions—one of the most known examples is the spontaneous parametric down conversion (SPDC), where pump light is being dissected into a pair of actual photons that were not present in the system primary to the interaction. At the same year mentioned above (1961), the paper by W. H. Louisell, A. Yariv and A. E. Siegman on quantum fluctuations of parametric processes [5] is considered as the founding year of the quantum nonlinear optics as well.

The integration between nonlinear optics and the very fast evolving field of plasmonics (metal-optics) is of special interest and is the subject of this chapter.

Active Plasmonics and Tuneable Plasmonic Metamaterials, First Edition. Edited by Anatoly V. Zayats and Stefan A. Maier.

This interest stems from the following reasons: *metal nonlinearities*—metal exhibits both bound electrons nonlinearities (similar to other optical media) and free carrier (plasma) hydrodynamic-like nonlinearities; *surface nonlinearities*—plasmonics always occurs on metal–dielectric interfaces, thus interface nonlinearities, spatial symmetry breaking, and related phenomena are highly important; *field enhancement*—one of the merits of plasmonic structures, either in localized or in propagation forms, is substantial field concentration and enhancement, which makes optical nonlinearities much more accessible and applicable; *multipolar*—for example, electric quadrupole or magnetic dipole interactions can be designed by properly shaping the plasmonic structures and these higher order interactions are intrinsically nonlinear (and nonlocal); *subwavelength-scale nonlinear waves*—the ability to generate nonlinear optical features such as spatial solitons at the nanoscale, below the wavelength of light; *non-phase-matched parametric processes*—plasmonic nanostructures are sometimes effective ultra small cavities (slow-light devices), which are replacing the very cumbersome phase-matching configurations of nonlinear conversion schemes.

The very well known historical example of plasmonic-enhanced nonlinearity (although the term plasmonics was not yet coined) is surface-enhanced Raman scattering (SERS), taking place on the interface of rough noble metal, and was shown to enhance the magnitude of the scattering by 14 orders of magnitude compared to the unassisted phenomenon [6]. In the last decade or so, with the explosion in plasmonics research, many related optical nonlinearities were studied, demonstrated, and proposed for applications, for example, for active plasmonic circuitry (the latter is discussed in a different chapter of this book).

In this chapter we discuss some of the nonlinear phenomena closely related to the interplay of light with metallic objects, generally termed *nonlinear plasmonics*.

The electromagnetic wave equation in its general form is given by

$$\nabla \times \nabla \times \vec{E}(\vec{r},t) + \frac{1}{c^2}\frac{\partial^2}{\partial t^2}\vec{E}(\vec{r},t) = -\frac{1}{\varepsilon_0 c^2}\frac{\partial^2}{\partial t^2}\vec{P}(\vec{r},t), \tag{2.1}$$

where $\vec{E}(\vec{r},t)$ is the electrical field vector, c the speed of light in vacuum, ε_0 the vacuum permittivity, and $\vec{P}(\vec{r},t)$ a general polarization term, reflecting the medium response to the incident electromagnetic wave. $\vec{P}(\vec{r},t)$ encapsulates all the required material information, such as linear and nonlinear dispersive responses. The inclusion of spatial dispersion may be further introduced by more advanced techniques [7] and will not be discussed here in much detail. The polarization term may be decomposed in a series of electrical field, and in the frequency domain it is given by [8, 9]

$$P_i = \varepsilon_0 \chi_{ij} E_j + 2\chi_{ijk} E_j E_k + 4\chi_{ijkl} E_j E_k E_l, \tag{2.2}$$

where subscripts denote vector components at certain directions and χs are susceptibility tensors of appropriate rank. Enhancement of nonlinearity, as reflected from Equation 2.2, may be gained by manipulation of the intrinsic material susceptibilities or the enhancement of electrical fields. While quantum engineering of internal structures may be performed [10, 11], here we emphasize the contributions of the so-called local field enhancement and fast spatial fields variations, where plasmonics is one of the most promising approaches.

The chapter is organized as follows: In Section 2.2 we discuss physical mechanisms, responsible for nonlinearities in metal structures, emphasizing the difference between local and nonlocal nonlinearities, although they are intermix when the predominant effects are occurring near interfaces. In Section 2.3 we present nonlinear surface plasmon polariton (SPP) propagation, when the nonlinearity is either stemming from the metal or enhanced by the metallic waveguide structure. Nonlinear modes, plasmon solitons, and nonlinear waveguide devices are discussed. In Section 2.4 we introduce nonlinearities enhanced by local surface plasmons within or in the vicinity of nanoparticles and nanoantennas. Sometimes the actual source of nonlinearity is not completely defined under such highly localized enhanced fields, since the largest field and field gradients are occurring exactly at the metal–dielectric interfaces and the nonlinear polarizability can be in either of both the media.

2.2 METALLIC NONLINEARITIES—BASIC EFFECTS AND MODELS

A unified analytical model for metal nonlinearities at the extended optical regime (300–2000 nm) is still missing. The main theoretical challenge is to extract clear information from the complex solid-state structure of metallic objects, surfaces, and surrounding environment in order to isolate the main contributors to certain nonlinear effects. Moreover, nowadays experimental data from different sources are not always consistent mainly because of the large dependence of the mechanisms on the exact quality of the sample and sample surfaces. In general terms, nonlinearities of metal structures originate from two families of physical effects; the first includes bulk point contributions, related to the saturations of interband optical transitions [12], and carriers heating contribution [13, 14]. These types of nonlinear response are exhibited at frequencies exceeding the interband energy gap occurring at the visible spectral range for most of the noble metals. The second family of nonlinearities is attributed to the nonlinear dynamics of the free carriers' plasma, which may be exhibited at any frequency, even below the interband transition. The detailed discussion of metal nonlinearity deserves a book of its own, and here we only briefly elucidate the different characteristics of the two families of nonlinearity by two examples (Sections 2.1 and 2.2) and then discuss their combined effects as a source for parametric process (Section 2.3). Finally, we look into material damage induced by extreme light intensity, which limits the field of metal nonlinear optics and conversely serves as an important mechanism in metal processing (Section 2.4).

2.2.1 Local Nonlinearity—Transients by Carrier Heating

Metal nonlinearity based on collective carriers heating is predominantly observed in short pulse experiments and was employed in pump probe experiments to extract the fast dynamics of collective nonradiative relaxation processes in electron plasma.

Generally, femtosecond (fs) pulses are used for generation of nonequilibrium excitations of the metal electrons via interband transitions (e.g., pulses with ~400 nm central frequency for Au). The carriers' dynamics may be described by the so-called two-temperature model, where electrons and lattice are assumed to be in internal equilibrium but not in mutual equilibrium ([15] and references therein). In the frame of this model the nonequilibrium electron distribution, created by an fs laser pulse, is thermalized via very fast electron–electron scattering processes, taking place on the scale of hundreds of fs, and subsequently creates a modified Fermi–Dirac electron distribution of "hot" carriers. Electron–phonon scatterings are much slower processes, taking place on the timescale of tens of picoseconds, and they equalize temperatures of the lattice and electrons. Most of the interesting phenomena occur in the window between these two characteristic timescales. The effect is predominantly local because nonlocal mechanisms such as diffusion are much slower compared to the electron–electron thermalization process.

The modified electron distribution yields a broadening of the localized plasmon resonances, modifying the related absorption coefficient. The linear behavior of a sample is recovered after few electron–phonon scattering events. Modification of absorption in metal nanoparticles 10 and 100 nm in size was studied as the function of fs laser excitation intensity [16]. Figure 2.1 shows the recovery dynamics of the

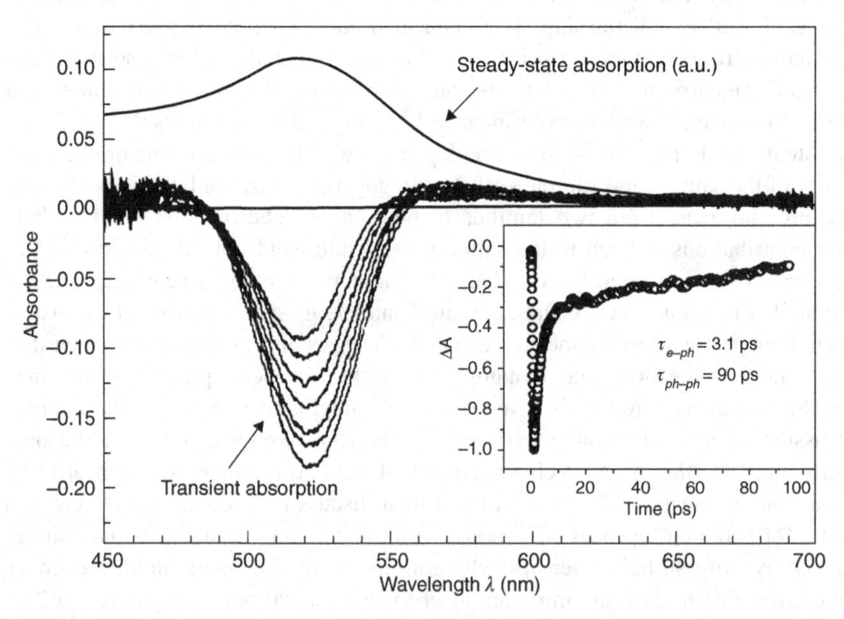

FIGURE 2.1 Independence of the electron–phonon and phonon–phonon relaxation times on the composition of gold and mixed gold–silver nanoparticles. The transient absorption spectra of 15 nm spherical gold nanoparticles after excitation at 400 nm with 100 fs laser pulses recorded at different delay times is shown. Also shown is the steady-state UV–vis absorption spectrum of the colloidal gold solution. The inset shows the decay of the transient bleach when the particles are monitored at the bleach maximum at 520 nm. Fitting of the decay curve gives electron–phonon and phonon–phonon relaxation times of 3.1 and 90 ps, respectively. (From Reference 16.)

absorbance. The simple model attributes the broadening to the temperature dependence of the electrons scattering rate. The Drude approximation with temperature-dependent damping of dielectric permittivity of a bulk metal $\varepsilon_M(\omega, T)$ is given by

$$\varepsilon_M(\omega, T) = \varepsilon_b - \frac{\omega_p^2}{\omega(\omega + i\hbar\gamma(\omega, T))}, \tag{2.3}$$

where ε_b is the background permittivity, which holds all the information on the interband transitions, ω_p the plasma frequency of conduction electron gas, and $\gamma(\omega, T)$ the scattering rate. In the frame of the simple model ([17 and references therein) this term is given by

$$\gamma(\omega, T) = \frac{\omega^2}{4\pi\omega_p^2}\left(1 + \left(\frac{k_B T_e}{\hbar\omega}\right)^2\right), \tag{2.4}$$

where k_B is the Boltzmann constant and T_e the electron gas temperature. The quality factor of the plasmonic resonance is roughly the ratio between the real and imaginary parts of the material permittivity and it is reduced with the increase in the temperature, as may be seen from Equation 2.4.

This intensity-dependent (pump) transient absorption (probe) may be enhanced by orders of magnitude, using metamaterial composition, based on nanorod arrays, and further more by nonlocal effects due to the fine features of such structures, as will be discussed in Section 2.4 in connection to parametric processes. The collective coherent mode of the structure overlaps with many nanorods, enhancing the effect and resulting in much faster collective response. Moreover, the steep modal dispersion of this collective excitation together with fast refractive index changes, initiated by the pump, provides much more flexibility and sensitivity for the modulation of the transmitted signal. Appropriate designs may provide a route to ultrafast, low-power all-optical information processing in the deep subwavelength scale [17].

2.2.2 Plasma Nonlinearity—The Ponderomotive Force

Nonlinearity of noble metals may originate from pure hydrodynamics of the free carrier plasma due to nonlocal effects and will also be viable at frequencies lower than those related to the interband transitions. The nonlocal effects of collective interactions of charged particles make the dielectric permittivity to be average-intensity dependent. Such nonlinearity originates from the ponderomotive force ($\vec{F}_{PM}(\vec{r})$) [18], repelling charged carriers from regions of high field intensities and thus effectively diluting locally the electron plasma in the conduction band within the high field intensity regions:

$$\vec{F}_{PM}(\vec{r}) = -\frac{1}{m}\left(\frac{e}{\omega}\right)^2 \left(\vec{E}(\vec{r}) \times \left(\nabla \times \vec{E}(\vec{r})\right) + \vec{E}(\vec{r}) \cdot \nabla\vec{E}(\vec{r})\right), \tag{2.5}$$

where e and m are the electron charge and mass, respectively, ω is the beam central frequency, and $\vec{E}(\vec{r})$ is the local electrical field amplitude. The conservative ponderomotive force may be recast into a potential that is subsequently added to the Fermi energy and results in effective change in electron concentration within the conduction band. The inclusion of fermion statistics in this hydrodynamic approach is required for metals, since the Boltzmann distribution typically used in plasma physics is inapplicable for metals, where the Fermi level is situated deep in the conduction band. The intensity-dependent (according to the shifted Fermi level) concentration of the charged carriers gives rise to nonlocal Kerr-like nonlinear electrical permittivity with a highly dispersive coefficient [19]:

$$\varepsilon_{PM}\left(\left|\vec{E}(\vec{r})\right|^2\right) = \varepsilon_M + \frac{3}{2}\left(\frac{\omega_p}{3\pi^2\varepsilon_0\hbar m e}\right)^{2/3}\left(\frac{e}{\omega}\right)^4\left|\vec{E}(\vec{r})\right|^2, \tag{2.6}$$

where ε_M is the linear part of the metal–dielectric constant and $\hbar$ the Planck's constant. This nonlinear Kerr coefficient is highly dispersive ($\sim 1/\omega^4$) and for the telecom wavelength of 1.5 μm is on the order of 10^{-18} (m^2/V^2)—comparable with that of typical nonlinear glasses. Detailed book chapter on the dynamical effects in metal plasma may be found in Reference 20.

A number of experimental studies present third-order nonlinearity with susceptibility of the same order of magnitude as predicted by the ponderomotive model. Four-wave mixing (FWM) experiments estimate $\chi^{(3)} = 0.2 \times 10^{-18}(m^2/V^2)$ at $\lambda = 633$ nm [21], which is two orders of magnitude larger than values reported for highly nonlinear crystals, for example, $LiNbO_3$.

2.2.3 Parametric Process in Metals

Parametric optical processes in metals will result from both local interband-based nonlinearity and plasma-based nonlinearity and in most cases from their combinations. Second-order interactions require broken spatial symmetry, resulting from either the quantum confining potential or macroscopic (but still subwavelength) geometric features, as may be shown from general considerations of symmetry and time reversal. Thus the second-order susceptibility $\chi^{(2)}$ cannot be exhibited by crystals with inversion symmetry and very poorly by amorphous/polycrystalline materials, which are the typical form of both bulk metals and substrates used conventionally in plasmonics. However, the required conditions may be simply fulfilled by geometrical violation of reflection symmetry, such as at abrupt distinct boundaries. Since plasmonics relies on the existence of such interfaces and the surface plasmons reside on them, "efficient" second-harmonic-related phenomena are expected to occur.

In early studies of metal second-order processes by Bloembergen et al. [22] and Jha [23], the conduction band response was modeled in terms of Sommerfeld free electrons, with an extension to a set of basic hydrodynamic equations for free plasma. It is important to note that the mere inclusion of Lorentz magnetic force makes these equations to be inherently nonlinear, coupling charge velocity with overall magnetic field—each of these quantities oscillates with the fundamental frequency and hence generates a source for frequency doubling. The presence of magnetic field violates the time-reversal conditions and results in the so-called magnetic dipole contribution to nonlinearity. Additional source for nonlinear polarization originates from large gradients of electrical field (inherent in plasmonics) that yield multipole contributions. This formalism describes the bulk properties of conduction carriers and does not include the contribution of "surface states" occurring in the region between the nominal interfaces to a depth of few Fermi wavelengths into the metal (a few nanometers for noble metals), where the carrier dynamics is modified. The hydrodynamic formalism was further extended by Sipe et al. ([24] and references therein along with citing articles) and the shortened version of nonlinear polarizability $\vec{P}^{(2\omega)}$ may be written as [25]

$$\vec{P}^{(2\omega)} = \beta \vec{E}(\omega)\left[\nabla \cdot \vec{E}(\omega)\right] + \gamma \vec{\nabla}\left[\vec{E}(\omega) \cdot \vec{E}(\omega)\right] + \delta'\left[\vec{E}(\omega) \cdot \vec{\nabla}\right]\vec{E}(\omega), \quad (2.7)$$

where β, γ, and δ' are parameters related to the magnetic dipole and electric quadrupole effects and may be estimated from specific material properties or alternatively measured. A comprehensive theoretical model, taking into account both bound and free-electron contribution to the linear response and nonlinear terms due to free electrons, was developed in Reference 26 and includes six terms, collapsing to Equation 2.7 with certain assumptions on material losses.

Experimental studies of SHG on reflection from flat metal mirrors were first reported in Reference 27. Advanced experimental techniques used in Reference 25 show that the surface contribution predominates the bulk; however, both should be taken into account for proper modeling.

Magnetic dipole interactions are relatively small at the high frequencies of the optical regime. However, their contribution to the SHG may be probed by creation of artificial materials (metamaterials) with significant magnetic responses at the optical frequencies. Arrays of split-ring resonators were shown to enhance the nonlinear phenomena at frequencies near the resonance of the artificial magnetic dipoles [28].

Another consequence of second-order nonlinearity is the photon drag effect, resulting from the zero frequency term of the nonlinear polarizability (related to what is known as optical rectification). The hydrodynamic theory of the effect is developed in Reference 29, the experimental demonstration was first performed for Germanium sample [30], while the plasmonic version was demonstrated in Reference 31 for Kretschmann geometry of SPP excitation and periodic dielectric grating on the metal surface [32]. The extreme nanoplasmonic confinement may also lead to giant photon drag effect and may result in the generation of THz source terms [33].

Theories for higher harmonic generation (third and above) on flat metal surfaces, based on Sommerfeld free-electron models, with subsequent solution of Schrödinger equation in the Kramers–Henneberger accelerating frame were developed [34] and are in good agreement with experimental data [35]. Coherence properties of higher harmonics were studied and it was shown that while the second harmonic is partially coherent, at higher harmonics incoherent emission is dominant [36]—meaning that the predominant effect that we are observing at "higher harmonic generation" is metal luminescence excited by multiphoton absorption. The reason is seemingly because of the very fast electron scattering, causing the loss of coherence of the carriers during the pump pulse duration. Since the efficiencies of second and third harmonics have different power dependence on the excitation, at high-enough intensities third-order process may overcome the second order. In the experimental observation of Pappadogiannis et al. [37] a 4 μm thick polycrystalline gold surface was exposed to 290 fs laser pulses. The duration of the pulses is shorter than the electron–phonon relaxation time, but larger than carrier–carrier scattering time, meaning that electrons are self-thermalized to an equilibrium temperature of thousands of Kelvin. Under these conditions the efficiency of the third harmonic process overcomes the second harmonic at 11 GW/cm^2 pump laser intensity.

The first observation of the third harmonic generation from individual colloidal gold nanoparticles was demonstrated in Reference 38. Pump laser at 1500 nm generated a 500 nm signal, in the vicinity of a localized plasmon resonance. The third harmonic signal intensity was shown to be proportional to the square of the particles' surface area.

2.2.4 Metal Damage and Ablation

Nonlinear experiments are typically performed with high-intensity lasers, operating at short pulsed regime to increase peak powers. Thus potential optical damage in nonlinear plasmonic experiments is an important limitation. One of the most pronounced bottlenecks is surface roughness increasing with downscale of noble metal experimental samples, which may cause undesired local field enhancements and effectively magnify the incident field by order of magnitudes, acting as primary sources of electromagnetic damage. The term "laser ablation" is sometimes associated with this type of damage, when materials are evaporated from host surfaces. Different models of laser ablation exist; among them are thermal and photochemical; we refer the readers to a couple of review papers on the subject—References 39 and 40. Briefly, there are four key mechanisms that may lead to the optical damage: linear absorption, resulting in heating of electrons and lattice; avalanche breakdown, where accelerated free electrons cause cascaded ionization; multiphoton ionization; and direct field ionization, initiated by very short energetic pulses [8].

In an experimental study [41], the damage of gold and nickel thin films induced by fs laser pulses was investigated by light-scattering measurements. Film thickness was found to be a key parameter for the measured thresholds. The main mechanism, involved in the process, is the carrier diffusion, occurring in the first picoseconds and strongly depends on electron motilities and phonon scattering cross sections. In the regime of long (ns) light pulses the energy dissipation and transport is defined by the heat diffusion in the lattice. In both regimes the melting is predicted to begin at the surface.

The damage thresholds, as the function of film thickness, are depicted in Figure 2.2.

For extra smooth surfaces the main damage mechanism is seemingly the creation of electron plasma in the air above the surface; for example, for a 4 μm thick

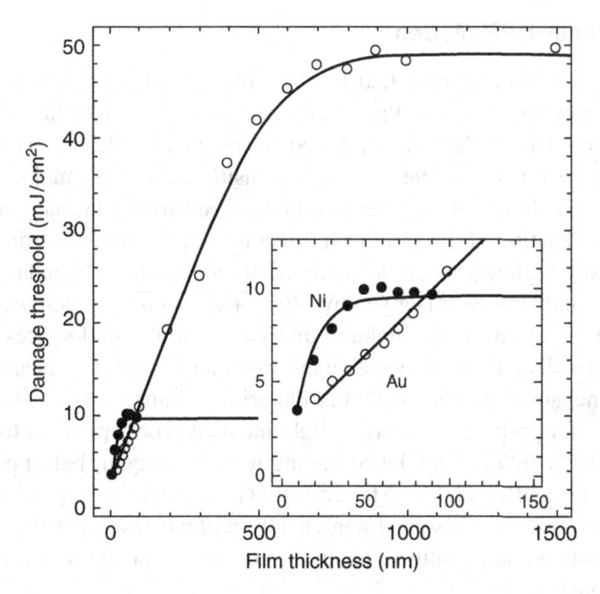

FIGURE 2.2 Film thickness dependence of the absorbed laser fluence at the damage threshold extrapolated to the single shot value for Au (open circles) and Ni (filled circles). The inset shows a magnification of the thin film range. The solid lines are calculated using the two-temperature model. (From Reference 41.)

polycrystalline gold surface, exposed to 290 fs laser pulses, plasma creation was observed at an excitation intensity of 120 GW/cm^2 [32]. The temperature of metal surfaces was analytically estimated by the solution of electromagnetic diffraction problem together with heat-transfer equations, distinguishing between different timescales and particle sizes [42].

Laser ablation may also be employed for metal processing, for example, hole drilling on metal surface. Pronko et al. [43] drilled 300 nm diameter holes in a 600 nm film of silver evaporated onto a glass substrate, using Ti:sapphire laser, producing 200 fs pulses with an overall energy of 40 nJ.

2.3 NONLINEAR PROPAGATION OF SURFACE PLASMON POLARITONS

Nonlinear phenomena in general planar and waveguiding structures were extensively studied and we refer the readers to a book chapter in Reference 44, where comprehensive studies of nonlinear waveguide structures are presented. Nonlinear phenomena may lead to contra-intuitive and unexpected results, such as TE surface wave over single interfaces, prohibited in the linear regime for nonmagnetic materials. In this section we discuss some nonlinear phenomena of propagating SPPs, where the nonlinearity stems either from the metal or from the dielectric constituents of the plasmonic waveguide structures.

2.3.1 Nonlinear SPP Modes

The propagation of SPP is modified by third-order nonlinearity even in the most basic SPP-supporting structure: the single metal–dielectric interface [19]. Specifically, SPP nonlinear modes are supported by metal–air interface at wavelengths longer than those related to the interband transition; thus the major nonlinearity stems from the ponderomotive forces within the free carrier plasma (Section 2.2.2). An intuitive explanation of this nonlinear effect is related to the electron depletion in the high-intensity regions that are located just at the metal–air boundary. The linear modal effective index of SPP is given by $\sqrt{(\varepsilon_M + \varepsilon_D)/\varepsilon_M\varepsilon_D}$, where ε_M and ε_D are the dielectric permittivities of metal and dielectric medium half-spaces. As a result from the carriers' depletion, the negative metal–dielectric constant near the interface becomes less negative, thus approaching the critical value of $\varepsilon_M = -\varepsilon_D$, which is the cutoff value for propagating SPP. At light intensity, corresponding to this critical value, the intensity-induced modal reshaping results in equal, but opposite, power flow in the metal and dielectric. The corresponding intensity-dependent nonlinear dispersion relations may be solved analytically without resolving the exact modal shape. The set of coupled nonlinear algebraic equations may be solved for the mode propagation constant β:

$$\left[\varepsilon_M + \frac{\varepsilon_D^2}{\beta^2 - \varepsilon_D}\right] E_z^2 - \varepsilon_D \beta \frac{E_z(0)\, E_x(0)}{\sqrt{\beta^2 - \varepsilon_D}} - \frac{1}{2}\chi_{PM} E_x^4(0) + \frac{1}{2}\chi_{PM} E_z^4(0) = 0,$$

$$\beta \varepsilon_D \frac{E_z(0)}{\sqrt{\beta^2 - \varepsilon_D}} = \left(\varepsilon_M + \chi_{PM} E_x^2(0) + \chi_{PM} E_z^2(0)\right) E_x(0), \tag{2.8}$$

where $E_x(0)$ and $E_z(0)$ are the x (perpendicular to the surface) and z (the direction of the propagation) components of the electrical field phasor amplitude at the interface, ε_M and ε_D the linear permittivities of the metal and dielectric substrate, respectively, and χ_{PM} the nonlinear ponderomotive susceptibility, defined in Equation 2.6. Solutions of the above equations for different field intensities on the surface are represented in Figure 2.3 [19] showing clearly the intensity-dependent cutoff wavelengths for the nonlinear modes:

2.3.2 Plasmon Solitons

Diffraction of electromagnetic (light) beams can be mitigated by nonlinearity, resulting in the generation of solitons—wave packets that maintain (roughly speaking) their envelopes as invariants during propagation. Solitons are one of the most explored features in nonlinear optics (and nonlinear physics in general). In this section we present solitons on planar metal–dielectric structures considering nonlinearities of one or both material components.

One of the prominent advantages of incorporating plasmons in nonlinear schemes is the capability to reduce modal volumes below classical limit of diffraction [45]. Tapered or corrugated plasmonic waveguides provided nanofocusing of optical fields

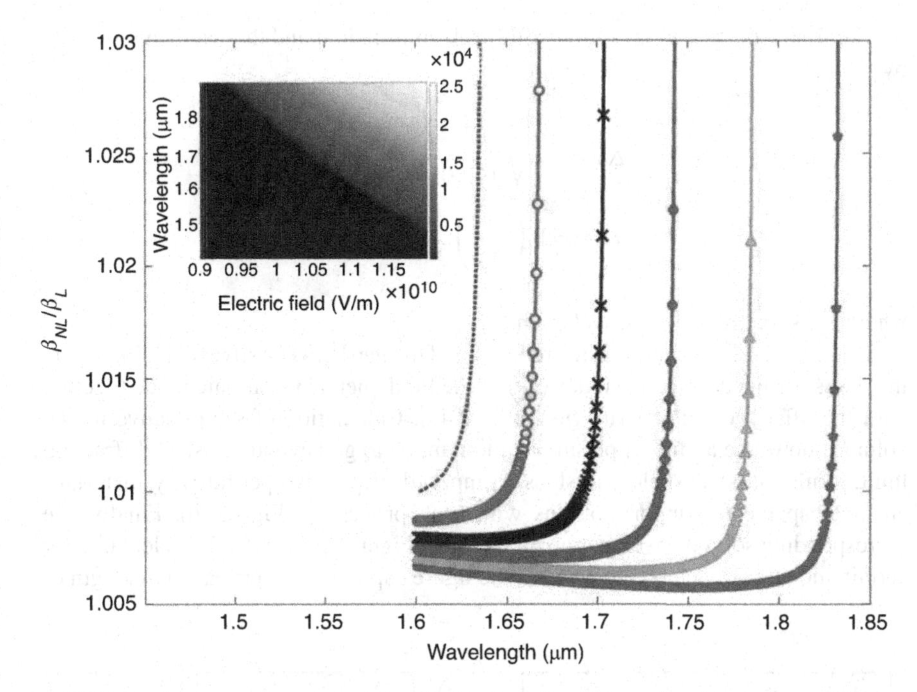

FIGURE 2.3 Nonlinear dispersion relation of a single-surface SPP on air–gold interface at different interface electric field amplitudes: dashed, 12 GV/m; circles, 11.5 GV/m; crosses, 11 GV/m; diamonds, 10.5 GV/m; triangles, 10 GV/m; stars, 9.5 GV/m. The inset is the nonlinear effective index normalized by the linear one versus the wavelength and field amplitude. (From Reference 19.)

[46] with apparent larger confinement in metal–insulator–metal (MIM) structures [47,48]. Generally, nonlinear dispersion relations for MIM waveguides with even-order nonlinearities (Kerr or higher) may be derived if lateral confinement is neglected [49]. When the confinement is taken into account (in Kerr-nonlinear slab, bounded by silver claddings) hybrid-vector spatial plasmon solitons may emerge. The most striking effect of plasmonics is that when the separation between the two silver layers of the MIM structure is reduced to increase the transverse confinement into the deep subwavelength regime, the field envelope in the lateral dimension (bound only by the nonlinearity) is reduced as well [50], which is an inverse effect compared to a nonlinear all-dielectric waveguide. The plasmon soliton scenario may be formulated in terms of nonlinear Schrödinger equation as follows:

$$ja_z + \frac{1}{2}\beta a_{yy} + \frac{2}{3}\beta k_0^2 n_0 n_2 \frac{I_3}{I} |a|^2 a = 0, \qquad (2.9)$$

where I and I_3 are certain field averages across the waveguide transversal direction, k_0 the free-space wavevector amplitude, and n_0 and n_2 the linear and nonlinear (Kerr) refractive indexes of the MIM core, respectively.

The first-order plasmon soliton with peak amplitude η and the width Δy is given by

$$\Delta y \cdot \eta = \sqrt{\frac{3I}{16\pi^2 n_0 n_2 I_3}} \cdot \lambda_0,$$
$$a \sim \operatorname{sech}\left(\frac{y}{\Delta y}\right), \tag{2.10}$$

where λ_0 is the free-space wavelength.

Simulation results show that the reduction of the gap between silver cladding layers increases the linear effective index (n_0) of the fundamental plasmonic mode together with the effective nonlinearity (n_2) (Fig. 2.4a). Calculation of the effective modal volume shows the ability of plasmon soliton mode to go beyond classical diffraction limit, while classical dielectric slabs, composed of positive permittivity materials, are not capable to support solitons with such properties (Fig. 2.4b). Finally, the corresponding soliton envelopes are depicted in Figures 2.4c and 2.4d, calculated for certain model parameters, indicated in the figure captions. The propagation length of

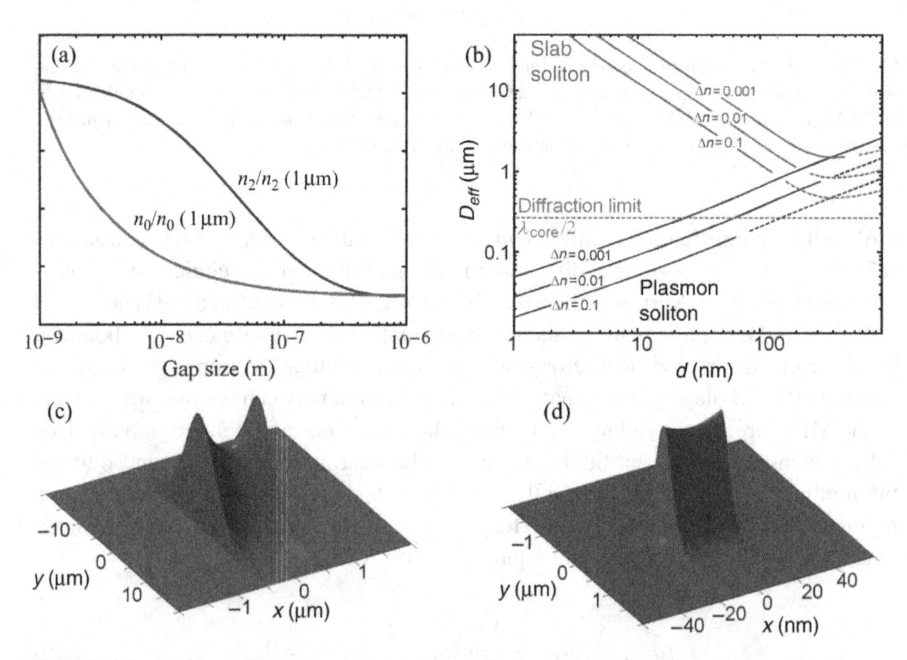

FIGURE 2.4 (a) Normalized linear and nonlinear effective indexes inside plasmonic gap at $\lambda_0 = 1.5$ μm. (b) Effective modal size D_{eff} versus gap thickness of the nonlinear Kerr slab embedded in (blue) metal and (red) air. $\lambda_0 = 820$ nm, $n_0 = 1.5$. Intensity distributions of plasmon soliton beam in Kerr medium embedded within silver layers. (c) $\Delta n = 0.005$, $\lambda_0 = 1550$ nm, $d = 1$ μm, $\varepsilon_m = -103.5 - i10$, $D_{eff} = 2.4$ μm. (d) $\Delta n = 0.1$, $\lambda_0 = 820$ nm, $d = 30$ nm, $\varepsilon_m = -30.2 - i1.6$, $D_{eff} = 95$ nm. (From Reference 51.)

the mode was shown to be larger than the typical soliton length as well as the decaying length due to metal absorption, clearing the possibility of soliton phenomenon in this structure.

Nonlinear propagation of SPPs on the boundary of a metal and a nonlinear Kerr dielectric in the presence of losses was further considered and a self-focusing phenomenon with the formation of slowly decaying spatial solitons was numerically approved in Reference 52. While References 50 and 52 used Kerr dielectric media, a similar phenomenon may also occur due to the ponderomotive nonlinearity of the metal film itself. Finite difference time domain (FDTD) simulation of SPP propagation in a plasmonic MIM waveguide comprising a narrow linear dielectric medium between two metallic layers showed nanofocusing features for high field intensities. MIM-plasmon has higher (compared to single surface) mode overlap with metal cladding. Figure 2.5a shows the diffraction pattern of MIM SPP for different initial powers. In the low-power regime the regular in-plane diffraction pattern combined with propagation losses is observed. Self-focusing of the SPP mode is achieved due to the ponderomotive nonlinearity at field intensities as low as ~kW/μm^2 making SPP self-focusing a practically achievable phenomenon and is depicted in Figure 2.5b.

An interesting approach for partially overcoming propagation losses of plasmon soliton in MIM structure was proposed in Reference 53, where additional tapering at a properly chosen angle generates additional field concentration thus enhancing the field intensity and enabling longer propagation (Fig. 2.6).

The propagation losses may be compensated by incorporation of an active medium—spatial surface plasmon solitons on the metal stripe surrounded by active and passive dielectric media were proposed [54]. Stable solitons are achieved by proper coupling between SPPs propagating along the active and passive interfaces.

Multilayered metallo-dielectric structures with embedded nonlinearities may lead to the creation of the so-called discrete solitons. Arrays of coupled waveguides are known to produce discrete diffraction patterns, where mode, launched into a central waveguide, exhibits maximal intensity in some edge waveguides at the output. This diffraction may be suppressed by introduction of nonlinearity into one or more of the

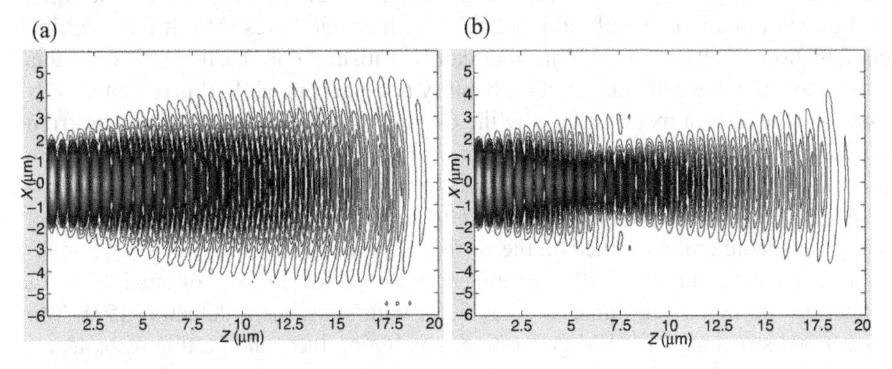

FIGURE 2.5 Nonlinear propagation in MIM: (a) beam diffraction at linear regime; (b) self-focusing due to ponderomotive nonlinearities of metal claddings.

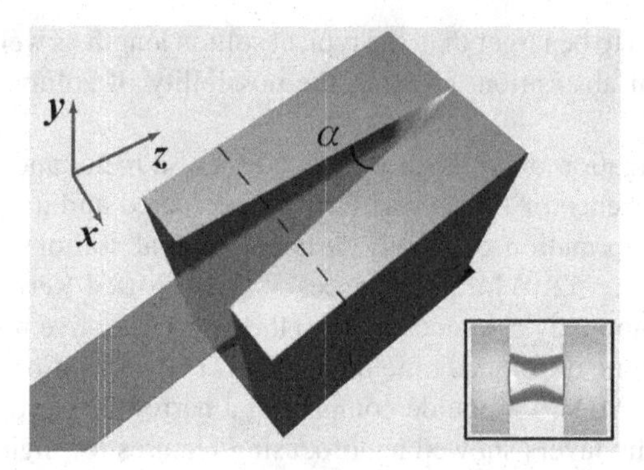

FIGURE 2.6 Schematic of the three-dimensional plasmon focusing in a tapered slot waveguide. Focusing occurs in the horizontal plane, due to a taper, and in the vertical plane, due to the nonlinear self-focusing effect. Inset shows the waveguide cross section with the magnetic field distribution of a symmetric plasmon mode. (From Reference 53.)

material components. Discrete solitons on nanoscale were studied in Reference 55 and were shown to be considerably different from conventional nonlinear dielectric waveguide arrays by combined interplay between periodicity, nonlinearity, and SPPs.

2.3.3 Nonlinear Plasmonic Waveguide Couplers

While many nonlinear SPP-based devices are feasible, we exemplify this notion by looking at an example of nonlinear couplers.

Plasmonic traveling modes (as well as any propagating modes) are "dark" since their propagation constants are always higher than those of the core dielectric media. This is the reason that the excitation of surface plasmons requires additional effort, such as implementation of Kretschmann/Otto configurations or etching of spatial gratings. However, free-space excitation of surface plasmons may also be assisted by nonlinear effects. One of the examples is nonlinear FWM, where fulfillment of phase-matching conditions involves the momentum of the resulting wave to be equal to the vector combination of momenta of three incident waves [56]. If the resulting combination results in momentum that can be satisfied only by the SPP, only this latter wave will have the maximal probability to be generated. In this scheme, SPPs may be excited on a metal surface by the overlap of two laser beams, incident from different angles to the normal. Two beams with $\lambda_1 = 707$ nm and $\lambda_2 = 800$ nm were used; fixing the angle of one beam, efficient excitation of surface plasmon occurs at a respective angle of the second beam. The generated SPP is monitored by outcoupling using the spatial grating etched on the surface (Fig. 2.7).

This principle may be further extended and used for imaging of small objects, scattering the nonlinear signal to a far field, with subsequent detection [57]. The conceptual structure for FWM generation, combining local and grating resonances, was shown to enhance the nonlinear phenomena by three orders of magnitude [58]. Directional coupler based on coupled nonlinear MIM waveguides was investigated

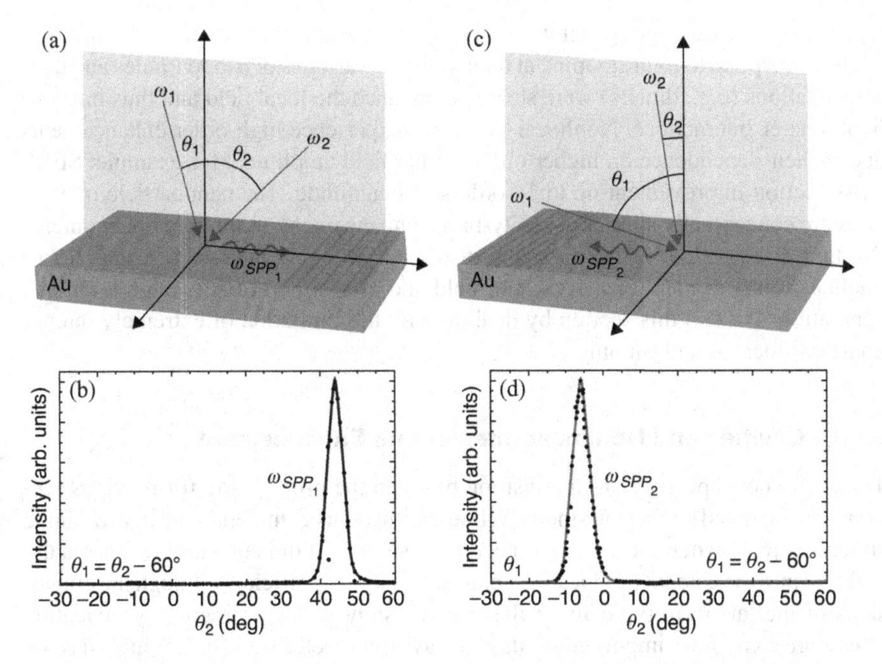

FIGURE 2.7 (a and c) Experimental configuration with overlaid scanning electron microscope images of the sample surface showing the gratings used for SPP outcoupling. The figures illustrate the resonant incident angles (arrows) and the direction of SPP propagation (wavy lines). (a) Configuration for plasmon $\lambda_{SPP_1} = 633$ nm excited by $\omega_{SPP_1} = 2\omega_1 - \omega_2$. (c) Configuration for plasmon $\lambda_{SPP_2} = 921$ nm excited by $\omega_{SPP_2} = 2\omega_2 - \omega_1$. (b and d) Experimental results showing the intensity of the outcoupled surface plasmons as a function of excitation angle Θ_2 using $\Theta_1 = \Theta_2 - 60°$. Black dots are data and the solid curves are Gaussian fitting functions. The peak positions agree with the theoretical predictions in Equation 2.4. (Taken from Reference 56.)

in Reference 59 and was shown to have substantial novel characteristics compared to pure dielectric nonlinear devices. Electrically controlled optical modulator creating modulated higher harmonics signal, originating from nonlinear dielectric material within the nanometric slit in a metal film, was demonstrated [60]. Power-dependent modal dispersion and switching was proposed in metallo-dielectric nanoparticle chain, embedded in nonlinear Kerr dielectric [61]. Two resonant modes are supported by the structure in the linear regime—longitudinal and transversal, which are different by 20 nm. Nonlinearities are shifting the mode resonances and mixing them up modifying the overall transmission.

2.4 LOCALIZED SURFACE PLASMON NONLINEARITY

Most of the current research on plasmonic enhancement of nonlinear interactions is related to localized surface plasmons generated near metallic nanoparticles, nanoantennas, and metamaterials based on such particles. Metallic nanoparticles at or near their plasmonic resonance generate highly localized electric field intensities that effectively correspond to light stopping at their vicinity. Varieties of nanoparticles and their constellations (e.g., dimers) were shown to enhance the local field and thus improve light–matter interactions. Nonlinear processes experience high-order enhancements due to their dependence on higher orders of the field amplitude, for example, SERS cross-section improvement up to 14 orders of magnitude. The nanoparticle interactions are conceptually similar to cavity-based enhancement of nonlinear phenomena. We first discuss this similarity and further look at a few interesting examples of nonlinear enhancement of the vacuum field and of non-perturbative high harmonic generation. We end this section by dealing with the limitation of extremely intense nonlinear localized plasmonics.

2.4.1 Cavities and Nonlinear Interactions Enhancement

The cavity concept serves as a transition between the propagating (of previous section) and localized surface plasmons. While cavities can be implemented in waveguide structures [62], when the latter are becoming shorter in dimension, they eventually collapse into the category of localized plasmon of a nanoparticle. Although the dimensions of the latter are typically in the extreme subwavelength regime, propagation effects are sometimes important for its nanocavity characteristics [63]. While all types of cavities are predominantly "stopped light" devices, the plasmonic nanoparticles enable virtually close to zero volume electric field intensity confinement. In terms of linear phenomena, the enhancement of the spontaneous emission rates in cavities relative to the free space is described in terms of the so-called Purcell factor [64], given by

$$P = \frac{3Q\left(\lambda_c/n\right)^3}{4\pi^2 V_{eff}} \tag{2.11}$$

where P is the Purcell enhancement factor, depending on the cavity quality factor Q, central wavelength λ_c, average modal refractive index n, and effective cavity modal volume V_{eff}. Quality factors of ultra small plasmonic cavities are limited by the ratio of real-to-imaginary parts of electric permittivity (generally dozens at optical and near-infrared range) and cannot be improved much. On the other hand, the modal volumes may go far below the diffraction limit, approaching zero values in theoretical estimations [63]. Relying on the above, the linear Purcell factor in plasmonic environment may reach extremely high values. In terms of nonlinear phenomena, the cavity enhancement will improve the interaction, multiplying the effect by Purcell factor in power, equal to the order of nonlinear process. Thus SHG may be improved by a factor of P^2 if the cavity is resonant at the doubled frequency.

In addition to local enhancement of the field (reduction of the mode volume), a short cavity may assist the generation of macroscopic nonlinear signal, by overcoming the required phase-matching condition. In parametric processes, for example, SHG, the pump frequency (ω) and the resulting doubled frequency (2ω) are generally not propagating with the same phase velocity due to dispersion of the nonlinear structure. Various phase-matching techniques may compensate this phase mismatch resulting in constructive interference for the nonlinear process making the whole structure to act as a macroscopic source of nonlinearity. In cavity nonlinear optics, however, the standing waves (for a cavity based on a waveguide structure) or the quasistatic fields (in localized plasmonic cavity) are generating different mechanism replacing the phase-matching. The proper theory, accompanied by experimental demonstration, was reported in Reference 65. The plane wave expansion, commonly used for propagating nonlinear optics, is replaced by eigen mode expansion in terms of cavity modes:

$$\tilde{E}_{\omega_i}(\vec{r},t) = E_{\omega_i}(t)\, e^{i\omega_i t} F_{\omega_i}(t), \tag{2.12}$$

where $\tilde{E}_{\omega_i}(\vec{r},t)$ is the total electrical field with central frequency ω_i, $E_{\omega_i}(t)$ the slowly varying time-dependent amplitude, and $F_{\omega_i}(t)$ the spatial mode dependence. The wave equation with nonlinear polarization as the source term may be written in terms of the above mode expansion, resulting in the following equation for the second harmonic field:

$$\begin{aligned} \frac{dE_2(t)}{dt} &= i\gamma\chi^{(2)}\frac{\omega_2}{4\varepsilon_2}E_1^2(t) - \frac{\alpha}{2}E_2(t), \\ \gamma &= \int d^3\vec{r}\, F_2^*(\vec{r})\, F_1^2(\vec{r}), \end{aligned} \tag{2.13}$$

where α is a linear cavity loss, ε_2 the cavity material permittivity at ω_2, $\chi^{(2)}$ the nonlinear susceptibility of the cavity, and γ the nonlinear mode overlap. It may be clearly seen that the classical phase-matching condition in cavity nonlinear optics should be replaced by the modal overlap. Hence, properly designed cavities, resonant at both fundamental and second harmonics, with spatially overlapping fields may dramatically enhance the nonlinear conversion efficiency. The first photonic example of such a design is presented in Figure 2.8.

The advantage of doubly resonant plasmonic structure was demonstrated for surface-enhanced Raman spectroscopy, where a pair of particles provides the desired resonant structure for both pump and Stokes frequencies [66].

SHG from core–shell nanocavities, containing nonlinear material, was demonstrated to enhance the process by two to three orders of magnitude owing to the nonlinear core [67]. Surface patterning of core–shell geometry with nanogaps filled by GaAs was shown to enhance second harmonic signal at the spectral window collocated with minimal linear transmission [68]. Grooves, etched in a metal surface, carefully organized into grating geometry, were shown to enhance FWM efficiency by two orders of magnitude [58]. Improvement of FWM by four orders of magnitude was demonstrated during the creation of plasmonic dimer configuration, when one particle was brought in proximity to another attached to an NSOM tip [69]. Dark

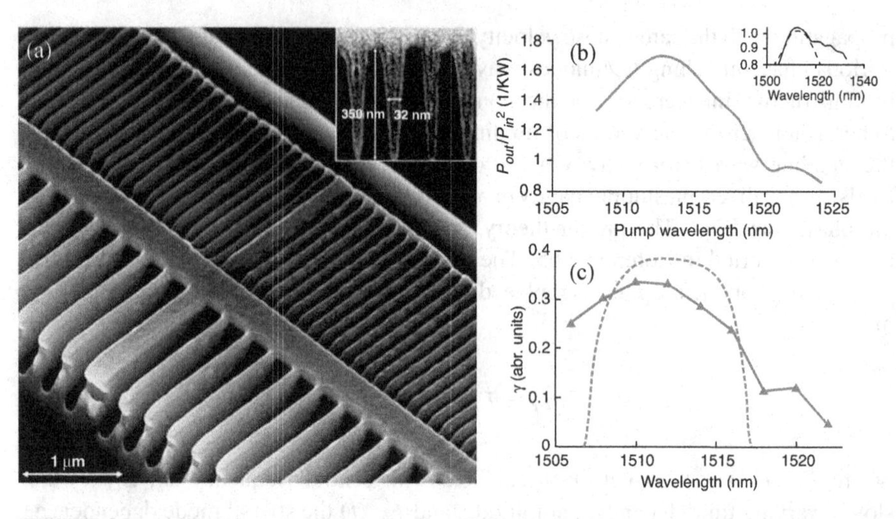

FIGURE 2.8 (a) Scanning electron microscopy (SEM) image of the fabricated integrated double-resonance microcavity. The inset is a SEM cross-section image of the shortest period grating. (b) Measured SH generation efficiency P_{out}/P_{in}^2 wavelength dependence. The inset is the microcavity transmission spectrum around the 1512 nm resonance. The solid curve is the measured and the dashed curve is the calculated (FDTD) spectrum. (c) Nonlinear mode overlap γ wavelength dependence. The solid curve is the measured and the dashed curve is the calculated spectrum. (From Reference 65.)

plasmonic modes (weakly coupled to far-field radiation) may be excited via second harmonic generation [70].

2.4.2 Enhancement of Nonlinear Vacuum Effects

Plasmonic cavities may also influence the properties of nonlinear emitters. It may be shown that emitters with initially low radiation efficiency may be substantially improved by plasmonic structures [71]. One of the very exciting processes with potential applications in quantum computing, transmission, and imaging is spontaneous two-photon emission (STPE), where electron–hole recombination process is accompanied by spontaneous emission of a photon pair. The process was recently observed using semiconductor structures [72] and subsequently enhanced by almost three orders of magnitude by depositing plasmonic array of bow-tie antennas on the top of semiconductor bulk [73]. STPE is an inherently weak second-order quantum-mechanical process and hence requires significant enhancement for competing with other candidates (such as SPDC) for various applications. The emission spectrum of STPE is very broadband since each pair of photons may be emitted with any energy combination with the only restriction on the overall energy conservation of the whole process. Dielectric cavities with high quality factors (Q) have very narrow spectral bandwidths. Such cavities fail to enhance the broadband STPE process. Contrarily, plasmonic cavities may exhibit significant Purcell factors by the reduction of modal volumes, while keeping quality factors to be relatively low. The relative nonlinear enhancement by a cavity may be roughly estimated by the ratio of the modified radiative density of states to the free-space values:

$$\frac{R_{cav}(\omega_1)}{R_{free}(\omega_1)} = F_p(\omega_1) F_p(\omega_0 - \omega_1) \frac{D_{cav}(\omega_1) D_{cav}(\omega_0 - \omega_1)}{\omega_1^2 (\omega_0 - \omega_1)^2},$$
$$F_p(\omega_1) = \frac{\pi^2 c^3}{n^2(\omega_1) n_g(\omega_1) V}, \tag{2.14}$$

where R_{cav} and R_{free} are the STPE rate densities (at ω_1) within the cavity and free space, respectively, D_{cav} the cavity spectral response (assumed to have a Lorentzian shape), and ω_0 the bandgap transition frequency. $F_p(\omega_1)$ is the usual density of states in homogeneous material with refractive and group indexes of $n(\omega_1)$ and $n_g(\omega_1)$, respectively, and effective volume V. For a single-resonance wideband cavity, the maximal enhancement is obtained for a resonance, centered at $\omega_0/2$—when the two emitted photons are degenerate in frequency. The experimental structure was comprised of a 200 nm AlGaAs active layer, with one photon photoluminescence centered at 826 nm, and STPE peak at 1653 nm. A 60 nm thick gold bow-tie antenna array was fabricated on the surface of the active layer using e-beam lithography and was designed to have its plasmonic resonance at $\omega_0/2$ (Fig. 2.9a). The active layer was pumped optically.

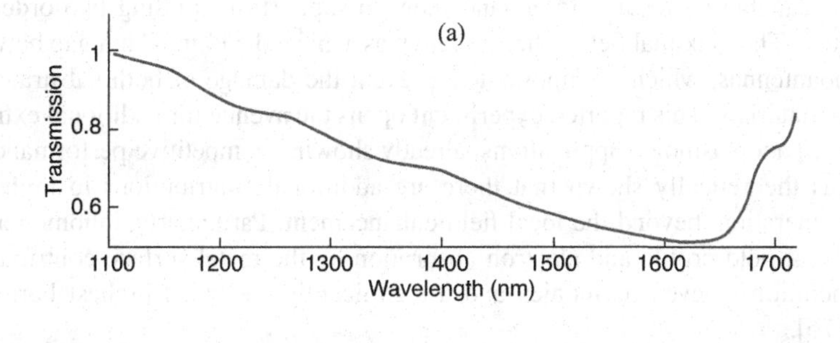

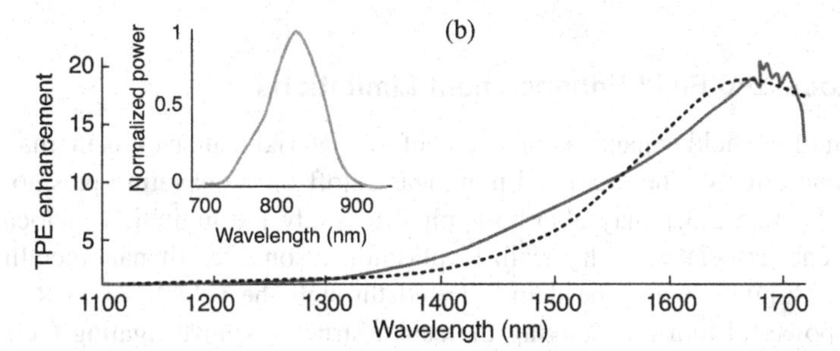

FIGURE 2.9 (a) Transmission spectrum of the passive bow-tie particle array. (b) Spectrum of the enhancement in TPE by the bow-tie array: measurement, solid blue line; FDTD calculation, dashed black line. The inset is the one-photon emission spectrum. (From Reference 73.)

The resulting enhancement (Fig. 2.9b) of STPE was achieved over a relatively wide spectral range and by the directly measured value of enhancement factor of ~20, and subtracting the background emission that was not enhanced, effective enhancement of about three orders of magnitude was exhibited—making this interesting nonlinear emission process to be close in intensity to regular LED emission.

2.4.3 High Harmonic Generation

The generation of high harmonics is applicable for coherent x-rays sources, laser processing, and attosecond spectroscopy. First experimental realizations considered flat metallic mirrors, while harmonics were generated by reflection. For example, 15th harmonic order, corresponding to a wavelength of 55.3 nm, was observed [74].

High harmonic generation was demonstrated recently by using localized plasmonic structures—specifically an array of bow-tie antennas, deposited on sapphire substrate [75]. The source of the nonlinear generation was a jet of argon gas, which was ionized by short femtosecond pulses (10 fs). Intense laser pulse causes ionization of core electrons, which are recaptured by the core atom, when the electrical field is reversed in direction. Acceleration of these electrons against the positively charged core produces "Bremsstrahlung" radiation resulting in a comb of higher harmonics. The required laser intensity for such ionization processes is very high (the threshold in this particular case is 10^{13} W/cm^2), while the initial laser intensity provided 10^{11} W/cm^2—two orders of magnitude lower than required. The plasmonic structure was capable by local field enhancement to support the missing two orders of magnitude. The maximal field enhancement was achieved within 20 nm gap between the nanoantennas, which is important to prevent the damage of both substrate and antenna materials. This reported experiment opens the avenue for additional extreme nonlinear nanoplasmonic applications, already showing competitive performances.

It was theoretically shown that there are additional contributions to high harmonic generation, beyond the local field enhancement. Particularly, inhomogeneity of the local field profile and electron absorption by the metal surface contribute to the generation of even harmonics and a significantly enhanced highest harmonic cutoff [76].

2.4.4 Localized Field Enhancement Limitations

Limitation of the field enhancement, even before material damage sets in, may result from plasma effects. The predicted nonlinear cutoff on propagating plasmons discussed in Section 2.3.1 may also be applied to study the limitation of local field enhancement, associated with localized plasmon resonances of nanometallic particles. This limitation may be demonstrated through the vehicle of SERs. SERS is a very powerful tool for studying molecular structures, investigating their internal vibrational degrees of freedom. Raman scattering is a third-order process in quantum-mechanical description and inherently very weak and inefficient. However, by attaching a molecule to a rough metallic surface, or a metallic nanoparticle and relying on their local field enhancement, it is possible to increase the process efficiency by factors as high as 10^{14} relative to the scattering process in regular dielectric media. Electromagnetic theory explains this giant scattering rate enhancement by excitation of collective oscillations of electrons on the surface, coupled to external light field, namely plasmons, that modifies the local density of radiation states, resulting in two-order enhancement of the scattering efficiency. The overall scattering enhancement factor due to plasmonic cavity is proportional to E^4, where E is the local field enhancement factor. This local field enhancement on metal surfaces strongly depends on the detailed surface geometry as well as on the chromatic dispersion of the metal itself. Complete theoretical model of SERS is very involved and still remains a subject of intensive research. Additional theoretical considerations (beyond electromagnetic), such as chemical considerations on formation of charge-transfer complexes, were shown to affect SERS efficiency as well [77]. Nevertheless, most of the experimental observations of enhanced nonlinear phenomena were explained by theory. Some of the reported SERS experiments were performed with excitation fields with low intensities, for example, 10 mW on 100 μm^2 [6, 78] where noble metals can be treated as a linear dielectric. However, for higher excitation intensities the emergence of significant nonlinear ponderomotive effects may cause the field localization to disappear. In order to demonstrate this limitation, cylindrically shaped roughness on a flat metal surface may be investigated—in a similar way as is done for calculation of SERS enhancement [79]. The linear solution of such a quasi-static scenario (in principle, may be analytically solved) is depicted in Figure 2.10a for touching silver bumps of 15 nm radius. This specific geometry generates resonant conditions at a frequency where $\varepsilon = -5.4 + 0.4j$, including the material losses [80]. Incident beam with electrical field amplitude of 10^9 V/m and 460 nm free-space wavelength and polarization, parallel to the surface, will generate local fields, repelling the free electrons out of

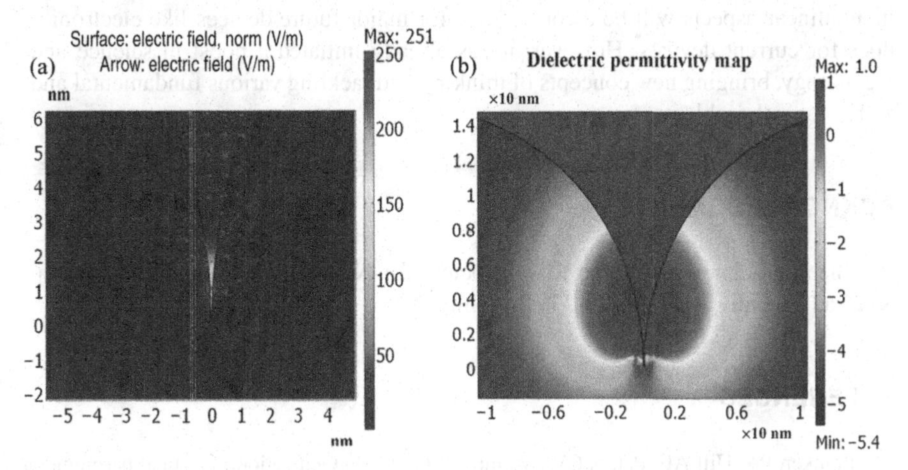

FIGURE 2.10 (a) Linear solution of cylindrical bumps—E/E_0. Field enhancement factor is about 251 at resonant condition. (b) Dielectric constant map at an external exciting field of 10^9 V/m. In the metal regions near the narrowest gap, the dielectric constant is now very close to the vacuum value. (From Reference 79.)

the high-intensity regions of the structure. The dielectric permittivity—depending on the local carrier concentration—is thus modified and recalculated accordingly to the linear solution and Equation 2.5, and the resulting permittivity distribution is shown in the color map depicted in Figure 2.10b. It can be clearly seen that the free carriers have been almost completely depleted from the region of interest, thus destroying the geometry of the confining potential. The complete self-consistent solution of such a nonlinear problem is involved, since the effect cannot be solved perturbatively and linear solution is not a good first estimate anymore.

2.5 SUMMARY

In this chapter we presented some major activities in the field of nonlinear nanoplasmonics. A large number of recent scientific publications on this topic are emphasizing its significance and, of course, we could not cover all the facets of this subject in this chapter. We emphasized the uniqueness of metal and its support of nonlinear effects associated with free-electron plasma. Nonlinear devices, based on nonlinear optical modes and spatial solitons in waveguide structures, were discussed. Nonlinear effects in nanoplasmonic cavities were presented—unifying propagating and localized plasmons for distinctive improvement of nonlinear light–matter interactions at deep subwavelength scale.

Progress in technological, computational, and experimental capabilities of last decades in the field of nanoscience made numerous theoretical propositions of the past to be experimentally relevant in the present and near future and is already pushing reconsideration and investigation of new nanoscale phenomena. In this sense the field of nonlinear nanoplasmonics is situated at this special interface between "old school" and recent "technological edge." It is still too premature to claim that plasmonics with its nonlinear aspects will be a cornerstone for major future devices like electronics does for current devices. However, it has already initiated a boost in science and technology, bringing new concepts of thinking and tackling various fundamental and technological problems.

ACKNOWLEDGMENTS

P. Ginzburg acknowledges the Royal Society for Newton International Fellowship and Yad Hanadiv for Rothschild Fellowship.

REFERENCES

1. Franken PA, Hill AE, Peters CW, Weinreich G (1961) Generation of optical harmonics. *Phys. Rev. Lett.* 7: 118–119.
2. Burnett NH, Baldis HA, Richardson MC, Enright GD (1977) Harmonic generation in CO_2 laser target interaction. *Appl. Phys. Lett.* 31: 172.

3. Gerstner E (2007) Laser physics: extreme light. *Nature* 446: 16–18.

4. Steen WM, Mazumder J, Watkins KG (2010) *Laser Material Processing*. 4th ed. Springer.

5. Louisell WH, Yariv A, Siegman AE (1961) Quantum fluctuations and noise in parametric processes. I. *Phys. Rev.* 124: 1646–1654.

6. Nie S, Emory SR (1997) Probing single molecules and single nanoparticles by surface-enhanced Raman scattering. *Science* 275: 1102–1106.

7. Agranovich VM, Ginzburg VL (1984) *Crystal Optics with Spatial Dispersion, and Excitons*. Springer.

8. Boyd RW (2008) *Nonlinear Optics*. 3rd ed. Academic Press.

9. Yariv A, Yeh P (2006) *Photonics: Optical Electronics in Modern Communications*. 6th ed. Oxford University Press.

10. Khurgin J (1988) Second-order nonlinear effects in asymmetric quantum-well structures. *Phys. Rev. B* 38: 4056–4066.

11. Ginzburg P, Hayat A, Orenstein M (2007) Nonlinear optics with local phasematching by quantum based meta-material. *J. Opt. A Pure Appl. Opt.* 9: S350.

12. Piredda G, Smith DD, Wendling B, Boyd RW (2008) Nonlinear optical properties of a gold-silica composite with high gold fill fraction and the sign change of its nonlinear absorption coefficient. *J. Opt. Soc. Am. B* 25: 945.

13. Lepeshkin NN, Schweinsberg A, Piredda G, Bennink RS, Boyd RW (2004) Enhanced nonlinear optical response of one-dimensional metal-dielectric photonic crystals. *Phys. Rev. Lett.* 93: 123902.

14. Perner M, Bost P, Lemmer U, von Plessen G, Feldmann J, Becker U, Mennig M, Schmitt M, Schmidt H (1997) Optically induced damping of the surface plasmon resonance in gold colloids. *Phys. Rev. Lett.* 78: 2192.

15. Jiang L, Tsai H (2005) Improved two-temperature model and its application in ultrashort laser heating of metal films. *J. Heat Transfer* 127: 1167.

16. Link S, El-Sayed MA (1999) Spectral properties and relaxation dynamics of surface plasmon electronic oscillations in gold and silver nanodots and nanorods. *J. Phys. Chem. B* 103(40): 8410–8426.

17. Wurtz GA, Pollard R, Hendren W, Wiederrecht GP, Gosztola DJ, Podolskiy VA, Zayats AV (2011) Designed ultrafast optical nonlinearity in a plasmonic nanorod metamaterial enhanced by nonlocality. *Nat. Nanotechnol.* 6: 107–111.

18. Nicholson DR (1983) *Introduction to Plasma Theory*. 2nd ed. John Wiley & Sons.

19. Ginzburg P, Hayat A, Berkovitch N, Orenstein M (2010) Nonlocal ponderomotive nonlinearity in plasmonics. *Opt. Lett.* 35: 1551.

20. Bennemann KH (1998) *Non-linear Optics in Metals*. Clarendon Press.

21. Palomba S, Harutyunyan H, Renger J, Quidant R, van Hulst NF, Novotny L (2011) Nonlinear plasmonics at planar metal surfaces. *Phil. Trans. R. Soc. A* 369: 3497–3509.

22. Bloembergen N, Chang RK, Jha SS, Lee CH (1968) Optical second-harmonic generation in reflection from media with inversion symmetry. *Phys. Rev.* 174: 813–822.

23. Jha SS (1965) Theory of optical harmonic generation at a metal surface. *Phys. Rev.* 140: A2020–A2030.

24. Sipe JE, So VCY, Fukui M, Stegeman GI (1980) Analysis of second-harmonic generation at metal surfaces. *Phys. Rev. B* 21: 4389–4402.

25. Xiang Wang F, Rodríguez FJ, Albers WM, Ahorinta R, Sipe JE, Kauranen M (2009) Surface and bulk contributions to the second-order nonlinear optical response of a gold film. *Phys. Rev. B* 80: 233402.

26. Centini M, Benedetti A, Sibilia C, Bertolotti M, Scalora M (2009) Second harmonic generation from metallic 2D scatterers. *Proc. SPIE* 7354: 73540F.

27. Brown F, Parks RE, Sleeper AM (1965) Nonlinear optical reflection from a metallic boundary. *Phys. Rev. Lett.* 14: 1029–1031.

28. Klein MW, Enkrich C, Wegener M, Linden S (2006) Second-harmonic generation from magnetic metamaterials. *Science* 313: 502–504.

29. Goff JE, Schaich WL (1997) Hydrodynamic theory of photon drag. *Phys. Rev. B* 56: 15421–15430.

30. Gibson AF, Kimmit MF, Walker AC (1970) Photon drag in germanium. *Appl. Phys. Lett.* 17: 75–77.

31. Vengurlekar AS, Ishihara T (2005) Surface plasmon enhanced photon drag in metals. *Appl. Phys. Lett.* 87: 091118.

32. Kurosawa H, Ishihara T (2012) Surface plasmon drag effect in a dielectrically modulated metallic thin film. *Opt. Express* 20: 1561–1574.

33. Durach M, Rusina A, Stockman MI (2009) Giant surface plasmon induced drag effect (SPIDER) in metal nanowires. *Phys. Rev. Lett.* 103 186801-1-4.

34. Varró S, Ehlotzky F (1994) Higher-harmonic generation from a metal surface in a powerful laser field. *Phys. Rev. A* 49: 3106–3109.

35. Farkas Gy, Tóth Cs, Moustaizis SD, Papadogiannis NA, Fotakis C (1992) Observation of multiple-harmonic radiation induced from a gold surface by picosecond neodymium-doped yttrium aluminum garnet laser pulses. *Phys. Rev. A* 46: R3605.

36. Georges AT (1996) Coherent and incoherent multiple-harmonic generation from metal surfaces. *Phys. Rev. A* 54: 2412–2418.

37. Papadogiannis NA, Loukakos PA, Moustaizis SD (1999) Observation of the inversion of second and third harmonic generation efficiencies on a gold surface in the femtosecond regime. *Opt. Commun.* 166: 133–139.

38. Lippitz M, van Dijk MA, Orrit M (2005) Third-harmonic generation from single gold nanoparticles. *Nano Lett.* 5(4): 799–802.

39. Ho JR, Grigoropoulos CP, Humphrey JAC (1995) Computational study of heat transfer and gas dynamics in the pulsed laser evaporation of metals. *J. Appl. Phys.* 78: 4696.

40. Marla D, Bhandarkar UV, Joshi SS (2011) Critical assessment of the issues in the modeling of ablation and plasma expansion processes in the pulsed laser deposition of metals. *J. Appl. Phys.* 109: 021101.

41. Güdde J, Hohlfeld J, Müller JG, Matthias E (1998) Damage threshold dependence on electron-phonon coupling in Au and Ni films. *Appl. Surf. Sci.* 127–129: 40–45.

42. Tribelsky MI, Miroshnichenko AE, Kivshar YS, Luk'yanchuk BS, Khokhlov AR (2011) Laser pulse heating of spherical metal particles. *Phys. Rev. X* 1: 021024.

43. Pronko PP, Dutta SK, Squier J, Rudd JV, Du D, Mourou G (1995) Machining of submicron holes using a femtosecond laser at 800 nm. *Opt. Commun.* 114: 106–110.

44. Mihalache D, Bertolotti M, Sibilia C (1989) Nonlinear wave propagation in planar structures. In: Wolf E, editor. *Progress in Optics*. Vol. 27. Elsevier. Chapter 4.

45. Gramotnev DK, Bozhevolnyi SI (2010) Plasmonics beyond the diffraction limit. *Nat. Photonics* 4: 83–91.

46. Stockman MI (2004) Nanofocusing of optical energy in tapered plasmonic waveguides. *Phys. Rev. Lett.* 93: 137404.

47. Ginzburg P, Arbel D, Orenstein M (2006) Gap plasmon polariton structure for very efficient microscale to nanoscale interfacing. *Opt. Lett.* 31: 3288–3290.

48. Ginzburg P, Orenstein M (2007) Plasmonic transmission lines: from micro to nano scale with $\lambda/4$ impedance matching. *Opt. Express* 15: 6762–6767.

49. Rukhlenko ID, Pannipitiya A, Premaratne M, Agrawal GP (2011) Exact dispersion relation for nonlinear plasmonic waveguides. *Phys. Rev. B* 84: 113409.

50. Feigenbaum E, Orenstein M (2007) Plasmon soliton. *Opt. Lett.* 32: 674–676.

51. Feigenbaum E, Orenstein M (2007) Modeling of complementary (void) plasmon waveguiding. *J. Lightw. Technol.* 25(9): 2547–2562.

52. Davoyan AR, Shadrivov IV, Kivshar YS (2009) Self-focusing and spatial plasmon-polariton solitons. *Opt. Express* 17: 21732–21737.

53. Davoyan AR, Shadrivov IV, Zharov AA, Gramotnev DK, Kivshar YS (2010) Nonlinear nanofocusing in tapered plasmonic waveguides. *Phys. Rev. Lett.* 105: 116804.

54. Marini A, Skryabin DV, Malomed B (2011) Stable spatial plasmon solitons in a dielectric-metal-dielectric geometry with gain and loss. *Opt. Express* 19: 6616–6622.

55. Liu Y, Bartal G, Genov DA, Xiang Z (2007) Subwavelength discrete solitons in nonlinear metamaterials. *Phys. Rev. Lett.* 99: 153901.

56. Renger J, Quidant R, van Hulst N, Palomba S, Novotny L (2009) Free-space excitation of propagating surface plasmon polaritons by nonlinear four-wave mixing. *Phys. Rev. Lett.* 103: 266802.

57. Harutyunyan H, Palomba S, Renger J, Quidant R, Novotny L (2010) Nonlinear dark-field microscopy. *Nano Lett.* 10(12): 5076–5079.

58. Genevet P, Tetienne J, Gatzogiannis E, Blanchard R, Kats MA, Scully MO, Capasso F (2010) Large enhancement of nonlinear optical phenomena by plasmonic nanocavity gratings. *Nano Lett.* 10: 4880–4883.

59. Salgueiro JR, Kivshar YS (2010) Nonlinear plasmonic directional couplers. *Appl. Phys. Lett.* 97: 081106.

60. Cai W, Vasudev AP, Brongersma ML (2011) Electrically controlled nonlinear generation of light with plasmonics. *Science* 333: 1720–1723.

61. Panoiu N-C, Osgood RM, Jr (2004) Subwavelength nonlinear plasmonic nanowire. *Nano Lett.* 4(12): 2427–2430.

62. Maier SA, Kik PG, Atwater HA (2003) Optical pulse propagation in metal nanoparticle chain waveguides. *Phys. Rev. B* 67: 205402.

63. Feigenbaum E, Orenstein M (2008) Ultrasmall volume plasmons, yet with complete retardation effects. *Phys. Rev. Lett.* 101(16): 163902.

64. Purcell EM, Torrey HC, Pound RV (1946) Resonance absorption by nuclear magnetic moments in a solid. *Phys. Rev.* 69: 37.

65. Hayat A, Orenstein M (2007) Standing-wave nonlinear optics in an integrated semiconductor microcavity. *Opt. Lett.* 32: 2864–2866.

66. Banaee MG, Crozier KB (2011) Mixed dimer double-resonance substrates for surface-enhanced Raman spectroscopy. *ACS Nano* 5(1): 307–314.

67. Pu Y, Grange R, Hsieh C, Psaltis D (2010) Nonlinear optical properties of core-shell nanocavities for enhanced second-harmonic generation. *Phys. Rev. Lett.* 104: 207402.

68. Fan W, Zhang S, Panoiu N-C, Abdenour A, Krishna S, Osgood RM, Malloy KJ, Brueck SRJ (2006) Second harmonic generation from a nanopatterned isotropic nonlinear material. *Nano Lett.* 6(5): 1027–1030.

69. Danckwerts M, Novotny L (2007) Optical frequency mixing at coupled gold nanoparticles. *Phys. Rev. Lett.* 98: 026104.

70. Biris CG, Panoiu NC (2011) Excitation of dark plasmonic cavity modes via nonlinearly induced dipoles: applications to near-infrared plasmonic sensing. *Nanotechnology* 22: 235502.

71. Khurgin JB, Sun G, Soref RA (2007) Enhancement of luminescence efficiency using surface plasmon polaritons: figures of merit. *J. Opt. Soc. Am. B* 24: 1968–1980.

72. Hayat A, Ginzburg P, Orenstein M (2008) Observation of two-photon emission from semiconductors. *Nat. Photonics* 2: 238.

73. Nevet A, Berkovitch N, Hayat A, Ginzburg P, Ginzach S, Sorias O, Orenstein M (2010) Plasmonic nanoantennas for broad-band enhancement of two-photon emission from semiconductors. *Nano Lett.* 10: 1848–1852.

74. von der Linde D, Engers T, Jenke G, Agostini P, Grillon G, Nibbering E, Mysyrowicz A, Antonetti A (1995) Generation of high-order harmonics from solid surfaces by intense femtosecond laser pulses. *Phys. Rev. A* 52: R25–R27.

75. Kim S, Jin J, Kim Y, Park I, Kim Y, Kim S (2008) High-harmonic generation by resonant plasmon field enhancement. *Nature* 453: 757–760.

76. Husakou A, Im S-J, Herrmann J (2011) Theory of plasmon-enhanced high-order harmonic generation in the vicinity of metal nanostructures in noble gases. *Phys. Rev. A* 83: 043839.

77. Lombardi JR, Birke RL, Lu T, Xu J (1986) Charge-transfer theory of surface enhanced Raman spectroscopy: Herzberg–Teller contributions. *J. Chem. Phys.* 84: 4174.

78. Kneipp K, Wang Y, Kneipp H, Perelman LT, Itzkan I, Dasari RR, Feld MS (1997) Single molecule detection using surface-enhanced Raman scattering (SERS). *Phys. Rev. Lett.* 78: 1667.

79. Ginzburg P, Hayat A, Orenstein M (2008) Breakdown of surface plasmon enhancement due to ponderomotive forces. *Plasmonics and Metamaterials*, October, Rochester, NY. OSA Technical Digest (CD), Optical Society of America, paper MThC8.

80. Pendry JB (2000) In: Soukoulis CM, editor. *Intense Focusing of Light Using Metals*, NATO ASI Series. Springer http://link.springer.com/chapter/10.1007%2F978-94-010-0738-2_24.

3

Plasmonic Nanorod Metamaterials as a Platform for Active Nanophotonics

GREGORY A. WURTZ, WAYNE DICKSON, AND ANATOLY V. ZAYATS
Department of Physics, King's College London, Strand, London, UK

ANTONY MURPHY AND ROBERT J. POLLARD
Centre for Nanostructured Media, Queen's University of Belfast, Belfast, UK

3.1 INTRODUCTION

Metamaterials consist of ensembles of subwavelength structures with an engineered electromagnetic response, so that their electromagnetic properties are determined not by the properties of the material constituents but rather by the structural geometry [1–3]. In the optical and near-infrared spectral ranges, metallic nanostructures support surface plasmon excitations associated with the free-electron motion near the interface between a metal and a dielectric. This behavior is central in achieving the required resonant properties of "meta-atoms" forming metamaterials. The plasmonic metamaterials are based on various combinations of metallic nanostructures, such as split-ring resonators of different geometries, fishnets, or nanorods. They provide the opportunity to flexibly design optical properties by engineering the optical response of individual components as well as the electromagnetic interaction between them. Metamaterials may exhibit unique properties such as negative permeability and negative refractive index and may be used to design super- and hyper-lenses for diffraction-unlimited high-resolution imaging and cloaking [1–11]. The most important practical advantage of metamaterials in active nanophotonic

Active Plasmonics and Tuneable Plasmonic Metamaterials, First Edition. Edited by Anatoly V. Zayats and Stefan A. Maier.

applications, where one needs to efficiently control optical signals with external stimuli, is in the possibility to readily design the dispersion of plasmonic modes and related active response. This can be electric or magnetic-field-tuned optical properties, mechanically or all-optically tuneable metamaterials, and photonic switches and modulators [12–17]. These enhanced functionalities can be optimized by engineering the required spectral response, the electromagnetic mode localization and field enhancement beneficial for many applications.

Dielectric, semiconductor, and metallic nanowires and nanorods have recently attracted significant attention due to their remarkable optical and electronic properties [18–21]. Among others, these include light guiding in subwavelength diameter metallic wires, negative permeability of pairs of closely spaced metallic rods, and optical plasmonic antennas [22–24]. These effects rely on surface plasmon modes supported by a metallic nanowire or nanorod. Progress in nanofabrication techniques has opened up the possibility to fabricate arranged metallic nanorods in macroscopic size arrays with their long axes aligned perpendicular to the substrate [25]. This geometry of oriented and closely packed nanorods, with a much reduced influence of the underlying substrate, impacts both the spectral optical properties and the spatial field distribution in the array, compared to isolated or weakly interacting nanorods studied in colloidal solutions [26, 27] or placed on substrates [23, 28].

Optical properties of such nanorod arrays featuring deep-subwavelength sizes in lateral direction can be described by effective permittivity parameters. Maxwell-Garnett (MG) calculations of the optical properties of these metamaterials have shown that the strongly anisotropic optical behavior can be beneficial for spectroscopic, sensing, as well as near-field optical imaging applications [29–33]. The important property of nanorod metamaterials is the possibility of wide range tuning of the geometrical parameters of the structure so as to control the electromagnetic interaction between the rods. This can be done, for example, by changing the inter-rod distance and is crucial for determining the reflection, transmission, and extinction of the array, and associated linear and nonlinear properties. This type of anisotropic plasmonic metamaterial exhibits unique optical properties such as hyperbolic dispersion and related strong Purcell effect, high-resolution imaging capabilities, high sensitivity to refractive index variations and biosensing, as well as active optical functionalities including enhanced nonlinear properties [3, 34, 36, 37].

This chapter presents the review of plasmonic metamaterials based on aligned nanorod assemblies. After a brief discussion of the fabrication, microscopic and effective medium description of the nanorod metamaterial will be detailed. The role of hyperbolic dispersion and epsilon-near-zero (ENZ) response will be discussed. We will then focus on the designs of active metamaterials. Nonlinear optical properties arising from the hybridization of plasmonic nanorod metamaterial and nonlinear dielectric as well as due to intrinsic metal nonlinearity will be presented. The role of weak and strong coupling regime of molecular excitons and plasmonic modes of the metamaterial on the optical properties will also be described. Finally, the electro-optical response of nanorod metamaterial hybridized with liquid crystals (LCs) will be discussed.

3.2 NANOROD METAMATERIAL GEOMETRY

The nanorod metamaterials consist of aligned nanorods oriented perpendicularly to a supporting substrate (Fig. 3.1). The nanorods are generally embedded in a dielectric matrix but can also be free-standing in air. Typically, they are grown electrochemically in a substrate-supported, porous, anodized aluminum oxide template (AAO) [25]. The diameter, separation, and ordering of the metallic rods in the assembly are determined by the geometry of the AAO template, which is in turn controlled by the anodization conditions and post-chemical etching processes. The rod diameter and spacing can vary in the range of 20–80 nm with separation between rods in the range of 40–100 nm. The rod diameter and separation are difficult to control independently on as-grown Al films, but they can be completely decoupled using pre-structured Al films [38, 39]. One-step anodization process provides a weakly disordered hexagonal periodic lattice of pores. Perfect hexagonal lattice ordering can be achieved with a two-step anodization process. By pre-structuring the Al film before anodization, arbitrary lattice geometry can be achieved as well [40, 41].

The length of the rods can be chosen by controlling the electrodeposition time and can be varied from about 20 nm to the limiting value dictated by the thickness of the AAO template, typically a few hundred nanometers. All these parameters—material of the rod, rod aspect ratio, rod length, and separation in the array—impact the optical response of the metamaterial. The rod material can be any metal suitable for electrodeposition including but not limited to Au, Ag, Ni, and Co.

In contrast to geometries involving nanorods on substrates or suspended in solutions, the nanorods forming a metamaterial have typically a much higher aspect ratio, are standing on a substrate with their long axes oriented perpendicularly to it, and

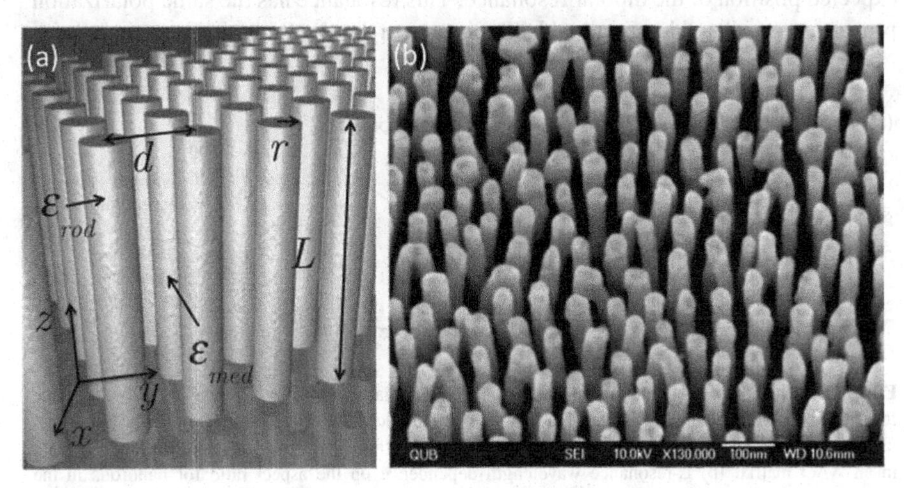

FIGURE 3.1 (a) Schematic representation of the nanorod metamaterial with parameter notations used throughout the chapter. (b) Scanning electron microscopy (SEM) image from a typical nanorod metamaterial made of Au nanorods. The AAO matrix was removed prior to SEM imaging, and the sample tilted with a viewing angle of 45°.

are placed in a quasi-regular lattice. An example of nanorod assembly is shown in Figure 3.1 presenting an SEM image of the assembly of Au rods after removal of the AAO matrix to enable their observation. The metamaterial can be additionally structured to create laterally confined geometries. This can be done using post-growth processing, for example, laser ablation [42], structuring of the underlying electrodes, or controlling AAO pore openings by focused ion beam [38, 39, 43].

3.3 OPTICAL PROPERTIES

3.3.1 Microscopic Description of the Metamaterial Electromagnetic Modes

The optical properties of nanorod assemblies (Fig. 3.2a) are reminiscent of those of individual plasmonic nanorods. For the nanorod metamaterial, plane wave-coupled modes are determined by two dipolar resonances associated with localized surface plasmon (LSP) excitation parallel and perpendicular to the nanorod long axis. The resonance associated with electron motion perpendicular to the nanorod long axis is excited with light having electric field component along this direction and is approximately situated at 520 nm in wavelength. The resonance associated with the electron motion parallel to the long axis is in the infrared for the nanorods with aspect ratios larger than 5 (Fig. 3.2b). For example, an isolated nanorod with a diameter of 30 nm and an aspect ratio of 10 embedded in AAO exhibits a longitudinal dipolar resonance at around 2 μm (Fig. 3.2b). The resonance demonstrating similar polarization properties in the array of nanorods is observed in the visible region of the spectrum at around 650 nm, therefore, strongly blue-shifted compared to the expected position of the dipolar resonance. This resonance has the same polarization properties as isolated longitudinal LSP related to a dipolar excitation associated with

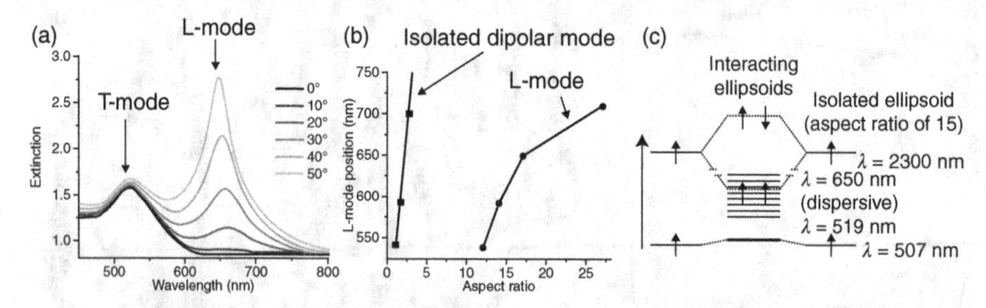

FIGURE 3.2 (a) Optical extinction spectra of the metamaterial measured at different angles of incidence for p-polarized light having an electric field component parallel to the nanorod long axis. The nanorods' length is 300 nm, diameter is 30 nm, their and the inter-rod distance is 60 nm. The nanorods are embedded in an AAO matrix. (b) L-resonance wavelength dependence on the aspect ratio for nanorods in the metamaterial (circles) and isolated ellipsoids (squares). (c) Schematic energy diagram of the plasmonic resonances in the metamaterial relative to the modes of individual nanorods. Strong electromagnetic coupling of the plasmonic resonances supported by individual nanorods leads to the formation of the collective plasmonic modes observed in (a).

a long nanorod axis and will be referred to as L-resonance in the metamaterial. In analogy, the complementary transverse mode of the metamaterial will be referred to as T-resonance as it is excited with an electric field component perpendicular to the nanorod long axis.

The observed behavior of the L-resonance can be explained by the formation of a spatially extended (collective) plasmonic resonance in the nanorod array due to the interaction between the plasmonic modes supported by individual nanorods in the assembly. In a simple consideration, this extended mode can be treated as a linear combination of these modes experiencing sensitive electromagnetic interaction from neighboring nanorods when $k_p l < 2\pi$, where k_p is the field extension of the LSP outside metal and l is the average distance between nanorods in the array. Considering the interaction between two nanorods only, coherent coupling between the longitudinal dipolar modes supported by these rods would result in the formation of two eigenstates representing the symmetric and antisymmetric configurations of the plasmonic dipole moments of the coupled rod system [51]. Including more interacting nanorods will result in the formation of a band of plasmonic states formed in the nanorod assembly.

This interaction is schematically shown in the energy diagram of Figure 3.2c where strong coupling between the nanorods shifts the symmetric state (parallel orientation of the dipole moments of the nanorods) in the short-wavelength spectral range compared to the longitudinal dipolar mode of the isolated nanorods. The antisymmetric state corresponds to the antiparallel arrangement of the dipole moments, is not optically active, and can be excited only in the near field.

The experimental extinction spectra $-\log_{10}(T)$, where T is the transmittance, measured for an array of Au nanorods embedded in an AAO matrix ($n \approx 1.6$) are shown in Figure 3.2a for different angles of incidence. At normal incidence or also with s-polarized light, where the incident electric field vibrates perpendicular to the nanorod long axis, the spectra reveal the T-mode at around 520 nm in wavelength [29]. At oblique incidence and with p-polarized light, where the incident electric field has a component both along and perpendicular to the nanorod long axes, two peaks are observed in the spectra: the above-mentioned T-resonance as well as the L-resonance at a longer wavelength. This long-wavelength peak becomes more pronounced at larger angles of incidence for which the longitudinal LSP is excited more effectively. The angular sensitivity of the spectra of Figure 3.2a reflects the strong anisotropy of the structure defined by the orientation of the nanorods in the metamaterial. The general behavior of these two observed resonances is consistent with the response of the dipolar modes supported by isolated or weakly interacting rods [29].

The L-resonance wavelength depends on both the rod aspect ratio and the distance between the rods in the array. In accordance with the dipolar plasmonic response of nanorods, an increase in the nanorod aspect ratio causes the two resonances to split further apart with the T-resonance undergoing a blue-shift while the L-resonance moves toward longer wavelengths. This behavior is illustrated in Figure 3.3 for either varying rod length (Fig. 3.3a) or varying rod diameter (Fig. 3.3b). The inter-rod distance can also be varied in order to tune the spectral position of the resonances in the array throughout the visible spectrum [45, 46]. Thus, this metamaterial provides

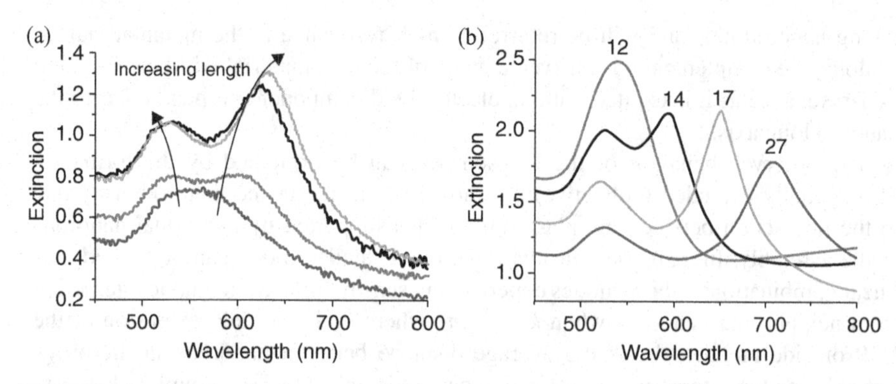

FIGURE 3.3 Optical extinction spectra of the metamaterial as a function of rod aspect ratio. (a) The nanorod length is varied from 100 to 250 nm for an inter-rod distance of 60 nm and a rod diameter of 30 nm. (b) The nanorod length is 400 nm and the inter-rod distance is about 60 nm. The curves are labeled according to the rod's aspect ratio corresponding to rods with diameter ranging from about 15 to 30 nm. The incident light is p-polarized. The angle of incidence is 40°.

excellent opportunities to control plasmonic resonances by controlling geometrical parameters of the nanorod arrays, such as the nanorod aspect ratio (and, thus, LSP wavelength for individual rods) as well as their separation (and, thus, the interaction between nanorods) resulting in the tunability of the metamaterial L-resonance from visible to near-infrared frequencies.

The field distribution in the metamaterial associated with the collective plasmonic L-resonance evolves from LSP fields of individual nanorods (Fig. 3.4). For a larger inter-rod distance of 500 nm, the field distribution around the rods, modeled using finite-element method, is dominated by the dipolar mode of the longitudinal plasmon resonance occurring at a wavelength of around 1950 nm. In this case, the field is

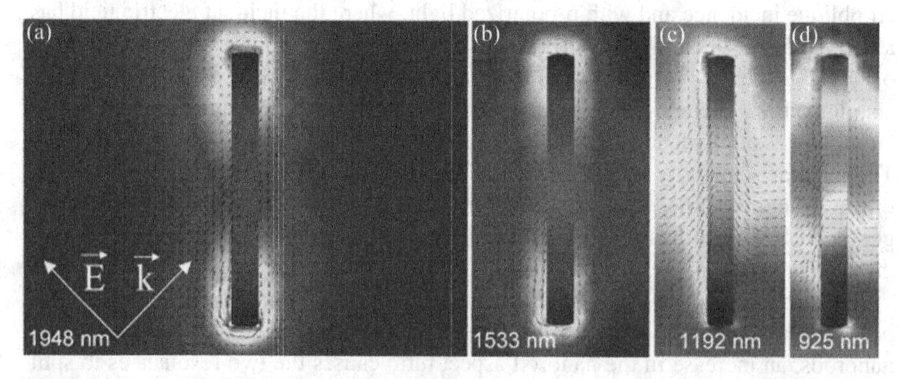

FIGURE 3.4 Electric field distribution in L-resonance plotted in the primitive cell of the nanorod array for different inter-rod distances: (a) 500 nm, (b) 200 nm, (c) 150 nm, and (d) 100 nm, calculated using the finite-element method. The nanorod length and diameter are 300 and 30 nm, respectively. The nanorods are embedded in AAO (refractive index $n = 1.6$) and supported by a glass substrate ($n = 1.5$). The superstrate is air ($n = 1$). The angle of incidence of TM-polarized probe light is 45°. The arrows in (a)–(d) show the direction of the Poynting vector.

localized at the rods' extremities with a small breaking of symmetry due to the presence of the substrate and illumination conditions. The effect of a reduction in the inter-rod distance leads to a blue-shift of the L-resonance, consistent with both the results from the effective medium calculations, detailed in the next section, and experimental observations. With the increase of the inter-rod coupling observed for smaller inter-rod spacings, the spatial distribution of the electric field and the power flow in the array are significantly modified compared to weakly interacting rods. In fact, the electric field in this geometry concentrates within the metamaterial slab, in the middle part of the rods. This evolution corresponds to a simultaneous decrease of the field amplitude at the rods' extremities. This reconfiguration of the spatial distribution of the field in the array is supported by experimental observations that showed the lack of sensitivity in the spectral position of the array's extinction, and the L-resonance position in particular, when the index of refraction of the superstrate is modified, that is, when only the rods' extremities are subjected to a change in refractive index [47].

It is instructive to examine the behavior of the Poynting vector within the assembly as a function of inter-rod coupling strength (Figs. 3.4a–3.4d). Going from the larger inter-rod distances to the strongly coupled regime, the electromagnetic energy flow around the rods evolves from being localized at the rods' extremities to a net energy flux in the layer containing the nanorods in a direction perpendicular to their long axes. Thus, the electromagnetic energy coupled to the array can effectively propagate from nanorod to nanorod within the metamaterial. In the geometry considered in the calculations, the maximum bandwidth of the L-resonance defined as $B_L = 2|\omega_0 - \omega_L|$, where ω_0 and ω_L are the resonant frequencies of the dipolar longitudinal mode of the isolated nanorods and the L-resonance, respectively, is obtained for the smallest inter-rod distance and is estimated to be about 1.4 eV for an inter-rod distance of 100 nm. This corresponds to a maximum group velocity of about 1.7×10^7 m/s if a linear dependence of the group velocity with bandwidth is assumed. The rate at which energy is lost in the assembly would then correspond to 1.2×10^{13} Hz, considering the 100 meV homogeneous broadening of the L-resonance [44]. The damping rate follows a monotonous and decreasing function of the inter-rod coupling strength, supporting the observation that the formation of the L-resonance enables electromagnetic energy to propagate in the layer defined by the nanorod metamaterial.

These properties were also investigated experimentally by gradually tuning the inter-rod coupling strength varying the inter-rod distance from about 170 to about 100 nm. This results in a shift of the L-resonance position from 1.74 to 2.25 eV [46]. Estimating the longitudinal resonance frequency of the isolated rod to be located around 0.54 eV, the effect of decreasing the inter-rod distance from 170 to 100 nm is to increase the bandwidth from 2.4 to 3.4 eV, respectively. Accounting for the bandwidth, the inter-rod distance, as well as the L-resonance linewidth, we can estimate the group velocity and propagation length of electromagnetic energy in the nanorod assembly. These quantities are plotted in Figure 3.5 as a function of inter-rod distance. The behavior of group velocity is quite interesting as it undergoes a nonmonotonous variation as a function of inter-rod coupling strength with a maximum obtained for an inter-rod distance of about 155 nm. In this case, a group velocity of

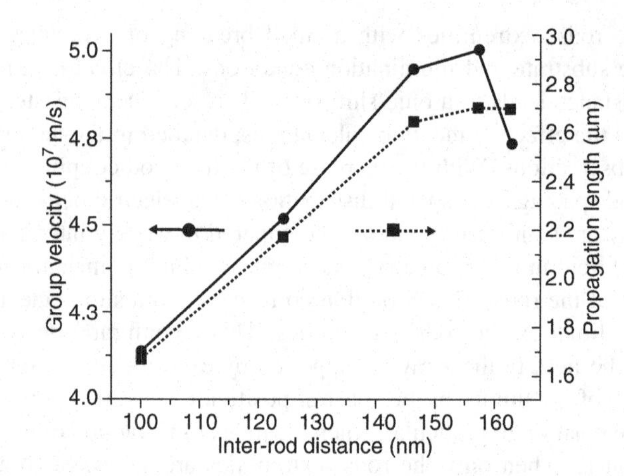

FIGURE 3.5 Group velocity (dots) and propagation length (squares) of electromagnetic energy of the L-resonance as a function of inter-rod distance; a decreasing distance corresponds to an increasing coupling strength.

about 5×10^7 m/s is reached, a value that is comparable to group velocities reported for other plasmonic guiding geometries [48, 49]. The origin of this maximum was associated to an increasing confinement of electromagnetic energy in the metamaterial as near-field interactions become predominant in the system with decreasing inter-rod distances [50].

Defining the inter-rod coupling strength for the L-mode as $J = \hbar|\omega_0 - \omega_L|$, coupling strengths on the order of 1 eV are typically observed in assemblies of nanorods but, more importantly, can be easily varied in order to control the degree of localization of electromagnetic energy within the assembly. For comparison, coupling strengths on the order of 60 to 200 meV, depending on the polarization, have been measured for other plasmonic structures [51, 52], while calculations show that coupling strength comparable to our observations can be reached in more complex plasmonic systems made of chains of interacting plasmonic nanoparticles [53].

The flexible geometry in assemblies of nanorods opens new opportunities to tailor the optical properties of plasmonic resonances both spectrally and spatially. The L-resonance obtained for strongly interacting nanorods shows particular potential in applications associated with guiding electromagnetic energy and optical manipulation applications on the nanometric scale. These applications are relevant since the recent demonstrations of the possibility to produce laterally confined assemblies of nanorods with single nanorod resolution [54], the optical properties of which should allow even further design of the field profiles and enhancement.

3.3.2 Effective Medium Theory of the Nanorod Metamaterial

Several analytical theories have been developed in the past to attempt describing the optical responses of nanoparticles such as nanorods. Mie [55] and Gans [56] solved Maxwell's equations explicitly for spheres and spheroids while others have made use of an effective medium approach such as the MG theory [57, 58]. In this section we will briefly present the results obtained from an effective medium theory (EMT) approach that allows one to accurately investigate the optical properties of an anisotropic medium made of plasmonic nanorods.

The geometry of the nanorod composites considered in the model is shown in Figure 3.1a. The nanorods of permittivity ε_{rod} are embedded into a host material with permittivity ε_{med} and are aligned along the z-direction of a Cartesian coordinate system. The nanorod cross section is assumed circular with the semiaxes r directed along the x- and y-coordinate axes, respectively, with a typical inter-rod separation of d. For an homogeneous metamaterial $rd \ll \lambda_0$, with λ_0 being the free-space wavelength, and the density of nanorod in the medium $N \ll 1$, the local fields in the composite are homogeneous across the rods. Under these conditions the effective permittivities along the nanorod length ε_{zz}^{eff} and diameter ε_{xx}^{eff} can be expressed as [59]

$$\begin{aligned}\varepsilon_{xx}^{eff} &= \varepsilon_{yy}^{eff} = +\frac{N\varepsilon_{rod}\varepsilon_{med} + (1-N)\varepsilon_{med}\left[\varepsilon_{med} + \frac{1}{2}(\varepsilon_{rod} - \varepsilon_{med})\right]}{N\varepsilon_{med} + (1-N)\left[\varepsilon_{med} + \frac{1}{2}(\varepsilon_{rod} - \varepsilon_{med})\right]},\\ \varepsilon_{zz}^{eff} &= N\varepsilon_{rod} + (1-N)\varepsilon_{med},\end{aligned} \tag{3.1}$$

with $N = \pi r^2/d^2$. Thus, the nanorod metamaterial behaves as a uniaxial anisotropic material with the effective permittivity tensor

$$\varepsilon_{eff} = \begin{bmatrix} \varepsilon_{xx} & 0 & 0 \\ 0 & \varepsilon_{yy} & 0 \\ 0 & 0 & \varepsilon_{zz} \end{bmatrix}.$$

Here, the effective permittivity does not depend on nanorod length but only on nanorod density. On the basis of this model, the calculated extinction for a typical nanorod assembly geometry deposited on a glass substrate is shown in Figure 3.6a. It is calculated using transfer matrix method [60] considering the effective material slab with the thickness equal to the nanorod length. Figures 3.6b and 3.6c show typical effective permittivities that can be calculated from the geometrical and material parameters of the nanorods. The model gives excellent quantitative results for both the spectral position and the amplitude of the resonances. However, in addition to the geometrical domain of validity, the model fails to account for nonlocal effects, which can become measurable when losses in the metal decrease. These nonlocal, spatial dispersion effects are important when electromagnetic fields in the nanorod medium fluctuate in space on the length scales on the order of the spatial variations of the local permittivity. This is discussed in more detail in the next section.

The validity domain of the above EMT approach depends significantly on the density of nanorods. In particular it fails to correctly describe the nanorod behavior in

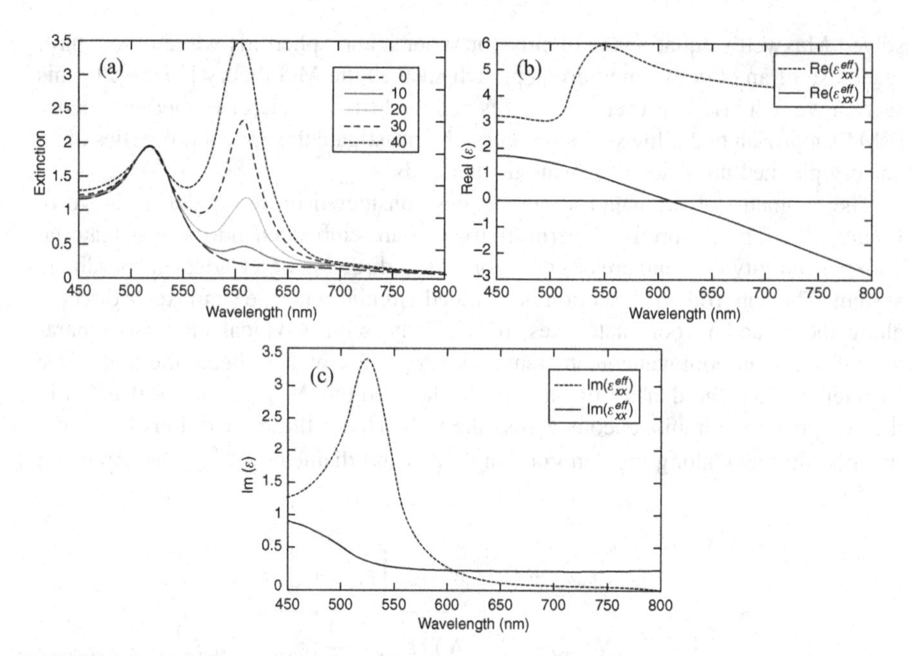

FIGURE 3.6 (a) Extinction spectrum for TM-polarized light calculated using the EMT. The rod length, diameter, and inter-rod distance are 300, 30, and 60 nm, respectively. The embedding medium is AAO. The associated complex-valued effective permittivity along and across the nanorod long axis is plotted in (b) and (c) for the real and imaginary parts, respectively.

the case where the interaction is weak and the optical properties become dependent on the individual nanorod parameters. In this case, EMT can be built from the dipolar response of individual nanorods as has been done by Atkinson *et al* [29]. In this model, the effective permittivity becomes dependent on the nanorod length. The behavior of the ε_{xx}^{eff} component is described by a single Lorentzian resonance related to the plasmonic mode associated with the short axis of the nanorods. This resonance red-shifts with decreased inter-rod distance but, more importantly, its strongly localized nature is weakly dependent on nanorod separation. The spectral behavior of ε_{zz}^{eff} is similar for larger inter-rod distances, when near-field interactions between the rods are negligible ($k_{p}l > 2\pi$). In this regime the rods are weakly interacting and the nature of the resonance is that of an isolated nanorod embedded in an effective medium made of AAO and neighboring nanorods as shown in Figure 3.7. With an increase in the nanorod concentration, the real part of the effective permittivity becomes negative indicating a metallic behavior of the metamaterial as a whole for light polarized along the nanorods' long axes. The frequency at which $\mathrm{Re}(\varepsilon_{xx}^{eff})$ changes sign microscopically corresponds to the excitation of the symmetric modes supported by the nanorod assembly with dipoles in adjacent nanorods excited in phase.

This regime in the anisotropic metamaterial with diagonal components of the permittivity tensor having opposite signs corresponds to the so-called hyperbolic

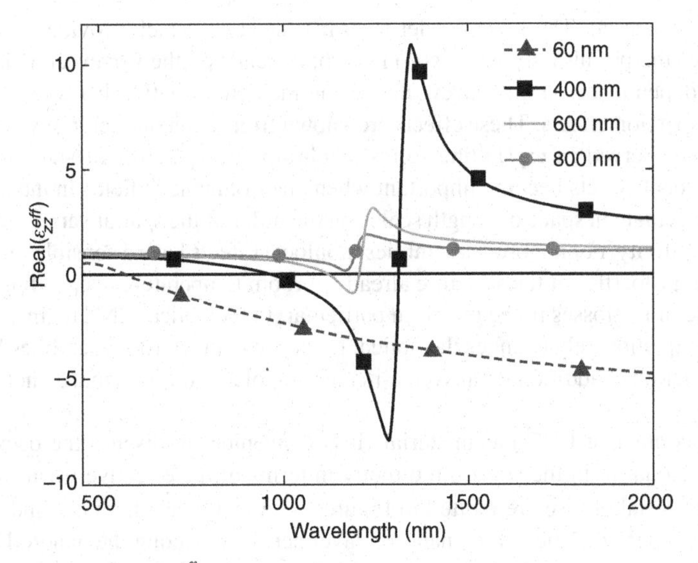

FIGURE 3.7 Spectra of Re(ε_{zz}^{eff}) calculated in the modified Maxwell-Garnet model for assemblies of Au nanorods with different periods.

dispersion regime, and metamaterials with these properties are sometimes called indefinite metamaterials. The consequences of hyperbolic dispersion are very significant for optical properties. In the hyperbolic regime, the metamaterial exhibits very strong Purcell factors leading to significant increase of the spontaneous emission rate due to very high density of the electromagnetic states [35, 61, 62]. In the regime of hyperbolic dispersion, the metamaterial slab behaves as an hyperlens, allowing superfocusing beyond the diffraction limit to be achieved. This is the result of the metamaterial supporting the propagation of electromagnetic modes with very long wavevectors [63].

The effective medium description breaks down and all the above effects smear out when the wavelength of the electromagnetic wave becomes comparable with the distance between the nanorods in the metamaterial. In this regime, the photonic crystal methodology should be applied for accurate description of the optical properties.

3.3.3 Epsilon-Near-Zero Metamaterials and Spatial Dispersion Effects

An ENZ electromagnetic regime occurs when the real part of the permittivity of the material vanishes or approaches zero. In the case of hyperbolic metamaterials, this takes place near the frequency where the transition from hyperbolic to elliptical dispersion regime occurs. In this situation, the real part of ε_{zz}^{eff} is close to zero. In order to describe the optical properties of the material in this ENZ regime, the higher order contributions to the permittivity series, which generally depend on the frequency and wavevector $\varepsilon(\vec{k}, \omega)$, should be taken into account to understand the optical properties of the metamaterial. The lowest order term in the series, which provides a nonzero real part of the permittivity when $\varepsilon(\omega)$ is zero, depends on the wavevector. The permittivity dependence on the wavevector of the incident radiation leads to nonlocal, spatial dispersion effects. These effects are known from semiconductor physics when they were observed in crystalline solids at ultralow temperatures [64, 65]. Practically, nonlocal effects become important when electromagnetic fields in the nanorod medium fluctuate in space on length scales on the order of the spatial variations of the local permittivity. For nanorod assemblies, nonlocal effects have a sensible impact on the optical properties of the structure already at room temperatures depending on the magnitude of the losses in the metal. In particular, in the optical ENZ regime, spatial dispersion qualitatively changes the optical properties of nanorod assemblies leading to the excitation of additional transverse-magnetic-polarized waves that do not exist in local EMT.

In the context of ENZ metamaterials it is convenient to discuss the occurrence of the L-resonance in the spectrum directly in terms of the effective permittivity of the metamaterial. These are plotted in Figures 3.6b and 3.6c where ε_{xx}^{eff} and ε_{zz}^{eff} are the complex permittivities of the nanorod layer across and along the nanorod length, respectively. In this description the L-mode corresponds to $\mathrm{Re}(\varepsilon_{zz}^{eff}) \to 0$. Both the oscillator strength and the width of the L-resonance are limited by losses in material components.

In metamaterials, as well as in crystalline media, spatial dispersion represents in general a relatively weak correction to the effective permittivity $\varepsilon_{zz}(\omega) = \varepsilon_{zz}^{0}(\omega) + \delta_{\varepsilon}(\vec{k})$ with $|\delta_{\varepsilon}(\vec{k})| \ll 1$, an effect that can no longer be neglected in the proximity to the ENZ condition where $\varepsilon_{zz}^{0}(\omega) = 0$. The presence of strong nonlocal terms dramatically changes the optical properties of materials and often leads to the appearance of the additional transverse or longitudinal waves [65].

The nanorod metamaterial layer can support the main wave and an additional transverse magnetic wave [34]. The main wave has a relatively smooth electric field profile and is responsible for the extinction observed in the local regime. In the same frequency range, the additional wave has a strongly nonuniform field distribution and can only weakly couple to incident plane waves. The strong interaction between these modes in the vicinity of the L-resonance leads to a substantial field mixing. As a result of this process, the profiles of both main and additional waves become almost uniform, and both waves contribute to the optical response of the structure, giving rise to two resonances in the optical extinction of the structure. These additional waves, present in most low-loss ENZ metamaterials, represent a new information channel in these systems and therefore may lead to new exciting applications of ENZ metamaterials [34].

These waves can be observed in nanorod metamaterials in the limit of small losses, as the nonlocal spatial dispersion effects are very sensitive to the presence of losses and cannot be detected if the loss is significant. In contrast to thin-film-based plasmonic metamaterials, the nanorods obtained through electrochemical means have poorer Au quality. Thus, the electron motion is limited to a path length on the order of a few nanometers. The effect can be quantitatively described by a restricted mean free

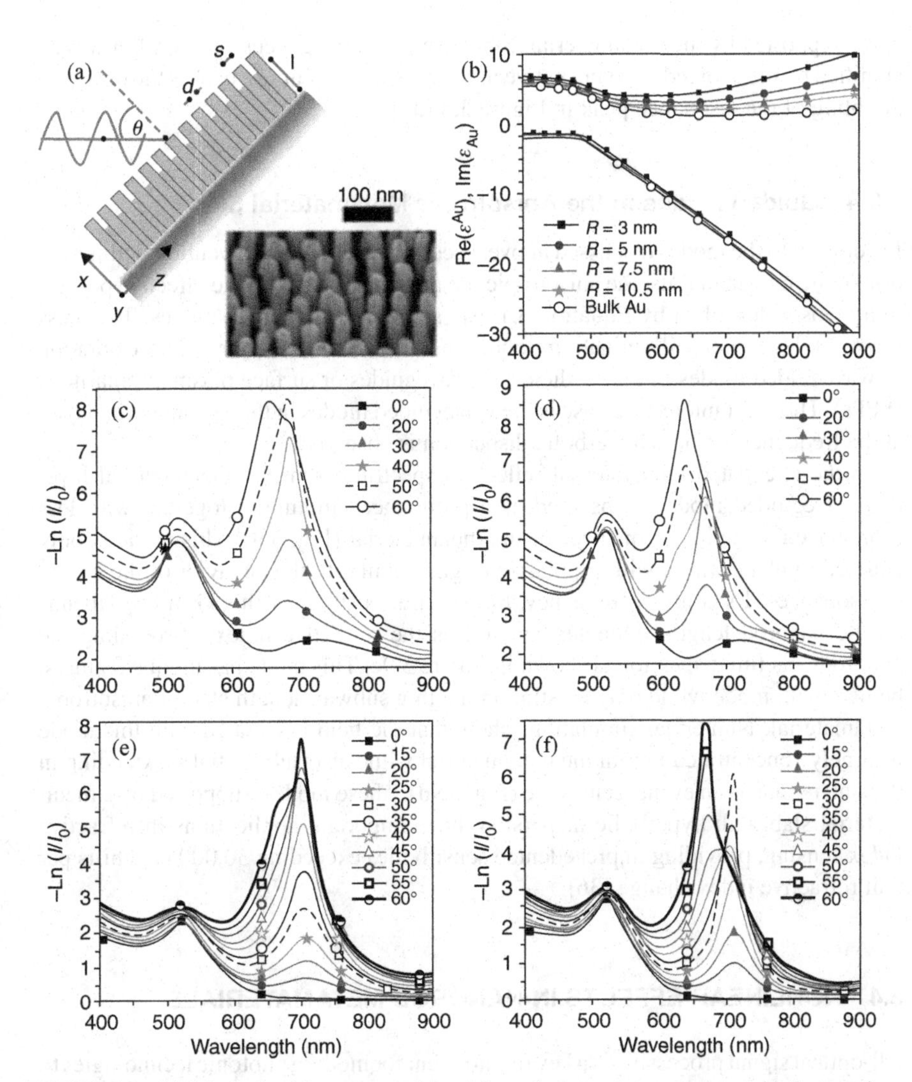

FIGURE 3.8 (a) Schematic of the metamaterial and the optical measurements along with the scanning electron microscopy image of the metamaterial after removal of the anodized aluminum oxide matrix. (b) Spectra of Au permittivity calculated for different values of parameter R. (c and d) Extinction spectra measured with p-polarized incident light for different angles of incidence with the nanorod assemblies (c) unannealed and (d) annealed at 300°C for 2 h. (e and f) Extinction spectra derived from full vectorial 3D simulations for the gold parameters corresponding to $R = 5$ nm (e) and $R = 10$ nm (f).

path R ($R \leq R_b$ where $R_b = 35.7$ nm in bulk gold) of the electrons on the permittivity of gold [34]. For values typical of R, which are on the order of 3 nm, local effects are predominant, and the optical extinction of the metamaterial presents a single L-resonance. However, as R is increased—this is achieved by way of thermal annealing of the nanorods—a splitting in the L-resonance, similar to that of Figure 3.8, is observed [29]. The origin of this splitting is the presence of a second TM-polarized wave supported by the metamaterial. Strong interaction between the two TM waves, signified by the avoided crossing between their dispersion curves, causes the observed doubling of the resonance peak in Figure 3.8 [34].

3.3.4 Guided Modes in the Anisotropic Metamaterial Slab

In addition to the modes discussed above, accessible under the conventional illumination of the metamaterials, the anisotropic metamaterial slab with the effective permittivity tensor described by Equation 3.1 also supports waveguided modes. To access these modes, special illumination conditions are needed, similar to the excitation of waveguided modes in planar dielectric waveguides or surface plasmon polaritons (SPPs). The most interesting case is the waveguided modes in the metamaterial layer, at the frequencies where hyperbolic dispersion is observed.

In the attenuated total internal reflection spectrum from the metamaterial layer this waveguided mode is observed as a pronounced minimum together with less pronounced T- and L-resonances of the metamaterial (Fig. 3.9). This mode is only coupled to plane waves with p-polarized light. Similar to the behavior of the T- and L-resonances, for a given frequency, the spectral position of this G-mode depends on the nanorod length, diameter, as well as the refractive index surrounding the nanorods. Its full dispersion is shown in Figure 3.9c. This mode has unique features, being a bulk mode while only existing in the thin subwavelength slab of anisotropic metamaterial. Numerical simulations show that the field associated with this mode is largely concentrated within the metamaterial while also substantially extending in the superstrate as an evanescent wave (Fig. 3.9d). These modes, supported by a metamaterial slab with hyperbolic dispersion, have important applications in refractive index sensing, providing unprecedented sensitivities exceeding 30,000 nm shifts per unit refractive index change [36].

3.4 NONLINEAR EFFECTS IN NANOROD METAMATERIALS

All-optical signal processing is a key requirement for modern photonic technologies to progress toward signal modulation and transmission speeds unsuitable for electronic handling [66]. Practical applications related to all-optical information processing are, however, severely limited by the inherently weak nonlinear effects in conventional materials that govern photon–photon interaction and diminish further as the switching speed increases. The increased photon–photon interaction and, thus, nonlinear optical response of nanostructured systems can be facilitated by the use of plasmonic metals [67]. It provides strong enhancement of the electromagnetic field, crucial for the observation of nonlinear interactions, and enhanced sensitivity to the adjacent dielectric medium and shows a temporal behavior of the optical properties at ultrafast timescales ranging from tens of femtoseconds to a few picoseconds depending on the electron plasma relaxation processes involved [68–70]. These characteristics make

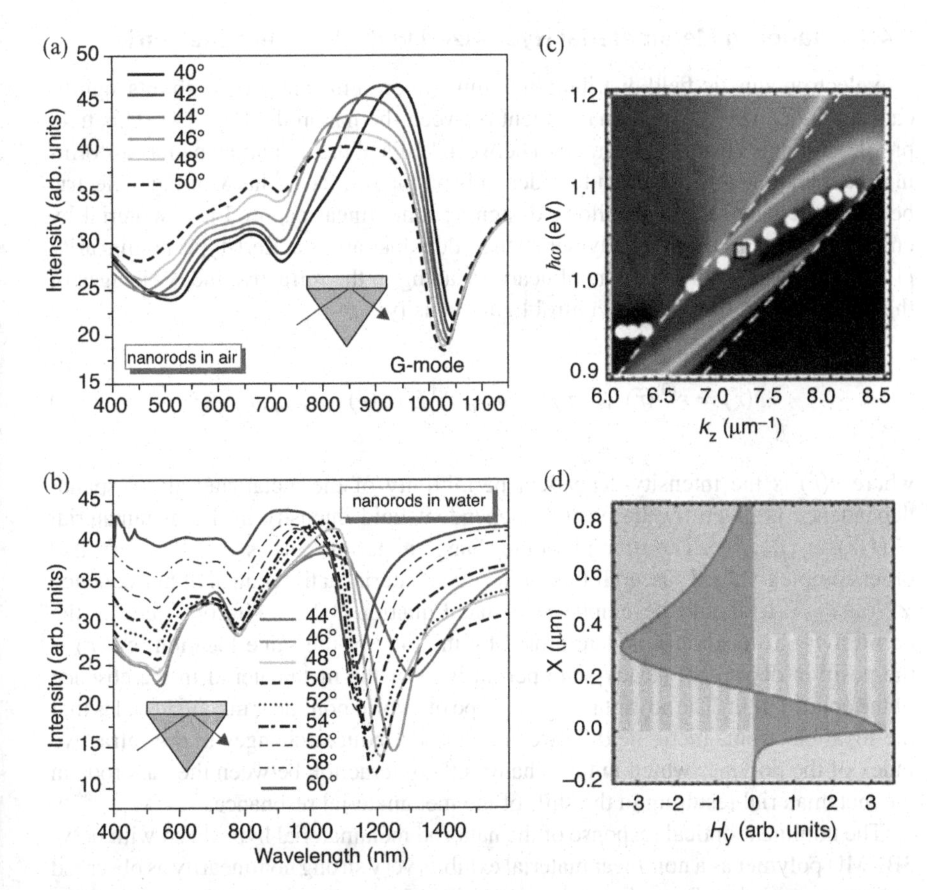

FIGURE 3.9 Reflection spectra of the metamaterials with the nanorods in (a) air ($n = 1$) and (b) water environment ($n = 1.33$) obtained in the ATR geometry with p-polarized light for different angles of incidence. The nanorod parameters are 380 nm length, 25 nm diameter, and 60 nm spacing between the rods (center-to-center). (c) Dispersion of the guided mode in the metamaterial/water system. Background: numerical simulations; circles: experimental dispersion; dashed lines: light lines in air (top left) and the substrate (bottom right). (d) Calculated electromagnetic field distribution of the guided mode in the metamaterial layer. The profile corresponds to the position indicated by the square in (c).

plasmonic structures very promising for ultrafast all-optical applications especially at low light intensities.

In this section, we present an overview of the characteristic nonlinear optical properties of nanorod metamaterials. We will start by considering the nonlinearities provided by a nonlinear dielectric incorporated within nanorods. Then, we will discuss the nonlinear effects due to metal nonlinearity, which can be enhanced by appropriate metamaterial design to achieve the ENZ regime. Both the ultrafast time response and the optical nonlinearity are enhanced due to the optically induced modification of the nonlocal behavior of the metamaterial.

3.4.1 Nanorod Metamaterial Hybridized with Nonlinear Dielectric

The electromagnetic field distribution within nanorod metamaterial exhibits significant field localization and enhancement between the nanorods (Fig. 3.4). This field profile originates from the interaction between LSPs of the individual nanorods forming the metamaterial, which in turn depends on the refractive index of the dielectric between the nanorods. This allows designing a nonlinear nanorod metamaterial by embedding nanorods in a polymer matrix demonstrating a Kerr-type nonlinearity [13, 45]. This is a third-order nonlinearity leading to the refractive index changes of the material with the use of a control light intensity:

$$\varepsilon(\vec{r}) = \varepsilon^{(0)}(\vec{r}) + 4\pi\chi^{(3)}(\omega_{pump})\left|\vec{E}\left(\varepsilon\left(\vec{r}\right), \omega_{pump}, \vec{r}\right)\right|^2, \qquad (3.2)$$

where $\varepsilon(\vec{r})$ is the intensity-dependent permittivity of the metamaterial; its spatial dependence is given by the position-dependent field intensity in the metamaterial $|\vec{E}(\varepsilon(\vec{r}), \omega_{pump}, \vec{r})|^2$. The strength of the change in permittivity is based on the third-order complex-valued susceptibility of the metamaterial at the control light frequency $\chi^{(3)}(\omega_{pump})$. It should be emphasized that Equation 3.2 is self-consistent as the permittivity at location $\vec{r}$ is determined by the field at the same location. $\varepsilon^{(0)}(\vec{r})$ is the position-dependent steady-state permittivity of the metamaterial in the absence of the control light. The advantage of this type of hybrid nonlinear metamaterial is that the local electromagnetic field between nanorods induces changes in the refractive index of the polymer which in turn changes the interaction between the nanorods in the metamaterial resulting in the shift of the metamaterial resonance.

The nonlinear optical response of the nanorod metamaterial hybridized with poly-3BCMU polymer as a nonlinear material exhibits very strong nonlinearity as observed in a pump–probe configuration. A laser light at 515 nm with a 10 ps pulse duration was used as the control light, therefore, interacting with the metamaterial via the transverse plasmonic resonance. A white light continuum was used as a signal beam across the visible spectrum. While the transverse resonance of the metamaterial provides a relatively modest field enhancement compared to the longitudinal resonance, the use of such a control light wavelength is required to induce sensible nonlinearities in the polymer [71,72].

With increasing control light intensity, the optical response of the nonlinear metamaterial exhibits a complex behavior (Fig. 3.10a) related to an interplay between the effects related to the intensity-dependent changes of the refractive index of the nanorod's surroundings and the structural modifications of the polymer under the control light illumination due to photopolymerization and/or photobleaching. The latter also leads to refractive index changes of the polymer matrix but while the former effect is reversible with respect to the variation of the control light intensity, the latter process is an irreversible change. The resultant effects of the inhomogeneous intensity distribution of the control field and the related spatial distribution of the induced refractive index profile result in a complex asymmetric change in the extinction peak associated with changes in both the real and imaginary parts of the

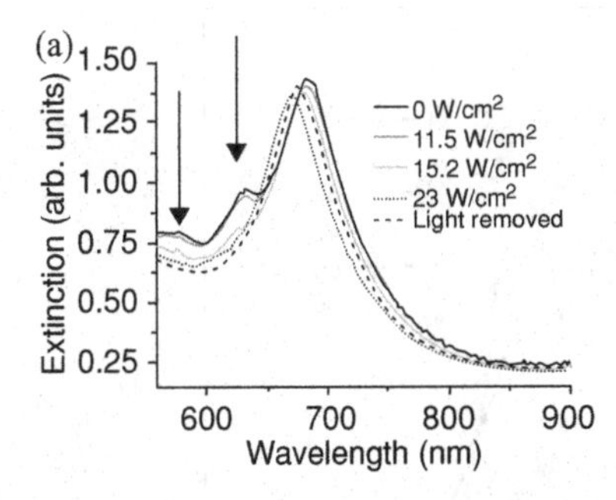

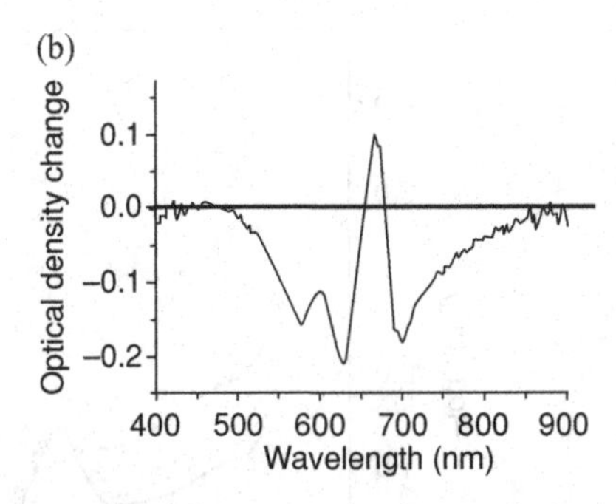

FIGURE 3.10 (a) Extinction spectra of the metamaterial with nanorods embedded in a nonlinear polymer after exposure to 515 nm control light of different intensities. (b) The change in the extinction spectrum after irradiation with the 515 nm light plotted from (a). The nanorod length, diameter, and separation are 300, 20, and 40 nm, respectively. The arrows indicate the bleaching of the absorption of the polymer.

refractive index of the polymer. A broadband nonlinear response, not related to irreversible polymer modifications, is observed in the spectral range from 625 to 800 nm (Fig. 3.10b). The associated tunability range of the L-resonance corresponds to the overall 20 nm wavelength shift with the 10 nm reversible range. The photochromic nonreversible effects in this type of nonlinear metamaterials can be useful for optical data storage applications, while reversible Kerr-type nonlinearities are required for nonlinear photonic applications involving switches and modulators.

3.4.2 Intrinsic Metal Nonlinearity of Nanorod Metamaterials

Metals exhibit intrinsic nonlinearities associated with the free-electron response determining their optical properties. Within the Drude–Lorentz approximation, the optical properties of Au originate from contributions of both free electrons in the conduction band and interband electronic transitions from the d-band to the conduction band [69, 70]. More subtle effects of nonlocality and quantum pressure can be taken into account in rigorous quantum mechanical treatment [73]. Nonlinear effects originate from the modification of the electron temperature and thus redistribution of the electrons in the conduction band [69, 74–76]. This nonlinear effect provides very fast switching times limited by the relaxation of the free electrons [77, 78]. While this type of metal nonlinearities has been studied for smooth metal films, nanoparticles, and nanostructures, including split-ring resonator-based metamaterials [79, 80], the use of nanorod metamaterials exhibiting ENZ regime allows to enhance the nonlinear response [37]. The nonlinearity occurs at the sub-picosecond timescale and with relatively weak peak pump intensity on the order of 10 GW/cm^2 resulting in changes in the optical density of up to $\Delta OD/OD = 0.44$, a significant increase over the generally observed values of $\Delta OD/OD \sim 0.1$ for low-concentration, non-interacting Au nanorods and smooth Au films [81–83].

The linear optical response of the plasmonic nanorod metamaterials used for intrinsic metal nonlinearities studies is shown in Figure 3.11. The transient extinction

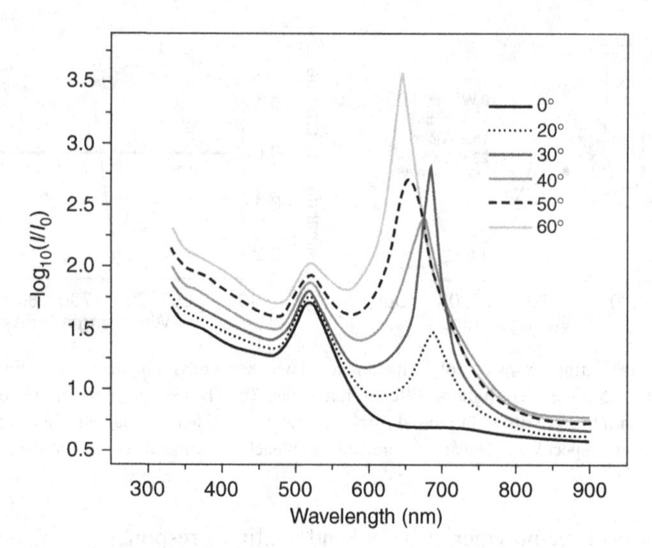

FIGURE 3.11 Angle-dependent linear extinction of the metamaterial. The rods are embedded in AAO and have an average length, diameter, and center-to-center spacing of 400, 20, and 70 nm, respectively. The illuminating light is TM-polarized.

spectrum $\Delta OD = -\log(T_{ON}/T_{OFF})$ for the TM-polarized probe light, with T_{ON} being the transmission of the sample at a finite delay time τ following excitation by the pump beam and T_{OFF} being the transmission of the sample in the ground state, that is, when $\tau \rightarrow \infty$, is mapped in Figure 3.12a for an angle of incidence AOI = 20°. Several cross sections of the transient response map are shown in Figure 3.12b as a function of time delay. In Figure 3.12c the angular dependence of the transient spectra is presented for the pump–probe time delay τ = 600 fs. The control light (λ = 465 nm, $\Delta\tau$ = 130 fs) has a peak intensity of about 1 GW/cm^2. This is well within the low excitation regime where the pump energy is mostly transferred to the electronic excitations of the rods and then coupled to available electron and phonon modes within the rods and to the dielectric environment [75, 81]. Remarkably, the ΔOD of 0.7 observed at 10 GW/cm^2 control intensity corresponds to an 80% change in transmission with a 130 fs pulse-width-limited rise time and a recovery time constant of less than 2 ps.

The dynamics of the spectral response observed in Figure 3.12a is specific to the plasmonic resonances of the assembly. While the behavior of the T-resonance is typical of a localized mode for which the plasmon resonance overlaps interband transitions, the behavior of the L-resonance is more complex. As a general trend, this resonance demonstrates a dispersive-type behavior resulting from modifications in the L-resonance width and spectral position, both of which are also typical of isolated plasmonic resonances in the absence of strong damping due to interband scattering [84]. However, while the latter would not show any angular dependency of the dispersive behavior, the L-resonance shows drastic changes in its transient spectral response with AOI as shown in Figure 3.12c. The divergence from the

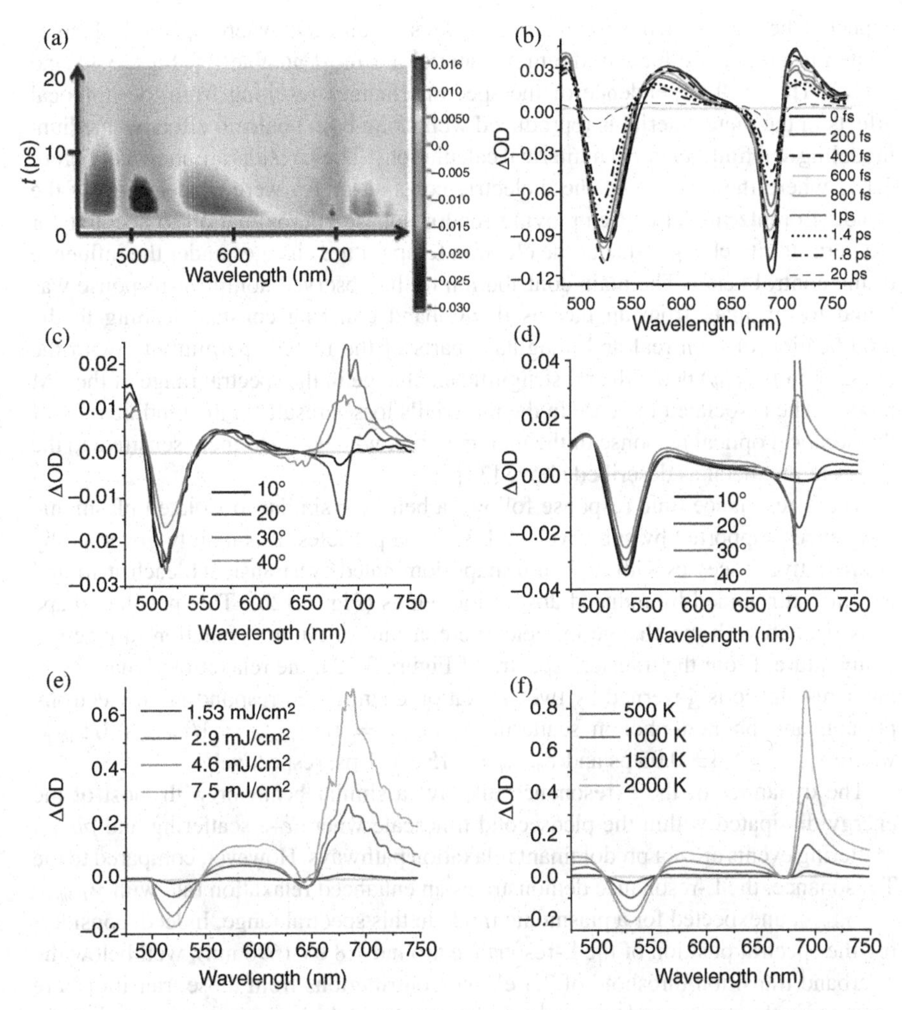

FIGURE 3.12 (a) Transient extinction map measured for the metamaterial in Figure 3.11 for TM-polarized probe light at an angle of incidence of 20°.The Pump wavelength is 465 nm; the fluence is 0.7 mJ/cm^2. (b) Transient extinction spectra at various time delays between pump and probe beams. Spectrum "0 fs" corresponds to the spectrum of the metamaterial measured directly after optical pumping. The other transient spectra correspond to increasing pump–probe time delays τ. (c) Transient extinction spectra at $\tau = 600$ fs measured for different angles of incidence. (d) Numerically modeled transient extinction spectra corresponding to the experimental situation in (c) with an electron temperature increase to 500 K following control light absorption, corresponding to $\Delta\text{Im}(\varepsilon)/\text{Im}(\varepsilon) \approx 0.01$; $\Delta\text{Re}(\varepsilon)/\text{Re}(\varepsilon)$ is negligible far from the interband transition. (e) Transient extinction measured for different pump fluences. (f) Transient extinction calculated for different electron temperatures. In both (e) and (f) the angle of incidence is 40° and the probe light is TM-polarized. Model spectra were calculated without any fitting parameters and do not account for the angular divergence of the probe beam and possible deviations in the angle of incidence.

expected behavior occurs for all time delays at an AOI where spectral changes related to nonlocal effects in the metamaterial are most notable (cf. Figs. 3.11 and 3.12c). This AOI dependence of the spectral changes resulting from the nonlocal effects in the metamaterial is reproduced well using both nonlocal effective medium modeling and finite-element numerical calculations. These results are shown in Figure 3.12d where the changes of the dielectric constant of Au were modeled using the Drude–Lorentz model and employing random phase approximation to account for the permittivity changes due to the electron temperature changes under the influence of the control beam. The main contribution to the observed nonlinear response was found to originate from an increased intraband damping constant leading to the modification of both real and imaginary parts of the metal's permittivity, with the increase in $\mathrm{Im}(\varepsilon_{Au})$ being the most significant change in the spectral range of the TM modes. The associated increase in the material's losses results in the modification of the nonlocal optical response of the metamaterial, which is extremely sensitive to the losses in the metal as described above [34].

The T-resonance time response follows a behavior similar to isolated plasmonic resonances supported by subwavelength-size Au particles. It is mainly governed by nonradiative processes with a spectral shape dominated by a transient bleach at around 515 nm surrounded by induced absorption wings (Fig. 3.12b). This peculiar shape is associated with the change in dielectric constant of Au as a function of electron temperature. From the transient spectra of Figure 3.12a, the relaxation dynamics of the T-resonance is governed by two relaxation channels corresponding to electron–phonon and phonon–phonon scattering: $\gamma_{T\text{-}mode} = 1/\tau_{T\text{-}mode} = 1/\tau_{e\text{–}ph} + 1/\tau_{ph\text{–}ph}$ where $\tau_{e\text{–}ph} = 1.6 \pm 0.05$ ps and $\tau_{ph\text{–}ph} = 115 \pm 0.1$ ps, respectively.

The dynamics of the L-resonance displays a similar behavior with most of the energy dissipated within the picosecond timescale where *e–e* scattering and *ph–ph* scattering events are not predominant relaxation pathways. However, compared to the T-resonance, the L-resonance demonstrates an enhanced relaxation rate with $\gamma_{L\text{-}mode} > \gamma_{T\text{-}mode}$, unexpected for a plasmonic mode in this spectral range. Indeed, considering the spectral position of the L-resonance around 1.8 eV (690 nm), well below the interband transition threshold of 2.4 eV, no contributions from those transitions are expected in the picosecond timescale. The dynamics of the L-resonance in the picosecond range must, therefore, comprise an additional relaxation channel identified as the waveguided mode and can be expressed as $\tau_{L\text{-}mode} = (\gamma_{L\text{-}mode})^{-1} = (\gamma_{e\text{–}ph} + \gamma_{Wg}.)^{-1}$ [34]. $\tau_{L\text{-}mode}$ was obtained from the spectra in Figure 3.2a for wavelengths spanning the resonance width. $\tau_{L\text{-}mode}$ values ranging from 0.76 ± 0.05 to 1.13 ± 0.05 ps were measured for different wavelengths around the resonance, resulting in a nonuniform modification of the resonance shape with delay time. A reduction in pump fluence to 0.7 mJ/cm^2 leads to a faster ~700 fs relaxation time due to a smaller increase of the electron temperature.

With their unusually strong intrinsic nonlinear optical response, nanorod metamaterials provide a new class of optical media for ultrafast strongly nonlinear processes, with numerous applications in optical communications, extraordinarily sensitive optical spectroscopies, and subwavelength imaging technologies. Coupled plasmonic nanoparticles have also been proposed as a means to propagate energy through linear chains of closely coupled nanoparticle chains, while still maintaining subwavelength lateral width. In this context, the nonlinear response of the delocalized L-resonance may provide ground for the comprehensive design of the next-generation nanophotonic circuitry. In all these applications the ability to switch and modulate optical properties, or indeed tune the properties of the nanostructures with the control optical beam, is of great importance.

3.5 MOLECULAR PLASMONICS IN METAMATERIALS

Controlling coherent electromagnetic interactions in molecular systems is a problem of both fundamental interest and important applicative potential in the development of photonic and optoelectronic devices. The strength of these interactions determines both the absorption and emission properties of molecules hybridized with nanostructures, effectively governing the optical properties of such a composite metamaterial. Here, we discuss strong coupling between a plasmonic mode of metamaterial and a molecular exciton. The coupling can be easily engineered and is deterministic as both spatial and spectral overlaps between the plasmonic structure and molecular aggregates are controlled [47]. These results, in conjunction with the flexible geometry of the assembly of gold nanorods, are of potential significance to the development of plasmonic molecular devices.

J-aggregates support excitonic states corresponding to electrically neutral electrons/holes pairs created by the absorption of photons. In photosynthetic processes, these states are used by plants to collect, store, and guide energy to the reaction center for energy conversion and could therefore drive artificial devices on identical principles. Excitonic states also show very strong nonlinear optical behavior that could be used to produce stimulated sources of photons and transistor-like action. Their coupling with plasmons appears as a particularly attractive approach to creating low-powered optical devices having demonstrated the possibility of using the strong scattering cross section and associated enhanced field concentrations of plasmons to generate stimulated emission of excitons of J-aggregates with very low excitation powers [85]. In addition, the coupling of J-aggregates to plasmons-supporting geometries presents genuine fundamental interest in the creation of mixed plasmon/exciton states [85–87]. These quasiparticles can be created if the energy of the excitonic mode is resonant with a plasmonic transition. Either weak or strong coupling can then be achieved depending on whether the transition probabilities associated to the system's eigenmodes are still governed by the Fermi golden rule. Tuning the coupling strength is therefore of particular interest to tailor the absorption and emission characteristics of the molecular semiconductor. Strong coupling was observed in different systems including microcavities where the quasiparticle comprises a cavity photon and a molecular exciton [88, 89] in plasmonic systems involving an SPP and a waveguide photon[90], and more recently between an SPP and an exciton [86, 91]. The flexible geometry of the nanorod metamaterial can be used to control the positioning of J-aggregates in the plasmonic structure as well as control the mixing of the hybrid system's eigenstates, offering a unique possibility in the designing of molecular plasmonic nanodevices with tailored optoelectronic functionalities.

The extinction spectra for both the assembly of nanorods and the J-aggregate when formed on a smooth Au film are shown in Figure 3.13. The metamaterial mode spectrum is dominated by the T- and L-resonances at 500 and 750 nm as discussed earlier. Because of the peculiar electric field distribution of the L-mode showing a maximum in the inter-rod spacing rather than around the rods' extremity, coupling of the excitonic state from the molecular aggregate to either or both the plasmonic modes of the assembly then requires creating a shell around the rods in which to introduce the dye. This opens up unique opportunities to controllably couple plasmonic structures with their environment. Indeed, the thickness of the shell represents an additional geometrical parameter allowing one to tune the optical properties of the plasmonic structure over the [500–900] nm range, while it simultaneously offers spatial selectivity to couple the plasmonic resonances with active molecules. Figures 3.14a and 3.14b show the extinction of the coupled system for two different shell thicknesses of 2.5 and 20 nm, respectively. Varying the shell thickness effectively tunes the plasmonic resonances with no measurable effect on the excitonic transition energy located around the wavelength of 622 nm. The shell thickness therefore can be used to control the spectral overlap between the plasmonic modes of the assembly of nanorods and the excitonic state of the J-aggregate.

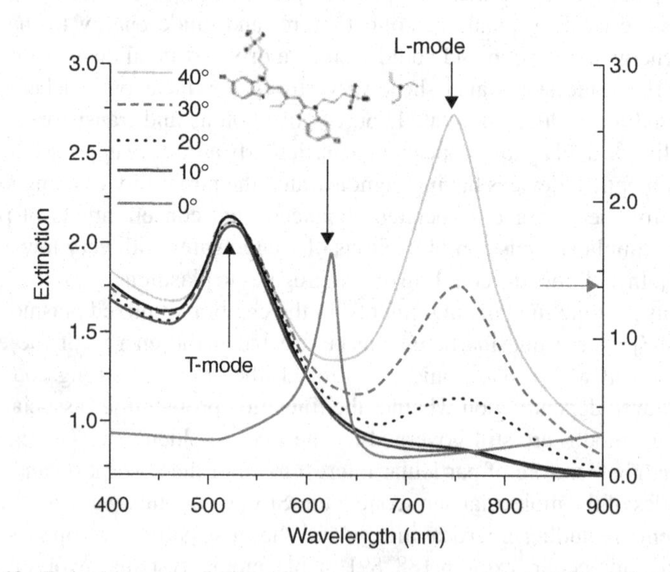

FIGURE 3.13 Extinction spectrum of the metamaterial with Au nanorods in AAO as a function of angle of incidence. The inset shows the molecular structure of the dye used to hybridize the metamaterial along with its extinction spectrum when J-aggregated on a 50 nm thick smooth Au film magnetron sputtered on a glass substrate. The rod's diameter, spacing, and length are 16, 50, and 250 nm, respectively.

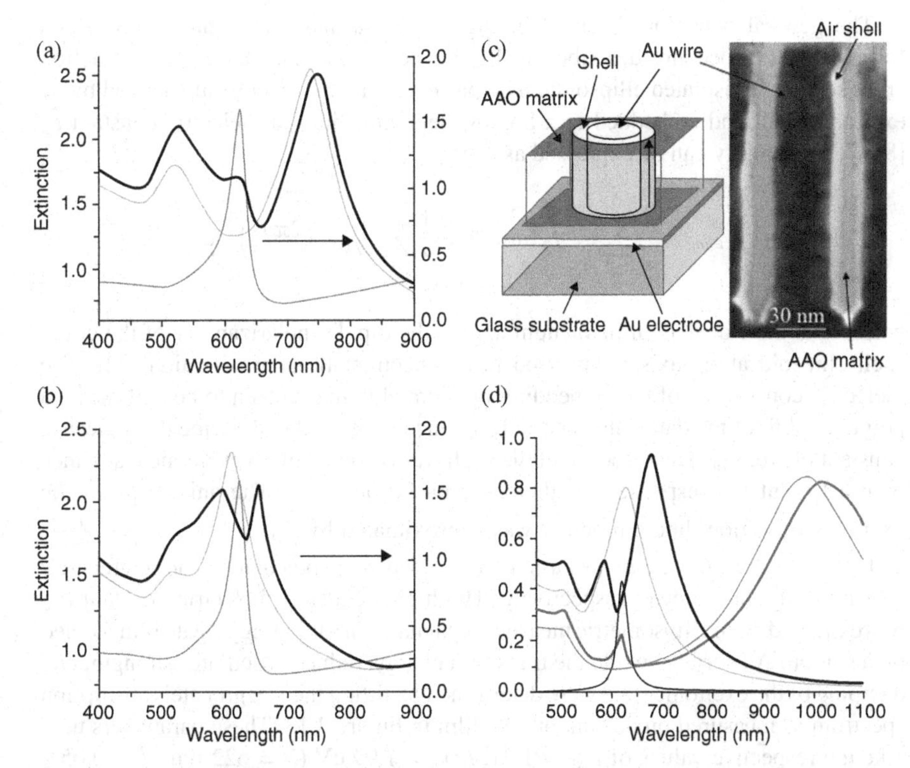

FIGURE 3.14 (a) Extinction spectrum of the active metamaterial with incorporated dye in the weak plasmon–exciton coupling regime along with the extinction of the metamaterial before dye incorporation and the extinction of the J-aggregate on an Au film. The shell thickness is 2.5 nm. (b) Extinction spectrum of the active metamaterial in the strong plasmon–exciton coupling regime along with the extinction of the passive metamaterial and the extinction of the J-aggregate on an Au film. The shell thickness is 20 nm. The spectra are taken for an angle of incidence of 40°. (c) TEM cross section of the core–shell structure and its schematics. The arrow shows the orientation of the molecular transition dipole moment with respect to the long axis of the Au nanorod when introduced in the shell. (d) Extinction from a coated Au ellipsoid calculated in the dipolar limit, illustrating the general behavior observed experimentally in (a) and (b). The thin lines correspond to the individual systems, while the thick lines describe the response of the hybrid system.

From Figure 3.14a it can be seen that when the overlap between the L-resonance ($\lambda_L = 737$ nm) and the excitonic transition ($\lambda_{JAgg} = 622$ nm) is small, then the spectral response of the extinction associated to the hybrid system ($\lambda_{hybrid+} = 745$ nm, $\lambda_{hybrid-} = 622$ nm) is essentially determined by the resonances of the isolated systems, that is, weak coupling is observed. Conversely, Figure 3.14b illustrates the regime where this overlap is strong ($\lambda_L = 610$ nm, $\lambda_{JAgg} = 622$ nm). The resonances of the hybrid system ($\lambda_{hybrid+} = 654$ nm and $\lambda_{hybrid-} = 593$ nm) then reflect the hybridization of the original resonances into mixed states with shared plasmon/exciton character, that is, strong coupling is observed.

The general behavior observed in Figures 3.14a and 3.14b can be illustrated following the procedure described in Reference 87 and calculating the extinction cross section for isolated ellipsoidal nanoparticles with an Au core surrounded by an excitonic shell and embedded in a homogeneous medium of dielectric constant ε_d [86]. This quantity can be expressed as

$$\sigma_{Extinction_i} = \frac{1}{3}\sum_i {}^{2\pi}/_{\lambda}\mathrm{Im}\{\alpha_i\} + \frac{1}{3}\sum_i {}^{1}/_{6\pi}\left({}^{2\pi}/_{\lambda}\right)^4 |\alpha_i|^2,$$

where λ is the wavelength in vacuum and α_i the dipolar polarizability of the core–shell ellipsoid along axis i expressed in the electrostatic approximation [92]. The dielectric constant ε_d of the embedding medium alumina is taken to be 2.56 while a polynomial fit of the data published in Reference 92 is used to describe the dielectric constant of Au, ε_{Au}. The response of the shell was modeled by an excitonic resonance with a Lorentzian response in a dielectric background ε_∞. The anisotropy of the exciton's transition dipole moment was approximated by $\varepsilon_{JAgg\parallel} = \varepsilon_\infty + \frac{\omega_0^2 \cdot f}{\omega_0^2 - \omega^2 - i\omega\gamma}$ and $\varepsilon_{JAgg\perp} = \varepsilon_\infty$, for the contributions along and perpendicular to the aggregate's transition dipole moment, respectively. The high-frequency dielectric constant ε_∞ was extracted from ellipsometric measurement made on a thin J-aggregate film formed on a smooth Au surface, while the transition energy, reduced oscillator strength, and damping of the excitonic state were determined by fitting the J-aggregate's absorption spectrum as measured on the smooth Au film in Figure 3.13. These parameters then take the respective values of $\varepsilon_\infty = 1.21$, $\hbar\omega_0 = 1.99$ eV ($\lambda = 622$ nm), $f = 0.054$, and $\hbar\gamma = 66$ meV. Figure 3.14d shows the calculated extinction cross section for core–shell ellipsoids with aspect ratios of 2.4 and 6 with a constant shell thickness of 2 nm. The thick lines correspond to the coupled plasmon/exciton system while the thin lines describe the extinction of the Au ellipsoids but surrounded by an air shell. The extinction of the J-aggregate is also shown. The aspect ratio of the ellipsoid was parameterized to modify the spectral overlap between the long-axis plasmonic resonance of the ellipsoid and the transition energy of the exciton, and therefore the coupling strength between the two transition dipole moments. In Figure 3.14d, only the long-axis plasmonic mode was coupled to the excitonic transition, reflecting the experimental observations. In a situation where both transverse and longitudinal modes from the ellipsoid were to be coupled with the excitonic transition, four modes would be observed in the hybrid's extinction spectrum: two transverse modes as well as two longitudinal modes.

In order to gain a better insight into the coupling mechanism let us consider a three-dimensional physical system whose Hamiltonian is H_0. The eigenstates of H_0 are $|\phi_T\rangle$, $|\phi_L\rangle$, and $|\phi_{JAgg}\rangle$ associated to the rods' dipolar transverse plasmonic resonance, the ANR's longitudinal resonance, and the J-aggregate excitonic state. The corresponding eigenvalues are E_T, E_L, and E_{JAgg} satisfying $H_0|\phi_T\rangle = E_T|\phi_T\rangle$, $H_0|\phi_L\rangle = E_L|\phi_L\rangle$, and $H_0|\phi_{JAgg}\rangle = E_{JAgg}|\phi_{JAgg}\rangle$, with $\langle\phi_i|\phi_j\rangle = \delta_{ij}$. Introducing coupling between these levels and searching for the eigenvalues of the coupled system in the $\{|\phi_T\rangle, |\phi_L\rangle, |\phi_{JAgg}\rangle\}$ basis, we rewrite the Hamiltonian $H = H_0 + V$, where V

is a time-independent perturbation accounting for the coupling between the different eigenstates of the isolated system. Assuming that the J-aggregate's transition dipole moment is oriented along the long axis of the nanorods only, the Hamiltonian takes the following simplified form, where V_{hybrid} couples $|\phi_L>$ and $|\phi_{JAgg}>$:

$$H = \begin{bmatrix} E_T + V_{11} & 0 & 0 \\ 0 & E_L + V_{22} & V_{hybrid} \\ 0 & V_{hybrid} & E_{JAgg} + V_{33} \end{bmatrix} \tag{3.3}$$

from which the eigenvalues are readily obtained:

$$E_{T'} = E_T + V_{11}, \tag{3.4}$$

$$E_{hybrid+} = \frac{1}{2}\left(E'_L + E'_{JAgg}\right) + \frac{1}{2}\sqrt{\left(E'_L - E'_{JAgg}\right)^2 + 4V^2_{hybrid}}, \tag{3.5}$$

$$E_{hybrid-} = \frac{1}{2}\left(E'_L + E'_{JAgg}\right) - \frac{1}{2}\sqrt{\left(E'_L - E'_{JAgg}\right)^2 + 4V^2_{hybrid}}, \tag{3.6}$$

where $E'_L = E_L + V_{22}$ and $E'_{JAgg} = E_{JAgg} + V_{33}$.

The eigenvector associated with E'_T is $|\psi_T> \sim |\phi_T>$ reflecting the fact that the J-aggregate's transition dipole moment and the T-mode were taken to be perpendicular to each other in our expression of the coupling Hamiltonian V. Consequently, for the transverse mode in the hybrid structure, the presence of the J-aggregate along the Au nanorod results in a net change in the energy E'_T through a change in the dielectric constant in the shell, with an eigenvector that is essentially unchanged from its original dipolar form. Neglecting the core–shell geometry as well as the presence of the substrate, an approximate expression for V_{11} can be found by assuming a loss-less Drude-like dielectric response of the form $\varepsilon(\omega) = 1 - \omega_p^2/\omega^2$ for an Au ellipsoid embedded in a homogeneous medium of effective dielectric constant ε_{eff}:

$$V_{11} = \frac{\hbar\sqrt{L}\omega_p(L-1)}{2\left[\varepsilon_{eff}(1-L)+L\right]^{3/2}}\Delta\varepsilon_{eff},$$

where ω_p is the bulk plasma frequency of Au, L is the geometric factor defined earlier, and $\Delta\varepsilon_{eff}$ reflects the change in dielectric constant induced by the introduction of the J-aggregate in the shell. V_{11} is an increasing function of the shell thickness through $\Delta\varepsilon_{eff}$ with an upper value of $V_{11} \sim -20$ meV measured for a complete removal of the AAO matrix ($\Delta\varepsilon_{eff} \sim \varepsilon_{air} - \varepsilon_{JAgg\perp} = -0.21$). The negative value of V_{11} reflects the increase in the index of refraction of the shell when introducing the molecule. Similar to V_{11}, V_{22} reflects the change in the L-resonance position upon J-aggregation in the shell via the off-resonance high-frequency dielectric constant of the J-aggregate.

However, V_{22} is more difficult to express analytically since the field associated to the L-mode within the assembly of nanorods is no longer following the dipolar response of the isolated rod's long-axis resonance. An estimated value of V_{22} can be made by measuring the sensitivity of the L-resonance as a function of shell thickness for the index of refraction in the shell $\sqrt{\varepsilon_{JAgg\perp}}$ of 1.1. This allows us to evaluate the variation ΔV_{22} of V_{22} over the shell thickness range considered in this study of about 65 meV while V_{22} at 736 nm is about –20 meV and ΔE_L is about 570 meV. V_{33} is the measure of the variation of the intermolecular coupling J within the aggregate when it adsorbs on the Au nanorods with respect to a reference state measured in water: $E'_{JAgg} = E_{JAgg} + V_{33}$. The loss in potential energy of the highest occupied molecular orbital (HOMO) upon J-aggregation determines the coupling energy J which can then be defined from the transition energy ΔE_{JAgg} of the aggregate as $2J = \Delta E - \Delta E_{JAgg}$, where ΔE is the transition energy of the monomer and J is the positive coupling energy. To evaluate V_{33}, J-aggregation has been studied on a smooth Au film, a borosilicate glass substrate, and a porous AAO from a dye solution in methanol. For all those situations the excitonic transition was observed at an identical wavelength of 622 nm, slightly blue-shifted from its value in water. This underlines that in the experimental configurations considered it is the interfacial geometry that governs the aggregation energy rather than the material properties of the substrate itself. Comparing Figure 3.13 with Figure 3.14a it is deduced that J-aggregation in the shell around the rods leads to an intermolecular coupling energy similar to the one measured on planar interfaces suggesting that the core–shell geometry does not strongly affect the value of J in the aggregate even for the smallest shell thicknesses studied ($\sim$2 nm). Consequently, the value of V_{33} is assumed to be shell thickness independent in the following. Considering the absorption frequency of the J-aggregate in both water ($\lambda = 650$ nm, i.e., $2J = 230$ meV) and when adsorbed on Au nanorods ($\lambda = 622$ nm, i.e., $2J = 150$ meV) we can deduce that $V_{33} = E'_{JAgg} - E_{JAgg} = \Delta E - 2J' - (\Delta E - 2J) = 80$ meV.

The eigenvectors $|\psi_+>$ and $|\psi_->$, associated to, respectively, $E_{hybrid+}$ and $E_{hybrid-}$, can be expressed as linear combinations of the longitudinal plasmonic orbital $|\phi_L>$ and the J-aggregate excitonic orbital $|\phi_{JAgg}>$ as $|\psi_+\rangle = \cos\frac{\alpha}{2}|\phi_L\rangle + \sin\frac{\alpha}{2}|\phi_{JAgg}\rangle$ and $|\psi_-\rangle = -\sin\frac{\alpha}{2}|\phi_L\rangle + \cos\frac{\alpha}{2}|\phi_{JAgg}\rangle$, where $\tan\alpha = \frac{V_{hybrid}}{\frac{1}{2}(E'_L - E'_{JAgg})}$. The effect of the coupling matrix V on H_0 is to hybridize the longitudinal and excitonic states into two new eigenstates with a mixed plasmon–exciton character. It follows that if $|E'_L - E'_{JAgg}| \gg 2V_{hybrid}$, then $\alpha \sim 0$, $\cos\alpha/2 \sim 1$, and $\sin\alpha/2 \sim V_{hybrid}/|E'_T - E'_{JAgg}|$, from which, if we assume in this instance that $E'_L > E'_{JAgg}$, we deduce the wavefunctions of the system as $|\psi_+\rangle = |\phi_L\rangle + \frac{V_{hybrid}}{E'_L - E'_{JAgg}}|\phi_{JAgg}\rangle \sim |\phi_L\rangle$ and $|\psi_-\rangle = -\frac{V_{hybrid}}{E'_L - E'_{JAgg}}|\phi_L\rangle + |\phi_{JAgg}\rangle \sim |\phi_{JAgg}\rangle$. These hybrid orbitals then resemble the plasmonic and excitonic orbitals from the uncoupled system in which case these orbitals represent a *weakly coupled* system as they retain their original character. This case is illustrated in Figure 3.14a for $E'_{JAgg} > E'_L$. Similarly, if we now consider the case for which $|E'_L - E'_{JAgg}| \ll 2V_{hybrid}$, then $\alpha \sim \pi/2$, $\cos\alpha = \sin\alpha \sim 1/\sqrt{2}$, and $|\psi_+\rangle = \frac{1}{\sqrt{2}}[|\phi_L\rangle \pm |\phi_{JAgg}\rangle]$. These functions are hybrid orbitals of the system of

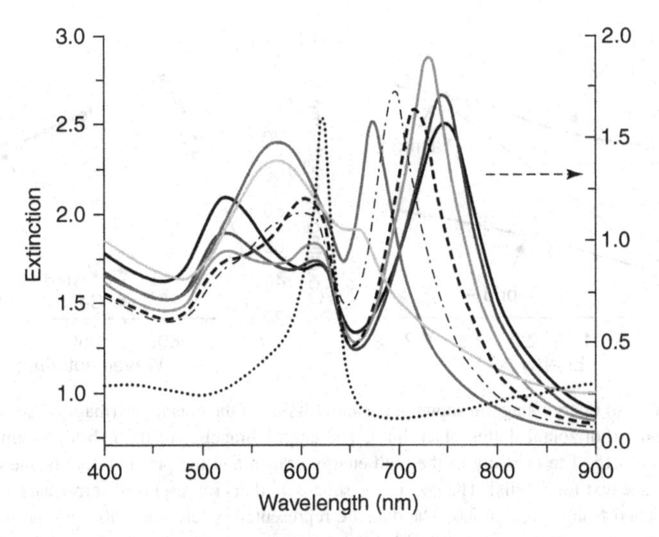

FIGURE 3.15 Extinction from the hybridized metamaterial with varying coupling strength between the plasmonic L-mode and the molecular excitonic transition. The spectra are taken at an angle of incidence of 40°. The plasmon–exciton coupling strength has been adjusted by tuning the spectral position of the L-resonance from 736 to 550 nm by increasing the thickness of the shell around the Au rods. The L-resonance position in air is shown for hybrids at 736 nm (2.5 nm), 733 nm (3.8 nm), 716 nm (5.6 nm), 700 nm (7.7 nm), 650 nm (8.9 nm), 600 nm (20 nm), and 550 nm (no shell, matrix completely removed). The black broken line shows the excitonic extinction on a 50 nm thick smooth Au film.

Figure 3.14b where the two original uncoupled orbitals overlap spectrally in which case strong coupling is achieved.

The unique characteristic of the core–shell nanorod geometry enables a continuous tuning of the plasmonic resonances by controlling the thickness of the shell created around the rods. The spectral position of the L-resonance can, therefore, be scanned through most of the visible spectrum and across the excitonic resonance to modulate the mixing of these states. The result is shown in Figure 3.15 where the spectrum of the hybrid assembly of nanorods is plotted for different positions of the L-resonance in air.

The occurrence of the strong coupling regime in the structure is evident since no resonances appear in the spectral range from 622 to 658 nm while the L-resonance resonance was continuously tuned from 736 to 550 nm. The anticrossing between the L-resonance and the excitonic transition is illustrated in Figure 3.16a. Deriving V_{hybrid} from Equations 3.5 and 3.6, we use Figure 3.15 to estimate the coupling strength within the hybrid as a function of the overlap between the L-resonance and the excitonic states as

$$V_{hybrid} = \sqrt{\frac{(E_{hybrid+} - E_{hybrid-})^2 - (E_L - E'_{JAgg})^2}{4}}, \tag{3.7}$$

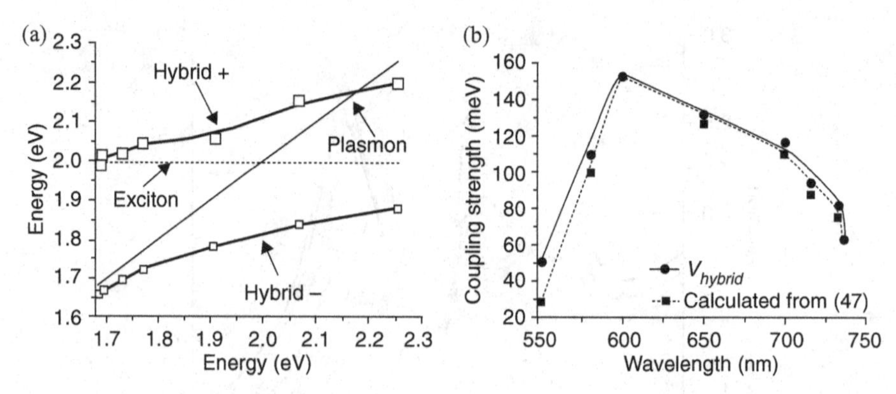

FIGURE 3.16 (a) Energy diagram showing the anticrossing of the plasmonic (diagonal solid black line) and the excitonic (horizontal dotted black line). The energy branches of the hybrid system are plotted based on Figure 3.15. The crossing of the high-energy branch and the plasmon line is due to V_{22} upon hybridization (see text for details). The size of the square markers represents the error bars. (b) Coupling strength calculated from Equation 3.6. The data are represented by circular dots, and the solid line is a guide for the eye. For comparison, the black square dots are obtained using the formula for the coupling strength as derived in Reference 47. The dotted line is a guide for the eye.

where the assumption $E'_L \sim E_L$ was made. The result is plotted in Figure 3.16 as a function of the original position of the L-resonance. A similar relationship derived from an analytical theory for the optical properties of ellipsoidal plasmonic particles covered by anisotropic molecular layers leads to very similar values for V_{hybrid} as a function of the L-resonance wavelength with a small overestimation of the coupling strength when using Equation 3.7. This overestimation reflects the effect V_{22} has on E'_L: The absolute value of V_{22} increases with shell thickness and is negligible at maximum coupling strength. This maximum is reached for an L-resonance in air located at 600 nm and corresponds to a Rabi splitting of 155 meV. This value is commensurate with the largest Rabi splitting observed in planar organic microcavities [93] and similar to recent observations for an excitonic state mixed with a planar SPP [94].

The ability to control the coupling strength between a metamaterial mode and a molecular exciton both spatially and spectrally is a major step toward implementing active and passive plasmonic devices. Based on the hybrid nature of the coupled states, the nonlinear optical properties of such metamaterial should demonstrate extreme sensitivity over a large spectral range; that is, by addressing one of the hybrid resonances one would effectively affect both coupled resonances and therefore affect the extinction of the device from 550 to 700 nm. This could be achieved optically in a pump–probe configuration, for example, and possibly also electrically by dynamically controlling the orientation/ordering of the molecular aggregate. Because of the coupled nature between plasmonic and excitonic states, future investigations on the emission properties of this hybrid structure should stimulate considerable interest as well [87].

3.6 ELECTRO-OPTICAL EFFECTS IN PLASMONIC NANOROD METAMATERIAL HYBRIDIZED WITH LIQUID CRYSTALS

Finally we will describe how the transmission through the metamaterial can be controlled electronically upon hybridization of the metamaterial with an electro-optic medium such as liquid crystals (LC) [12]. Plasmonic devices with electric field-tuneable resonances have been demonstrated for both plasmonic crystals and plasmonic waveguide-based devices [95–98]. In all these cases, the effect relies on the modification of the refractive index when the LC orientation is changed upon applied external electric field and the associated shift in the resonance position of the Bloch mode or waveguided mode.

In contrast, when the nanorod metamaterial is placed in proximity to the LC layer, reorientation of anisotropic molecules does not provide spectral tunability of the metamaterial resonances due to their field distribution: The field is concentrated inside the AAO matrix and is not sensitive to the refractive index changes of the superstrate. Nevertheless, polarization effects associated with the LC reorientation in external electric field combined with the polarization properties of the anisotropic metamaterials can be used for light switching and modulation applications [12].

Figure 3.17 shows the effect of the LC and electric applied field on the extinction spectra of the metamaterial. When the LC is present near the metamaterial, it completely extinguishes the L-resonance at 712 nm if no electric field is applied. The application of a small field leads to an initial increase in the extinction of the T-resonance and a decrease in the extinction at the L-mode amplitude. As the voltage is further increased, the extinction of the T-resonance decreases while the L-resonance starts to appear which then rapidly increases. A steady increase in the L-resonance amplitude with increasing applied voltage is observed up to saturation. This process is completely reversible and shows good reproducibility as the applied field is removed.

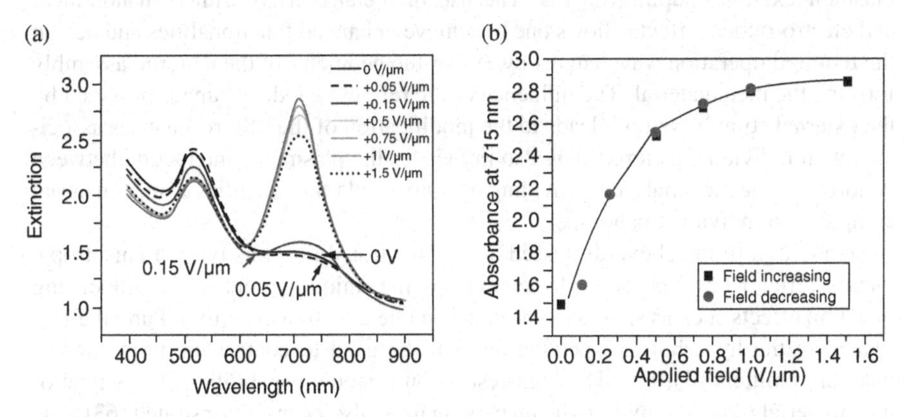

FIGURE 3.17 (a) Extinction spectra of the metamaterial (20 nm diameter, 400 nm long gold nanorods embedded in AAO) with a 100 mm top liquid crystal layer as a function of applied electric field. The incident light is TM-polarized and the angle of incidence is 40°. (b) Dependence of the metamaterial extinction at 715 nm as a function of applied electric field.

This effect is observed for illumination through the LC side. When the device is illuminated through the metamaterial, the L-resonance extinction is unchanged by the presence of the LC independently of the applied field. The observations of both electric field-dependent extinction and nonreciprocity with respect to the illumination side suggest that the polarization effects determined the observed electro-optical switching. The LC layer conformation has molecules oriented in the plane of the substrate (perpendicularly to the nanorod axes) when no electric field is applied. The TM-polarized incident light is then converted into TE-polarized light when transmitted through the LC molecules layer. In the configuration where light transmits first through the LC layer before interacting with the nanorod layer, the L-resonance cannot be excited and is not present in the extinction of the sample. Applying the voltage causes the LCs to orient along the field, that is, perpendicular to the substrate and along the direction of the nanorod long axes. In this orientation they no longer affect the polarization of the incident TM-polarized light and hence the L-resonance is observed again.

Comparing with previous work on gold nanorods [96], nanoparticles [97], and nanodots [98], it can be seen that the nanorod metamaterial allows observing much larger changes in the extinction spectra for smaller applied electric fields, and this without the need for a rubbing layer, but it does not provide tunability of the plasmonic resonances.

3.7 CONCLUSION

We have presented an overview of the optical properties of plasmonic nanorod metamaterials with an emphasis on their applications in active nanophotonics. In particular, we have focused on their nonlinear optical properties and ultrafast all-optical switching, electro-optical switching in LC-infiltrated metamaterials, and strong plasmon–exciton coupling effects. The use of metamaterials with both nonlinear and electro-optical effects allows one to achieve enhanced functionalities and design the required operation wavelengths by tuning the geometry of the nanorod assembly forming the metamaterial. The influence of the refractive index changes provided by the external stimuli not only leads to the modification of the LSP resonances associated with individual nanorods but also influences the plasmonic interaction between nanorods in the metamaterial, providing the improved functionalities of metamaterial compared to individual nanorods.

In addition to the above-discussed active functionalities, this type of anisotropic metamaterial with hyperbolic dispersion has the unique potential for influencing quantum effects such as spontaneous emission due to extremely strong Purcell effect related to the hyperbolic dispersion and can be used to control the properties of quantum emitters [35, 62]. The high-resolution imaging capabilities of this type of metamaterial related to hyperbolic dispersion have also been demonstrated [63]. The nanorod metamaterial has a record figure of merit for sensing, relying on the strong sensitivity of its resonant response to the plasmonic interaction between nanorods and, thus, on the refractive index of the analyte between nanorods [36, 99]. The ability to use different materials for building the nanorod metamaterial has been applied for designing novel ferromagnetic properties using assemblies of Ni and Co nanorods. Magnetic nanorod metamaterial exhibits peculiar magnetic properties as well as magneto-optical response with the prominent role of spin-waves in magnetic nanorods [100].

Further extension of the anisotropic metamaterial designs has been very recently demonstrated by replacing nanorods by plasmonic nanotubes. This type of nanostructures provides access to a wider range of plasmonic modes, which can be used for further designing optical properties [101]. This can be advantageous for both biosensing and nonlinear optical functionalities [102].

REFERENCES

1. Cai W, Shalaev VM (2009) *Optical Metamaterials: Fundamentals and Applications.* New York: Springer.
2. Zheludev NI (2010) The road ahead for metamaterials. *Science* 328: 582.
3. Soukoulis CM, Wegener M (2011) Past achievements and future challenges in the development of three-dimensional photonic metamaterials. *Nat. Photonics* 5: 523.
4. Smith DR, Pendry JB, and Wiltshire MCK (2004) Metamaterials and negative refractive index. *Science* 305: 788.
5. Tsakmakidis KL, Boardman AD, and Hess O (2007) 'Trapped rainbow' storage of light in metamaterials. *Nature (Lond.)* 450: 397.
6. Cai W, Chettiar UK, Kildishev AV, Shalaev VM (2007) Optical cloaking with metamaterials. *Nat. Photonics* 1: 224.
7. Ma H-F and Cui TJ (2010) Three-dimensional broadband ground-plane cloak made of metamaterials. *Nat. Commun.* 1: 1.
8. Gharghi M, Gladden C, Zentgraf T, Liu Y, Yin X, Valentine J, and Zhang X (2011) A carpet cloak for visible light. *Nano Lett.* 11: 2825.
9. Hoffman AJ, Alekseyev L, Howard SS, Franz KJ, Wasserman D, Podolskiy VA, Narimanov EE, Sivco DL, and Gmachl C (2007) Negative refraction in semiconductor metamaterials. *Nat. Mater.* 6: 946.
10. Zhang X and Liu Z (2008) Superlenses to overcome the diffraction limit. *Nat. Mater.* 7: 435.
11. Kundtz N and Smith DR (2010) Extreme-angle broadband metamaterial lens. *Nat. Mater.* 9: 129.
12. Evans PR, Wurtz GA, Hendren WR, Atkinson R, Dickson W, Zayats AV, and Pollard RJ (2007) Electrically switchable nonreciprocal transmission of plasmonic nanorods with liquid crystal. *Appl. Phys. Lett.* 91: 043101.
13. Dickson W, Evans P, Wurtz GA, Hendren W, Atkinson R, Pollard RJ, and Zayats AV (2008) Towards nonlinear plasmonic devices based on metallic nanorods. *J. Microsc.* 420: 415.
14. Lu Y, Rhee JY, Jang WH, and Lee YP (2010) Active manipulation of plasmonic electromagnetically-induced transparency based on magnetic plasmon resonance. *Opt. Express* 18: 20912.

15. Kang L, Zhao Q, Zhao H, and Zhou J (2008) Magnetically tuneable negative permeability metamaterial composed by split ring resonators and ferrite rods. *Opt. Express* 16: 8825.

16. Jun YC, Gonzales E, Reno JL, Shaner EA, Gabbay A, and Brener I (2012) Active tuning of mid-infrared metamaterials by electrical control of carrier densities. *Opt. Express* 20: 1903.

17. Lee SH, Choi M, Kim T-T, Lee S, Liu M, Yin X, Choi HK, Lee SS, Choi C-G, Choi S-Y, Zhang X, and Min B (2012) Switching terahertz waves with gate-controlled active graphene metamaterials. *Nat. Mater.* 11: 1476.

18. Friedman RS, McAlpine MC, Ricketts DS, Ham D, and Lieber CM (2005) High-speed integrated nanowire circuits. *Nature* 434: 1085.

19. Tong LM, Gattass RR, Ashcom JB, He SL, Lou JY, Shen MY, Maxwell I, and Mazur E (2003) Subwavelength-diameter silica wires for low-loss optical wave guiding. *Nature* 426: 816.

20. Xiang J, Lu W, Hu Y, Wu Y, Yan H, and Lieber CM (2006) Ge/Si nanowire heterostructures as high-performance field-effect transistors. *Nature* 441: 489.

21. Gudiksen MS, Lauhon LJ, Wang J, Smith DC, and Lieber CM (2002) Growth of nanowire superlattice structures for nanoscale photonics and electronics. *Nature* 415: 617.

22. Krenn JR, Lamprecht B, Ditlbacher H, Schider G, Salerno M, Leitner A, Aussenegg FR (2002) Nondiffraction-limited light transport by gold *nanowires. Europhys. Lett.* 60: 663.

23. Mühlschlegel P, Eisler H-J, Hecht B, and Pohl DW (2005) Resonant optical antennas. *Science* 308: 1607.

24. Shalaev VM, Cai WS, Chettiar UK, Yuan Hk, Sarychev AK, Drachev VP, and Kildishev AV (2005) Negative index of refraction in optical metamaterials. *Opt. Lett.* 30: 3356.

25. Evans P, Hendren WR, Atkinson R, Wurtz GA, Dickson W, Zayats AV, and Pollard RJ (2006) Growth and properties of gold and nickel nanorods in thin film alumina. *Nanotechnology* 17: 5746.

26. Lee K-S and El-Sayed MA (2006) Gold and silver nanoparticles in sensing and imaging: sensitivity of plasmon response to size, shape, and metal composition. *J. Phys. Chem. B* 110: 19220.

27. Pelton M, Liu M, Park S, Scherer NF, and Guyot-Sionnest P (2006) Ultrafast resonant optical scattering from single gold nanorods: large nonlinearities and plasmon saturation. *Phys. Rev. B* 73: 155419.

28. Schider G, Krenn JR, Hohenau A, Ditlbacher H, Leitner A, Aussenegg FR, Schaich WL, Puscasu I, Monacelli B, and Boreman G (2003) Plasmon dispersion relation of Au and Ag nanowires. *Phys. Rev. B* 68: 155427.

29. Atkinson R, Hendren WR, Wurtz GA, Dickson W, Zayats AV, Evans P, and Pollard RJ (2006) Anisotropic optical properties of arrays of gold nanorods embedded in alumina. *Phys. Rev. B* 73: 235402.

30. Ono A, Kato J-i, and Kawata S (2005) Subwavelength optical imaging through a metallic nanorod array. *Phys. Rev. Lett.* 95: 265407.

31. Podolskiy VA, Sarychev AK, Narimanov EE, and Shalaev VM (2005) Resonant light interaction with plasmonic nanowire systems. *J. Opt. A Pure Appl. Opt.* 7: S32.

32. Rahachou AI and Zozulenko IV (2007) Light propagation in nanorod arrays. *J. Opt. A Pure Appl. Opt.* 9: 265.

33. Rahman A, Kosulnikov SY, Hao Y, Parini C, Belov PA (2011) Subwavelength optical imaging with an array of silver nanorods. *J. Nanophoton.* 5: 051601.

34. Pollard RJ, Murphy A, Hendren WR, Evans PR, Atkinson R, Wurtz GA, Zayats AV, and Podolskiy VA (2009) Optical nonlocalities and additional waves in epsilon-near-zero metamaterials. *Phys. Rev. Lett.* 102: 127405.

35. Poddubny AN, Belov PA, Ginzburg P, Zayats AV, and Kivshar YS (2012) Microscopic model of Purcell enhancement in hyperbolic metamaterials. *Phys. Rev. B* 86: 035148/1.

36. Kabashin AV, Evans P, Pastkovsky S, Hendren W, Wurtz GA, Atkinson R, Pollard R, Podolskiy VA, and Zayats AV (2009) Plasmonic nanorod metamaterials for biosensing. *Nat. Mater.* 8: 867.

37. Wurtz GA, Pollard R, Hendren W, Wiederrecht G, Gosztola D, Podolskiy VA, and Zayats AV (2011) Designed ultrafast optical nonlinearity in a plasmonic nanorod metamaterial enhanced by nonlocality. *Nat. Nanotechnol.* 6: 107.

38. Liu NW, Datta A, Liu CY, Peng CY, Wang HH, and Wang YL (2005) Fabrication of anodic-alumina films with custom-designed arrays of nanochannels. *Adv. Mater.* 17: 222.

39. Liu N-W, Liu C-Y, Wang H-H, Hsu C-F, Lai M-Y, Chuang T-H, and Wang Y-L (2008) Focused-ion-beam-based selective closing and opening of anodic alumina nanochannels for the growth of nanowire arrays comprising multiple elements. *Adv. Mater.* 20: 2547.

40. Kustandi TS, Loh WW, Gao H, and Low HY (2010) Wafer-scale near-perfect ordered porous alumina on substrates by step and flash imprint lithography. *ACS Nano* 4: 2561.

41. Wolfrum B, Mourzina Y, Mayer D, Schwaab D, and Offenhusser A (2006) Fabrication of large-scale patterned gold-nanopillar arrays on a silicon substrate using imprinted porous alumina templates. *Small* 2: 1256.

42. Reinhardt C, Passinger S, Chichkov BN, Dickson W, Wurtz GA, Evans P, Pollard R, Zayats AV (2006) Restructuring and modification of metallic nanorod arrays using femtosecond laser direct writing. *Appl. Phys. Lett.* 89: 231117.

43. Lillo M, Losic D (2009) Ion-beam pore opening of porous anodic alumina: the formation of single nanopore and nanopore arrays. *Mater. Lett.* 63: 457.

44. Wurtz GA, Dickson W, O'Connor D, Atkinson R, Hendren W, Evans P, Pollard R, Zayats AV (2008) Guided plasmonic modes in nanorod assemblies: strong electromagnetic coupling *regime. Opt. Express* 16: 7460.

45. Dickson W, Wurtz, GA, Evans P, O'Connor D, Atkinson R, Pollard R, Zayats AV (2007) Dielectric-loaded plasmonic nano-antenna arrays: a metamaterial with tuneable optical properties. *Phys. Rev. B* 76: 115411.

46. Evans PR, Wurtz GA, Atkinson R, Hendren W, O'Connor D, Dickson W, Pollard RJ, and Zayats AV (2007) Plasmonic core/shell nanorod arrays: subattoliter controlled geometry and tuneable optical properties. *J. Phys. Chem. C* 111: 12522.

47. Wurtz GA, Evans PR, Hendren W, Atkinson R, Dickson W, Pollard RJ, Harrison W, Bower C, and Zayats AV (2007) Molecular plasmonics with tuneable exciton-plasmon coupling strength in J-aggregate hybridized Au nanorod assemblies. *Nano Lett.* 7: 1297.

48. Maier SA, Kik PG, Atwater HA, Meltzer S, Requicha A, and Koel BE (2002) Observation of coupled plasmon-polariton modes of plasmon waveguides for electromagnetic energy transport below the diffraction limit. *Proc. SPIE* 4810.

49. Maier SA, Kik PG, and Atwater HA (2003) Optical pulse propagation in metal nanoparticle chain waveguides. *Phys. Rev. B* 67: 205402.

50. Park SY and Stroud D (2004) Surface-plasmon dispersion relations in chains of metallic nanoparticles: an exact quasistatic calculation. *Phys. Rev. B* 69: 125418.

51. Jain PK, Eustis S, El-Sayed MA (2006) Plasmon coupling in nanorod assemblies: optical absorption, discrete dipole approximation simulation, and exciton-coupling model. *J. Phys. Chem. B* 110: 18243.

52. Rechberger W, et al. (2003) Optical properties of two interacting gold nanoparticles. *Opt. Commun.* 220: 137.

53. Koenderink AF and Polman A (2006) Complex response and polariton-like dispersion splitting in periodic metal nanoparticle chains. *Phys. Rev. B* 74: 033402.

54. Vlad A, Matefi-Tempfli M, Faniel S, Bayot V, Melinte S, Piraux L, and Matefi-Tempfli S (2006) Controlled growth of single nanowires within a supported alumina template. *Nanotechnology* 17: 4873.

55. Mie G (1908) Beiträge zur Optik trüber Medien, speziell kolloidaler Metallösungen. *Ann. Phys.* 25: 377.

56. Gans R (1915) Über die Form ultramikroskopischer Silberteilchen. *Ann. Phys.* 352: 270.

57. Maxwell-Garnett JC (1904) Colours in metal glasses and in metallic films. *Phil. Trans. R. Soc. Lond.* 203: 385.

58. Maxwell-Garnett JC (1906) Colours in metal glasses, in metallic films, and in metallic solutions. II. *Phil. Trans. R. Soc. Lond.* 203: 385.

59. Elser J, Wangberg R, Podolskiy VA, and Narimanov EE (2006) Nanowire metamaterials with extreme optical anisotropy. *Appl. Phys. Lett.* 89: 261102.

60. Yeh P, Yariv A, and Hong C-S (1977) Electromagnetic propagation in periodic stratified media. I. General theory. *J. Opt. Soc. Am.* 67: 423.

61. Tumkur T, Zhu G, Black P, Barnakov YA, Bonner CE, and Noginov MA (2011) Control of spontaneous emission in a volume of functionalized hyperbolic metamaterial. *Appl. Phys. Lett.* 99: 151115.

62. Krishnamoorthy HNS, Jacob Z, Narimanov E, Kretzschmar I, and Menon VM (2012) Topological transitions in metamaterials. *Science* 336: 205.

63. Yao J, Liu Z, Liu Y, Wang Y, Sun C, Bartal G, Stacy AM, and Zhang X (2008) Optical negative refraction in bulk metamaterials of nanowires. *Science* 321: 930.

64. Agranovich VM, Ginzburg VL (1984) *Crystal Optics with Spatial Dispersion and Excitons*. Springer-Verlag Berlin.

65. Hopfield JJ and Thomas DG (1963) Theoretical and experimental effects of spatial dispersion on the optical properties of crystals. *Phys. Rev.* 132: 563.

66. Gibbs HM (1985) *Optical Bistability: Controlling Light with Light*. New York: Academic.

67. Kauranen M, Zayats AV (2012) Nonlinear plasmonics. *Nat. Phot.* 6: 737.

68. Boyd RW (2003) *Nonlinear Optics*. New York: Academic.

69. Fatti ND, Bouffanais R, Vallée F, and Flytzanis C (1998) Nonequilibrium electron interactions in metal films. *Phys. Rev. Lett.* 81: 922.

70. Sun CK, Vallée F, Acioli L, Ippen EP and Fujimoto JG (1993) Femtosecond investigation of electron thermalization in gold. *Phys. Rev. B* 48: 12365.

71. Wurtz GA, Pollard R, and Zayats AV (2006) Optical bistability in nonlinear surface-plasmon polaritonic crystals. *Phys. Rev. Lett.* 97: 057402.

72. Smolyaninov II, Davis CC, and Zayats AV (2002) Light-controlled photon tunneling. *Appl. Phys. Lett.* 81: 3314.

73. Milman P, Keller A, Charron E, and Atabek O (2007) Bell-Type inequalities for cold heteronuclear molecules. *Physical Review Letters* 99: 130405.

74. Sipe JE, So VCY, Fukui M, and Stegeman GI (1980) Analysis of second-harmonic generation at metal surfaces. *Phys. Rev. B* 21: 4389.

75. Bigot J-Y, Halte V, Merle J-C, and Daunois A (2000) Electron dynamics in metallic nanoparticles. *Chem. Phys.* 251: 181.

76. Shahbazyan TV, Perakis IE, and Bigot J-Y (1998) Size-dependent surface plasmon dynamics in metal nanoparticles. *Phys. Rev. Lett.* 81: 3120.

77. Inouye H, Tanaka K, Tanahashi I, and Hirao K (1998) Ultrafast dynamics of nonequilibrium electrons in a gold nanoparticle system. *Phys. Rev. B* 57: 11334.

78. Del Fatti N, Voisin C, Achermann M, Tzortzakis S, Christofilos D, and Vallee F (2000) Nonequilibrium electron dynamics in noble metals. *Phys. Rev. B* 61: 16956.

79. Rotenberg N, Betz M, and van Driel HM (2008) Ultrafast control of grating-assisted light coupling to surface plasmons. *Opt. Lett.* 33: 2137.

80. MacDonald KF, Samson ZL, Stockman MI, and Zheludev NI (2009) Ultrafast active plasmonics. *Nat. Photonics* 3: 55.

81. Link S, Burda C, Mohamed MB, Nikoobakht B, and El-Sayed MA (2000) Femtosecond transient-absorption dynamics of colloidal gold nanorods: shape independence of the electron-phonon relaxation time. *Phys. Rev. B* 61: 6086.

82. Wiederrecht GP, Wurtz GA, and Bouhelier A (2008) Ultrafast hybrid plasmonics. *Chem. Phys. Lett.* 461: 171.

83. Varnavski OP, III TG, Mohamed MB, and El-Sayed MA (2005) Femtosecond excitation dynamics in gold nanospheres and nanorods. *Phys. Rev. B* 72: 235405.

84. Sönnichsen C, Franzl T, Wilk T, Plessen Gv, Feldmann J, Wilson O, and Mulvaney P (2002) Drastic reduction of plasmon damping in gold nanorods. *Phys. Rev. Lett.* 88: 077402.

85. Ozcelik S, Ozcelik I, and Akins DL (1998) Superradiant lasing from J-aggregated molecules adsorbed onto colloidal silver. *Appl. Phys. Lett.* 73: 1949.

86. Wiederrecht GP, Wurtz GA, and Hranisavljevic J (2004) Coherent coupling of molecular excitons to electronic polarizations of noble metal nanoparticles. *Nano Lett.* 4: 2121.

87. Bellessa J, Bonnand C, Plenet JC, and Mugnier J (2004) Strong coupling between surface plasmons and excitons in an organic semiconductor. *Phys. Rev. Lett.* 93: 036404.

88. Dintinger J, Klein S, Bustos F, Barnes WL, and Ebbesen TW (2005) Strong coupling between surface plasmon-polaritons and organic molecules in subwavelength hole arrays. *Phys. Rev. B* 71: 035424.
89. Lidzey DG, Bradley DDC, Armitage A, Walker S, and Skolnick MS (2000) Photon-mediated hybridization of Frenkel excitons in organic semiconductor microcavities. *Science* 288: 1620.
90. Hobson PA, Barnes WL, Lidzey DG, Gehring GA, Whittaker DM, Skolnick MS, and Walker S (2002) Strong exciton–photon coupling in a low-Q all-metal mirror microcavity. *Appl. Phys. Lett.* 81: 3519.
91. Christ A, Tikhodeev SG, Gippius NA, Kuhl J, and Giessen H (2003) Waveguide-plasmon polaritons: strong coupling of photonic and electronic resonances in a metallic photonic crystal slab. *Phys. Rev. Lett.* 91: 183901.
92. Bohren CF, Huffmann DR (1983) *Absorption and Scattering of Light by Small Particles*. New York: Wiley Science.
93. Ambjörnsson T, Mukhopadhyay G, Apell SP, and Käll M (2006) Resonant coupling between localized plasmons and anisotropic molecular coatings in ellipsoidal metal nanoparticles. *Phys. Rev. B* 73: 085412.
94. Agranovich VM, Litinskaia M, and Lidzey DG (2003) Cavity polaritons in microcavities containing disordered organic semiconductors. *Phys. Rev. B* 67: 085311.
95. Dickson W, Wurtz GA, Evans PR, Pollard RJ, and Zayats AV (2008) Electronically controlled surface plasmon dispersion and optical transmission through metallic hole arrays using liquid crystal. *Nano Lett.* 8: 281.
96. Chu KC, Chao CY, Chen YF, Wu YC, and Chen CC (2006) Electrically controlled surface plasmon resonance frequency of gold nanorods. *Appl. Phys. Lett.* 89: 103107.
97. Müller J, Sönnichsen C, Poschinger Hv, Plessen Gv, Klar TA, and Feldmann J (2002) Electrically controlled light scattering with single metal nanoparticles. *Appl. Phys. Lett.* 81: 171.
98. Kossyrev PA, Yin A, Cloutier SG, Cardimona DA, Huang D, Alsing PM, and Xu JM (2005) Electric field tuning of plasmonic response of nanodot array in liquid crystal matrix. *Nano Lett.* 5: 1978.
99. Yakovlev VV, Dickson W, McPhillips J, Pollard RM, Podolskiy VA, Zayats AV (2013) Ultrasensitive nonresonant detection of ultrasound with plasmonic metamaterials. *Adv. Mat.* DOI: 10.1002/adma.201300314
100. Veniaminova Y, Stashkevich AA, Roussigne Y, Cherif SM, Murzina TV, Murphy AP, Atkinson R, Pollard RJ, and Zayats AV (2012) Brillouin light scattering by spin waves in magnetic metamaterials based on Co nanorods. *Opt. Mater. Express* 2: 1260.
101. Hendren WR, Murphy A, Evans P, O'Connor D, Wurtz GA, Zayats AV, Atkinson R, and Pollard RJ (2008) Fabrication and optical properties of gold nanotube arrays. *J. Phys. Condens. Matter* 20: 362203.
102. Murphy A, Mcphillips J, Hendren W, Mcclatchey C, Atkinson R, Wurtz G, Zayats AV, and Pollard RJ (2011) The controlled fabrication and geometry tuneable optics of gold nanotube arrays. *Nanotechnology* 22: 045705.

4

Transformation Optics for Plasmonics

ALEXANDRE AUBRY
Institut Langevin, ESPCI ParisTech, Paris, France

JOHN B. PENDRY
Blackett Laboratory, Department of Physics, Imperial College London, London, UK

4.1 INTRODUCTION

Since the pioneering work of Mie [1] and Ritchie [2], there has been a vast amount of research effort to investigate the electromagnetic (EM) properties of metal/dielectric interfaces (e.g., [3–6] and references therein). Surface plasmon polaritons (SPPs) can show strong coupling to light and have wavelength of only a few tens of nanometers, hence beating the classical diffraction limit [7,8]. In particular, localized surface plasmons (LSPs) can be excited in a metal particle and give rise to a local enhancement of the EM field. In the literature, the maximization of the field enhancement is generally obtained by combining the strong overall resonance of the plasmonic structure with very small and sharp geometric features where hot spots arise. Following this principle, several plasmonic structures have been investigated, such as triangles/squares with sharp corners [9,10], nanoparticle dimers [10–24], and crescent-shaped nanoparticles [25–32] or nanoshells [33–38]. The significant field enhancement that such structures may provide has drawn considerable attention in surface-enhanced Raman spectroscopy [16,25,39–42] or enhanced fluorescent emission [43,44]. Until now, the theoretical description of the optical response of such structures has remained a challenge: Numerical simulations are generally performed and few qualitative arguments are proposed to explain these numerical results. An elegant physical picture to describe the interaction of LSPs in dimers or nanoshells is the plasmon hybridization

Active Plasmonics and Tuneable Plasmonic Metamaterials, First Edition. Edited by Anatoly V. Zayats and Stefan A. Maier.

model [15,36]. In analogy with molecular orbital theory, the dimer plasmons can be viewed as bonding and antibonding combinations of the individual nanoparticle plasmons. However, albeit elegant, the plasmon hybridization picture is a limited tool: Numerical simulations are still needed to calculate the optical response of a dimer. Another severe challenge for potential applications lies in the spectrum bandwidth over which plasmonic particles can efficiently operate. Indeed, small devices tend to be efficient collectors at just a few resonant frequencies, contrary to an infinite structure that naturally shows a broadband spectrum [6]. Nevertheless, transformation optics (TO) can take up this challenge by leading to the design of broadband plasmonic nanostructures acting as strong field concentrators [45].

TO has been used as a tool in EM design for some time [46–48] and has drawn considerable attention in the last seven years sparked by the seminal works of Leonhardt [49] and Pendry [50]. Whereas the Leonhardt approach approximates Maxwell's equations by the Helmholtz equation which is restricted to the far field, the latter approach is exact at the level of Maxwell's equations and can be deployed to treat both near and far fields and has even been applied to static fields. In this review we are concerned with near-field applications.

The paths of EM waves can be controlled by devising a material whose constitutive parameters should vary spatially in a way prescribed by coordinate transformations. Various applications have been proposed and implemented experimentally, among which is the famous EM cloak [51–56]. This strategy has been recently extended to plasmonics [57–64]. At visible and infrared frequencies, most of the energy of SPPs is contained in the dielectric layer adjacent to the metal slab. Hence, by devising the optical parameters of the dielectric placed on top of a metal surface, one can design plasmonic devices capable of different functionalities such as beam shifting, waveguide bending, cloaking, or focusing. Following this strategy, plasmonic Luneburg and Eaton lenses have been implemented experimentally at optical frequencies [60]. The carpet-cloak concept [52] has also been extended to plasmonics and tested experimentally in the visible regime [59,64].

In a series of papers [45, 65–81], an alternative strategy based on TO has been proposed for plasmonics. It consists in designing and studying analytically plasmonic devices capable of

- an efficient harvesting of light over a broadband spectrum in both the visible and the near-infrared regimes and
- a strong far-field to near-field conversion of energy, leading to a considerable field confinement and enhancement.

The strategy is as follows: Start with an infinite plasmonic system that naturally shows a broadband spectrum and then apply a mathematical transformation that converts the infinite structure into a finite one while preserving the spectrum [45]. TO provides us with a very general set of transformations but here we shall concentrate on a specific subset: the conformal transformations (CT) [82]. Both the electrostatic potentials and the material permittivity are preserved under 2D conformal mapping. Hence, it does not require the delicate design of a metamaterial with a spatial variation in its constitutive parameters. Note that conformal mapping has also been used in the past to study the electrostatic interaction between dielectric or metallic particles in the context of effective medium theory [83–87]. Following the CT strategy, various broadband plasmonic structures have been designed and studied analytically, such as 2D crescents [65,71], kissing nanowires [70], touching spheres [67], groove/wedge-like plasmonic structures [65], nanoparticle dimers [66,73], core-shell nanostructures [81], or or rough surfaces[75]. All these structures have in common to display geometrical singularities that simultaneously allow a broadband interaction with the incoming light as well as a spectacular nanofocusing of its energy.

Despite the power and elegance of the CT approach, one could claim that these theoretical predictions will be difficult to retrieve experimentally. First, the singular structures derived from the CT approach will suffer from inevitable imperfections due to the nanofabrication process. This issue has been solved in a recent paper [68]. Blunt-ended nanostructures can also be derived via CT and the possibility of designing devices with an absorption property insensitive to the geometry bluntness has been demonstrated. Second, the CT approach relies on the quasi-static approximation. Its range of application should be restricted to plasmonic devices whose dimension is at least one order of magnitude smaller than the wavelength (typically 40 nm in the optical regime). However, a conformal mapping of radiative losses allows to go beyond the electrostatic approximation and extend the range of validity of the CT approach until a nanostructure dimension of 200 nm [72]. The third potential limitation of the CT approach is nonlocality. The spatial extent of the EM fields in singular structures is comparable to the Coulomb screening length in noble metals, hence nonlocal effects can be important in the vicinity of the structure singularity. Recent studies have shown how nonlocality can be taken into account by conformal mapping [69, 78–80]. A compromise can be found between radiative losses and nonlocal effects by adjusting the nanostructure dimension such that the absorption bandwidth and field enhancement capabilities of the nanostructure are maximized.

In this chapter, we review the main designs provided by the CT approach and present the corresponding analytical predictions. The first section presents the general strategy and the two families of nanostructures that one can derive via conformal mapping: the singular plasmonic devices and the resonant nanostructures. In Sections 4.3 and 4.4, the broadband light harvesting and nanofocusing properties of singular structures are investigated. The vertex angle, which characterizes the strength of the singularity, is shown to be a key parameter. The conformal picture also provides novel physical insights into the formation of hot spots in the vicinity of the structure singularities. The issue of the nanostructure bluntness will also be addressed. The fourth section is dedicated to resonant nanostructures. The plasmonic hybridization model is revisited with TO and a full-analytical solution is provided for 2D nanostructures. In Section 4.6, we show how to take into account radiative losses in the CT approach. Besides being capable to predict analytically the optical properties of nanostructures until a dimension of 200 nm, the fluorescence enhancement of a molecule as well as its quantum yield can be predicted analytically in the vicinity of complex nanostructures. At last, the CT strategy is refined to investigate the impact that nonlocality has on the optical properties of plasmonic nanostructures.

4.2 THE CONFORMAL TRANSFORMATION APPROACH

A conformal map is an analytic transformation $z' = f(z)$ that preserves local angles [82]. If we consider an analytic function $\phi(z)$ in the complex plane with $z = x + iy$, it will satisfy

$$\frac{\partial^2 \phi}{\partial x^2} + \frac{\partial^2 \phi}{\partial y^2} = 0. \tag{4.1}$$

If we now make a coordinate transformation $z' = x' + iy' = f(z)$, $\phi(z')$ also satisfies the Laplace equation in the new coordinate system:

$$\frac{\partial^2 \phi}{\partial x'^2} + \frac{\partial^2 \phi}{\partial y'^2} = 0, \tag{4.2}$$

provided that the transformation is analytic everywhere in the region under consideration. Hence, in 2D electrostatics, a CT preserves the potential in each coordinate system:

$$\phi(x, y) = \phi'(x', y'). \tag{4.3}$$

Moreover, the continuities of the tangential component of the electric field $E_{||}$ and of the normal component of the displacement current $D_\perp = \epsilon E_\perp$ across a boundary are also conserved due to the preservation of local angles. Thus, the permittivity of each material is the same in each geometry:

$$\epsilon(x, y) = \epsilon(x', y'). \tag{4.4}$$

On the one hand, the CT approach allows a tractable analytical study of complex structures, in contrast with full-wave coordinate transformations that generally imply strong spatial variations and anisotropy for the constitutive parameters. On the other hand, the CT approach is only valid in the quasi-static approximation: The plasmonic structures that we will deduce by conformal mapping should be one order of magnitude smaller than the wavelength. However, we will show in Section 4.6 how to go beyond the quasi-static limit and take into account radiative losses in the CT picture.

The strategy of our approach is as follows: Start with an infinite plasmonic system that generally shows a broadband spectrum and apply a mathematical transformation that converts the infinite structure into a finite one whilst preserving the spectrum.

We will start by describing a set of canonic structures easily tractable analytically on which we will then apply an ensemble of transformations.

4.2.1 A Set of Canonic Plasmonic Structures

The set of infinite plasmonic structures displayed in Figure 4.1 corresponds to the initial geometries used for the CT approach. Figures 4.1a and 4.1b correspond to the classical insulator–metal–insulator (IMI) and metal–insulator–metal (MIM) structures widely studied in plasmonics [5]. A 2D dipole pumps energy into the SP modes of the metallic slab(s) that transport the energy out to infinity. Although the SP spectrum of a plane surface is degenerate at ω_{sp}, the SP frequency, in a thin slab, hybridization of SPs on opposing surfaces creates a continuous spectrum with a lower bound at zero frequency and an upper bound at the bulk plasma frequency ω_p. Conformal mapping of such structures will give rise to a set of perfectly singular metallic structures (kissing nanowires, 2D crescent with touching tips, etc.) capable of an efficient harvesting and a strong nanofocusing of light over a broadband spectrum [45] (see Section 4.2.2).

The CT is not only dedicated to the design of broadband plasmonic nanostructures but can also generate resonant nanostructures. To that aim, an array of 2D dipoles

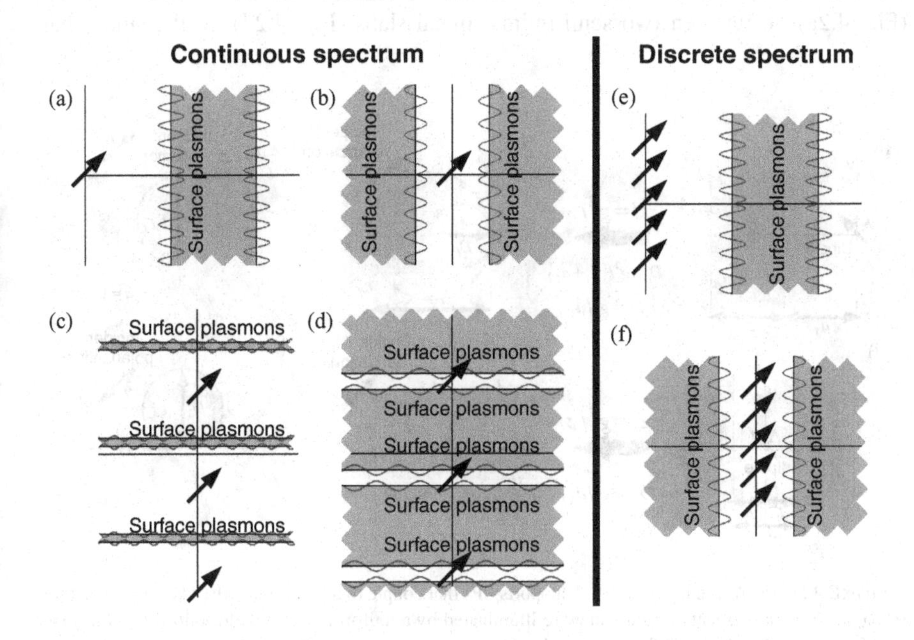

FIGURE 4.1 A set of canonic structures. A thin layer of metal (a) and two semi-infinite metallic slabs separated by a thin dielectric film (b) support SPs that couple to a 2D dipole source, transporting its energy to infinity. (c and d) An infinite and periodic stack of metallic slabs separated by dielectric layers can support hybridized SPs excited by an array of line dipoles. (e and f) The SP modes of the IMI and MIM structures are now excited by an array of 2D dipoles.

can be used to excite SPs in IMI and MIM geometries (see Figs. 4.1e and 4.1f). In that case, SPs can only be excited at certain resonant frequencies fixed by the pitch of the dipole array. Such a structure will give rise after conformal mapping to resonant nanostructures, such as metallic dimers [10–13, 15, 17, 18, 20–22, 24], 2D nanoshells [33–38], or a nanowire placed on top of a metal plate [39, 42, 88–90], that have been widely studied both numerically and experimentally in plasmonics (see Section 4.2.4).

At last, more exotic structures can be studied like the one shown in Figures 4.1c and 4.1d. A periodic and infinite stack of metallic slabs supports hybridized SPs over a broadband spectrum. The latter ones are excited by an infinite array of dipoles placed between each metallic slab. This geometry is unrealistic experimentally but can be transformed by conformal mapping into tips or wedge-like structures [91–93], common in plasmonics, but also to finite nanostructures such as 2D open crescents [25–32] or overlapping nanowires [14, 18, 19] (see Section 4.2.3). The propagation of SPs along rough metal surfaces [94–96] can also be investigated through a conformal map of the structures shown in Figures 4.1c and 4.1d.

4.2.2 *Perfect* Singular Structures

Our canonic system is a line dipole that can be located near a thin layer of metal (Fig. 4.2a) or between two semi-infinite metal slabs (Fig. 4.2d). In the study that

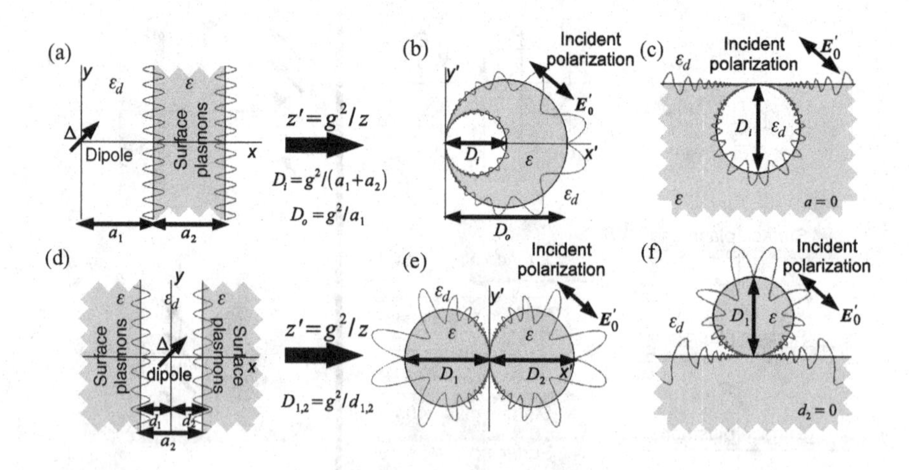

FIGURE 4.2 (a) A thin layer of metal supports SPs that couple to a line dipole. (b) The latter structure is mapped onto a crescent-shaped nanowire illuminated by a uniform electric field following an inverse transformation of Equation 4.5. (c) If the dipole is placed on the metallic surface in the original frame ($a = 0$), the inversion map of Equation 4.5 leads to a semi-infinite metal slab with a hole drilled just below its surface. (d) Two semi-infinite metallic slabs separated by a thin dielectric film support SPs excited by a 2D dipole source. (e) The latter structure is mapped with Equation 4.5 onto two kissing nanowires illuminated by a uniform electric field. (f) If the dipole is placed at the surface of one of the metal plates in the initial geometry, the transformed structure is a nanowire placed just on top of a metal plate.

follows, the metal permittivity is denoted as ϵ and the dielectric permittivity ϵ_d is fixed arbitrarily to 1. Now apply the following CT:

$$z' = \frac{g^2}{z}, \tag{4.5}$$

where $z = x + iy$ and $z' = x' + iy'$ are the usual complex number notations and g is an arbitrary length scale constant. Obviously all points at infinity in z translate to the origin in z' and planes translate into cylinders. Hence the resulting structures are a 2D crescent with touching tips (Fig. 4.2b) and two kissing cylinders (Fig. 4.2e).

For the crescent, we define a key parameter ρ which is the ratio between the inner and outer diameters of the crescent, D_i and D_o, respectively ($\rho = D_i/D_o$). Note that, if the line dipole is placed on the surface of the metal plate in the initial frames (Figs. 4.2a–4.2d), the resulting asymptotic structures are a cylindric hole drilled just below the surface of a metallic slab (Fig. 4.2e) and a nanowire placed on top of a metal film (Fig. 4.2f).

Transformation of the dipole is equally interesting – see Fig. 4.2. We have chosen the origin of our inversion at the center of the dipole. Thus the two charges comprising the dipole, very close to the origin in z, translate to near infinity in z', which gives rise to a uniform electric field $\mathbf{E}'_0$ in the transformed geometry. To prove it more rigorously, one can first express the incident potential due to the dipole $\Delta = \Delta_x \mathbf{u_x} + \Delta_y \mathbf{u_y}$ in the slab geometry,

$$\phi_0(x, y) = -\frac{1}{2\pi\epsilon_0}\frac{\Delta_x x + \Delta_y y}{x^2 + y^2} = -\frac{1}{2\pi\epsilon_0}\mathrm{Re}\left\{\frac{\overline{\Delta}}{z}\right\}, \tag{4.6}$$

with $\overline{\Delta} = \Delta_x + i\Delta_y$ the complex notation applied to the dipole moment Δ. The incident potential is preserved under the CT ($\phi_0(z) = \phi'_0(z')$), which yields the following expression for the incident potential in the transformed geometry:

$$\phi'_0(z') = -\frac{1}{2\pi\epsilon_0 g^2}\mathrm{Re}\left\{\overline{\Delta} z'\right\}. \tag{4.7}$$

This incident potential can be related to a uniform electric field, $\mathbf{E}'_0 = -\nabla\phi'_0$, such that

$$\mathbf{E}'_0 = \frac{1}{2\pi\epsilon_0}\frac{\Delta_x \mathbf{u_x} - \Delta_y \mathbf{u_y}}{g^2}, \tag{4.8}$$

with $\mathbf{u_{x'}}$ and $\mathbf{u_{y'}}$ the unitary vectors along the axis (Ox') and (Oy'). This confirms our previous intuition that the dipole Δ should map to a uniform electric field $\mathbf{E}'_0$. As we assume that the dimensions of the crescent and the kissing cylinders are small compared to the wavelength, the SP modes are well described in the near-field approximation. The uniform electric field can then be taken as due to an incident plane wave of polarization $\mathbf{E}'_0$.

The mathematics of CT closely links the physics at work in each of the very different geometries. Solving the relatively tractable slab problems solves the crescent and kissing cylinders problems. In particular, the dispersion relation of the SP modes supported by IMI and MIM structures under the near-field approximation can be transposed to all the transformed geometries. The dispersion of the excitations can be derived from the following condition [70,71]:

$$e^{2|k|a_2} - \left(\frac{\epsilon - 1}{\epsilon + 1}\right)^2 = 0. \tag{4.9}$$

The dispersion relation is thus given by,

$$ka_2 = \alpha = \begin{cases} \ln\left(\frac{\epsilon-1}{\epsilon+1}\right), \text{ if } \mathrm{Re}[\epsilon] < -1 \\ \ln\left(\frac{1-\epsilon}{\epsilon+1}\right), \text{ if } -1 < \mathrm{Re}[\epsilon] < 1 \end{cases} \tag{4.10}$$

with k the wavenumber of the SP modes. The dispersion relation is shown in Figure 4.3. The metal is assumed to be silver with an SP frequency $\omega_{sp} = 3.67$ eV and permittivity taken from Johnson and Christy [97].

The spectrum of modes is continuous with a lower bound at zero frequency and an upper bound at the bulk plasma frequency ω_p. At the SP frequency ω_{sp}, there is a singularity where the symmetry of the potential modes switches from antisymmetric to symmetric. Figure 4.3 indicates that the symmetric mode (namely, the SP excitation above the surface plasma frequency) is relatively narrowband compared to the antisymmetric mode. Therefore, in the following, we shall focus on the frequency range below the surface plasma frequency ω_{sp}.

In Figures 4.2a–4.2d, the dipole pumps energy into the SP modes of the metallic slab(s) which transport the energy out to infinity without reflection. The same modes

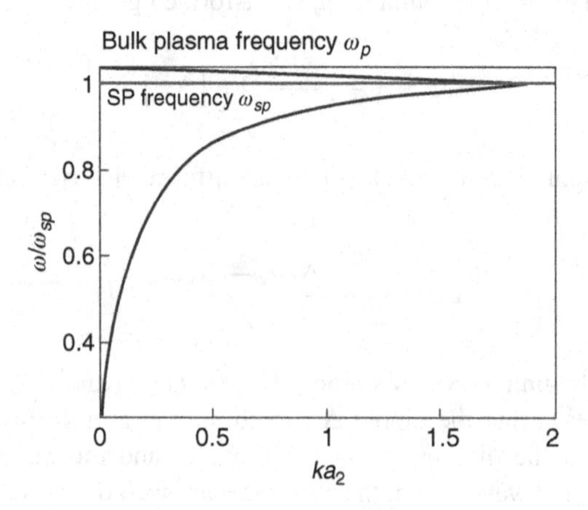

FIGURE 4.3 Dispersion relation for IMI and MIM plasmonic structures shown in Figures 4.2a and 4.2d.

are excited in the crescent and the kissing cylinders by the incident electric field $\mathbf{E}'_0$ (Eq. 4.8). Since we make the near-field approximation all dimensions are less than the wavelength but the changed geometry means that the excited modes have a net dipole moment. This provides coupling to the external field. Transformation of the modes tells us that in the crescent, modes are excited mainly at the fat part of the crescent and propagate around to the claws in an adiabatic fashion (Fig. 4.2c). Similarly, the plasmon modes propagate along the surface of the kissing cylinders toward the touching point (Fig. 4.2e). As SPs propagate toward the structure singularity, their wavelength shortens and velocity decreases in proportion. Just as modes excited in the original slab never reach infinity in a finite time, modes excited in the crescent or in kissing cylinders never reach the tips or the touching point. In an ideal lossless metal, energy accumulates toward the singularity, its density increasing with time without bound. In practice finite loss will resolve the situation leading to a balance between energy accumulation and dissipation. The harvesting and nanofocusing properties of the 2D crescent and kissing nanowires will be described in detail in Section 4.3. The physics described until now is similar for the two asymptotic structures shown in Figures 4.2c and 4.2f. The system consisting in a nanowire placed on top of a metal plate is investigated in detail in Reference 70.

Solving the 2D problem is interesting but quite restrictive in terms of applications. One can wonder if similar broadband and nanofocusing properties can be observed with 3D nanostructures like kissing spheres or a 3D crescent. This extension to 3D has been performed by Fernández-Domínguez et al., using coordinate transformation [67,78,79]. Starting this time from the desired geometry, a dimer of kissing spheres (see Fig. 4.4), the following 3D inversion can be applied:

$$\mathbf{r}' = \frac{g^2}{|\mathbf{r}|^2}\mathbf{r}, \tag{4.11}$$

where $\mathbf{r}$ and $\mathbf{r}'$ are the position vectors in the transformed and original space, respectively. Equation 4.11 transforms the spheres into two semi-infinite slabs. In contrast to 2D conformal mapping, the 3D inversion also acts on the material properties of the system. TO determines that the transformed slabs are not filled with a conventional metal, but with a modified material having a spatially dependent dielectric function of the form $\epsilon'(r') = \epsilon g^2/r'^2$ and that the relative permittivity between them is $\epsilon'_v(r') = g^2/r'^2$. Figure 4.4b describes the geometry and the permittivity distribution obtained from the inversion of Figure 4.4a.

Albeit more difficult to solve analytically, the 3D problem is actually very close to the 2D problem in terms of physics [67]. The SPs propagate along the surface of the two kissing spheres toward the touching point. As they approach the structure singularity, their velocity decreases, which leads to an accumulation of energy. Actually, the nanofocusing performance is even better since the energy is focused onto a single point in 3D instead of a line in 2D. The system also displays a continuous spectrum with a lower bound at zero frequency and an upper bound at the bulk plasma frequency. The light absorption and nanofocusing properties of the kissing spheres will be described in detail in Section 4.3.3.

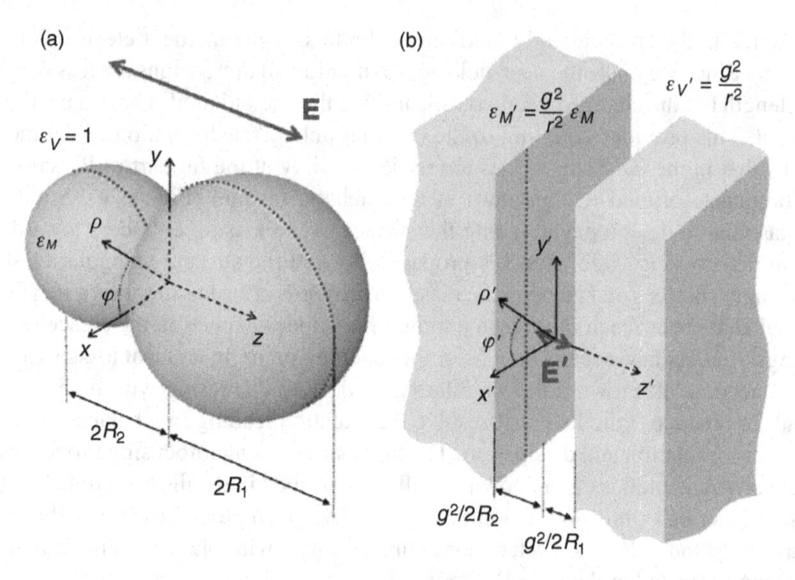

FIGURE 4.4 Transformation used to solve the kissing spheres problem. (a) Two metal nanospheres of radii R_1 and R_2 touching at a single point. (b) Slab geometry with the illumination and permittivities obtained from the inversion of the kissing nanospheres. Reprinted figure with permission from [67]. Copyright (2010) by the American Physical Society.

4.2.3 Singular Plasmonic Structures

4.2.3.1 *Conformal Mapping of Singular Structures*

A more systematic study of singularities in plasmonics is now presented and relies on the work made by Luo et al. [65]. The *mother* system is more complex than in the previous subsection (Figs. 4.5a and 4.5b). It consists in an array of 2D dipoles located at $z = i2\pi md$ (with m an integer) embedded into a periodic stack of infinite metallic films. Each element of the dipole array is assumed to have a dipole moment Δ. The total incident field ϕ_0 is given by

$$\phi_0(z) = -\frac{1}{2\pi\epsilon_0}\text{Re}\left\{\sum_{m=-\infty}^{m=+\infty}\frac{\overline{\Delta}}{z - i2\pi md}\right\}. \tag{4.12}$$

Now apply the following exponential transformation:

$$w = d\exp\left(\frac{z}{d}\right), \tag{4.13}$$

where $w = u + iv$ is the complex number notation. This CT maps lines parallel to the y-axis of the original coordinate system to circles centered around the origin in the w-plane. A metallic wedge (Fig. 4.5c) and a V-shaped metallic groove (Fig. 4.5d) are

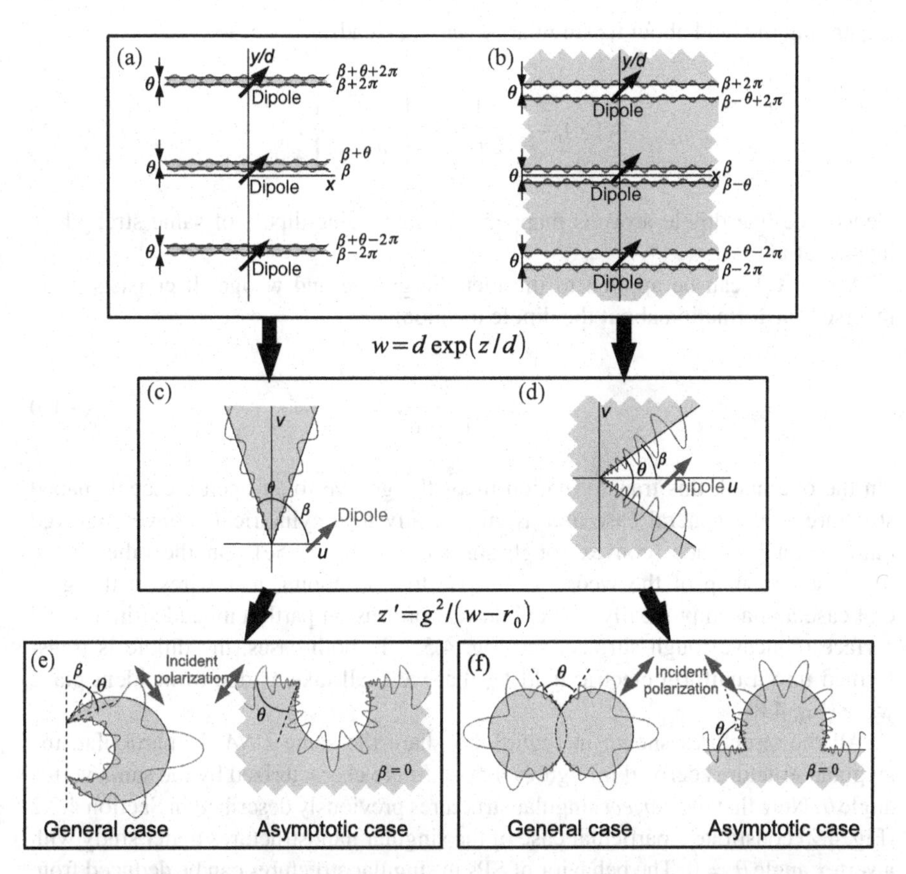

FIGURE 4.5 (a and b) Periodic metallic thin films support SPs that couple to a dipole array. The exponential map (Eq. 4.13) transforms these two structures into a metallic wedge (c) and a V-shaped metallic groove (d). The array of dipoles is transformed into a single dipole of same strength Δ located at $w = d$ (Eq. 4.15). Finally, under the inversion mapping (Eq. 4.16), the wedge/groove-like structures are converted to a metallic open crescent (e) and to overlapping nanowires (f). The dipole is mapped to a uniformed electric field $\mathbf{E}_0'$. In the particular case of a dipole placed along the metal wedge or groove, the transformed structures become a cylindrical groove engraved onto a metal surface (e) and a cylindrical protrusion partly embedded in a metal surface (f).

thus obtained. A geometric singularity is created at the origin by the exponential map. The vertex angle of the wedge and the groove is denoted as θ. As to the transformation of the source, the conservation of the incident potential ($\phi_0(z) = \phi_0'(w)$) leads to

$$\phi_0'(w) = -\frac{1}{2\pi\epsilon_0}\text{Re}\left\{\sum_{m=-\infty}^{m=+\infty} \frac{\overline{\Delta}}{d\ln(w/d)}\right\}. \tag{4.14}$$

Expanding the field about its singularity ($w = d$) leads to

$$\phi_0'(w) = -\frac{1}{2\pi\epsilon_0}\text{Re}\left\{\frac{\overline{\Delta}}{w-d}\right\}. \tag{4.15}$$

Hence, the line dipole array is mapped to a single line dipole of same strength Δ located at $w = d$.

A new CT can be applied to the metallic groove and wedge. It consists in an inverse transformation about the dipole location:

$$z' = \frac{g^2}{w-d}. \tag{4.16}$$

On the one hand, this transformation maps the groove to an open crescent-shaped structure in the general case and asymptotically to a cylindrical groove engraved onto a metal surface (convex rough surface, see Fig. 4.5e). On the other hand, the inversion map of the wedge gives rise to overlapping nanowires in the general case and asymptotically to a cylindrical protrusion partly embedded in a metal surface (concave rough surface, see Fig. 4.5f). In both cases, the dipole is transformed into a uniform electric field $\mathbf{E}_0'$ that we shall take as due to incident plane wave (Eq. 4.8).

All the structures shown in Figure 4.5 share the same *DNA*. In particular, the singular structures derived in Figures 4.5c–4.5f are characterized by the same vertex angle θ. Note that the *perfect* singular structures previously described in Section 4.2.2 (Fig. 4.2) constitute a particular case of the singular nanostructures under study with a vertex angle $\theta = 0$. The behavior of SPs in singular structures can be deduced from the *mother* systems (Figs. 4.5a and 4.5b). In particular, the dispersion relation of the SP modes supported by the initial system can be transposed to all the transformed geometries. This dispersion relation is defined by the following equation [65]:

$$\left(\frac{\epsilon-1}{\epsilon+1}\right)^2\left[e^{(2\pi-\theta)|k|d} - e^{\theta|k|d}\right]^2 - \left[e^{2\pi|k|d} - 1\right]^2 = 0. \tag{4.17}$$

Although, this relation is not tractable analytically, it can be solved recursively. The result is shown in Figure 4.6a. The metal is still assumed to be silver [97]. The plasmonic system is shown to support a continuous spectrum of bounded SP modes with symmetric and antisymmetric profiles of the tangential field E_x. The symmetric (or even) mode spans the frequency range $[\omega_c, \omega_{sp}]$, while the antisymmetric (or odd) mode spans the range $[\omega_{sp}, \omega_c']$. ω_c and ω_c' are the lower and upper bound cutoff frequencies determined by the following conditions:

$$\text{Re}\{\epsilon(\omega_c)\} = \frac{\theta - 2\pi}{\theta}, \quad \text{Re}\{\epsilon(\omega_c')\} = \frac{\theta}{\theta - 2\pi}. \tag{4.18}$$

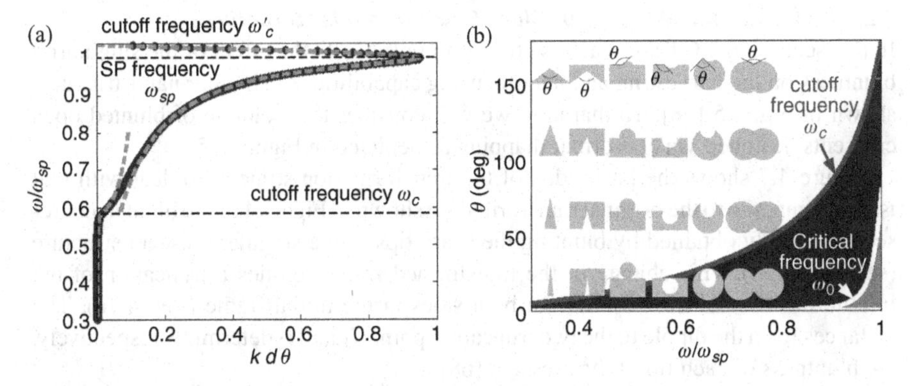

FIGURE 4.6 (a) Dispersion relation of the SP modes supported by the plasmonic structures shown in Figure 4.5. (b) Evolution of the lower bound cutoff frequency ω_c (red line, Eq. 4.18) and the critical frequency ω_0 (white line, Eq. 4.35) as a function of the wavelength λ and of the vertex angle θ. The spectrum of SPs is continuous from ω_c to ω_{sp} shaded and lightly-shaded. The field is divergent at the singularity of the nanostructure for $\omega < \omega_0$ shaded area. For both panels, the metal is assumed to be silver [97].

Although the dispersion relation cannot be solved analytically, two asymptotic solutions can be derived [73]:

$$kd\theta = \sqrt{\frac{3(\epsilon - \epsilon_c)}{\epsilon - \epsilon_c^3}}, \text{ for } \mathrm{Re}\{\epsilon\} \to \epsilon_c, \tag{4.19}$$

$$kd\theta = \frac{1}{2}\ln\left(\frac{\epsilon - 1}{\epsilon + 1}\right), \text{ for } \left|\ln\left(\frac{\epsilon - 1}{\epsilon + 1}\right)\right| \gg -\epsilon_c^{-1}. \tag{4.20}$$

These two asymptotes are superimposed to the dispersion relation in Figure 4.6a.

The fact that all the plasmonic structures shown in Figure 4.5 share the same dispersion relation implies that they display the same broadband properties from ω_c to ω_c', as summarized in Figure 4.6b. The bandwidth of these structures only depends on the vertex angle θ. The stronger the singularity is (i.e., smaller θ), the broader the bandwidth of the light harvesting process is. As shown by Figure 4.5, they also share the same nanofocusing mechanism with an accumulation of energy about their geometrical singularity due to the distortion of the space under conformal mapping. The field is even divergent at the singularity below a certain critical frequency ω_0 that we will define in Section 4.4.1 (see Fig. 4.6b). However, the nonlocal properties of ϵ encountered at small length scales will prevent the electric field from diverging at the singularity (see Section 4.7). Moreover, due to nanofabrication imperfections, the singularity of these structures cannot be perfectly reproduced in practice, which reduces the accumulation of energy at the tip of the singular structures.

4.2.3.2 *Conformal Mapping of Blunt-Ended Singular Structures*

In this section, the CT approach is applied to investigate the effect of the singularity bluntness on the harvesting and nanofocusing capabilities of singular nanostructures shown in Fig. 4.5 [68]. To that aim, we will consider the example of blunted open crescents designed via conformal mapping as depicted in Figure 4.5e.

Figure 4.7 shows the basic idea of the transformation strategy to deal with that issue. Compared to the original transformation shown in Figure 4.5, the blunt nanocrescent is simply obtained by blunting the sharp tips of the singular crescent structure (see Fig. 4.7b). This change in the transformed frame implies a truncation of the infinite metallodielectric system on both sides in the initial frame (Fig. 4.7a). The distances from the dipole to the two truncation points l_1 and l_2 determine, respectively, the bluntness of each tip of the crescent [68]:

$$\delta_j = \frac{D \sin\beta \tan(\theta/2)}{\left|\sin_h[(l_j + i\beta)/2]\sin_h[(l_j + i\beta + i\theta)/2]\right|}. \tag{4.21}$$

δ_1 and δ_2 denote the bluntness dimensions (or the diameters at the two blunt tips). The remaining geometrical parameters D, θ, and β are defined in the caption of Figure 4.7.

Since the whole system under investigation has now a finite size along x, the SP modes are now distributed at several discrete spatial frequencies $k_n = n\pi/(l_1 + l_2)$, with n an integer. Replacing the continuous spatial frequency k by the discrete spatial frequencies k_n into the previous dispersion relation (Eq. 4.17) leads to the following resonance condition:

$$\left(\frac{\epsilon-1}{\epsilon+1}\right)^2 \left[e^{nd\pi(2\pi-\theta)/(l_1+l_2)} - e^{nd\pi\theta/(l_1+l_2)}\right]^2 - \left[e^{2nd\pi^2/(l_1+l_2)} - 1\right]^2 = 0 \quad (4.22)$$

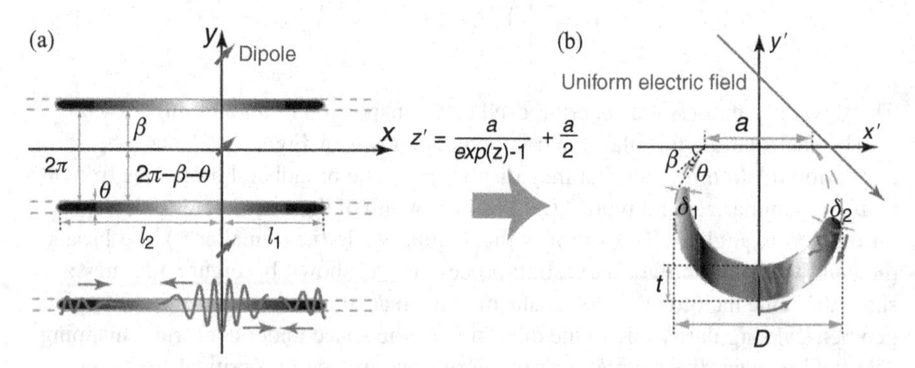

FIGURE 4.7 The coordinate transformation which bridges (a) the truncated periodic metallodielectric system and (b) the blunt crescent. D is the total dimension of the crescent, a denotes the distance between the two crescent tips, t represents the maximum distance between the inner and outer crescent boundaries (referred to as the crescent thickness), θ is the vertex angle, and β denotes the angle between the x-axis and the outer boundary of the crescent. Reprinted figure with permission from [68]. Copyright (2012) by the American Physical Society.

The bluntness imposed on the geometrical singularity makes the spectrum of modes become discrete. As shown in Reference 68, the number of SP modes decreases exponentially as the tip bluntness increases. The SP modes are characterized by a linear momentum k_n in the original frame that maps onto an angular moment n in the transformed geometries. In Section 4.4.2, we will show in detail the effect of the tip bluntness on the light harvesting and nanofocusing capabilities of singular nanostructures.

4.2.4 Resonant Plasmonic Structures

Until now, only singular structures have been studied within the CT approach. Resonant plasmonic structures are now considered since they do not rely on any geometrical singularity and thus will be easier to implement experimentally. To that aim, the initial geometry we consider are the MIM and IMI structures excited by an array of dipoles shown in Figures 4.8a and 4.8b. Unlike in Section 4.2.2, the modes supported

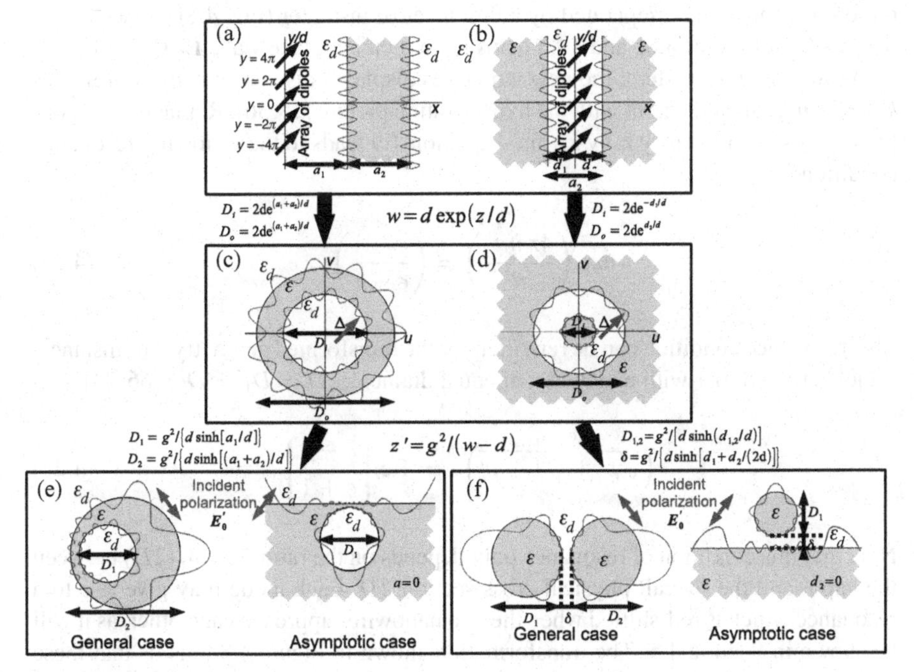

FIGURE 4.8 An array of dipoles Δ with a pitch $2\pi d$ excites the SP modes supported by an IMI (a) and MIM (b) structure. Upon exponential mapping (Eq. 4.13), these two materials are transformed into a metallic nanotube (c) and its complementary structure (d) excited by a single dipole of same strength Δ. Finally, an inverse transformation of (c) and (d) about the dipole location (Eq. 4.16) leads to a nanotube (e) and a dimer of nanowires (f) illuminated by a uniform electric field $\mathbf{E}'_0$. When the dipole is placed at the metal surface in (c) and (d), these two structures become asymptotically an insulator hole drilled into a semi-infinite metal plate (e) and a nanowire placed on top of a metal plate with a thin layer of insulator between them (f).

by these systems are discrete due to the presence of the dipole array and characterized by a linear momentum $k_n = 2\pi n/d$, with n an integer.

Now apply the exponential transformation already defined in Equation 4.13. The IMI structure (Fig. 4.8a) leads to a metallic nanotube (see Fig. 4.8c), whereas the MIM structure (Fig. 4.8b) gives rise to the complementary structure, that is, an insulator nanotube drilled into a block of metal (see Fig. 4.8d). The transformation of the source is the same as before: The array of dipoles Δ maps into a single dipole Δ located at $w = d$ (Eq. 4.15). The linear momentum of the SP modes supported by the original systems now maps to the angular momentum of SPs that propagate along the insulator–metal surfaces. The configurations described in Figures 4.8c and 4.8d can correspond to the experimental situation of a molecule of transition dipole moment Δ interacting with LSPs supported by a nanotube [98, 99], but we can go beyond these simple configurations by applying a new inverse transformation at the dipole position (Eq. 4.16). On the one hand, the metallic tube (Fig. 4.8c) is transformed into an off-axis nanotube in the general case and asymptotically to a hole drilled below the surface of a metal plate (Fig. 4.8e). On the other hand, the insulator tube (Fig. 4.8d) maps to a dimer of nanowires in the general case and asymptotically to a nanowire on top of a metal plate separated by a thin layer of insulator (Fig. 4.8f). Once again, the dipole of Figures 4.8c and 4.8d maps to an incident plane wave $\mathbf{E}'_0$ (Eq. 4.8).

As their *mother* system, the new structures display a discrete spectrum of modes $k_n = 2\pi n/d$, in agreement with the hybridization picture [15,36]. Replacing the continuous spatial frequency k by k_n into Equation 4.9 leads to the following resonance condition:

$$\exp\left(\frac{4\pi n a_2}{d}\right) = \left(\frac{\epsilon - 1}{\epsilon + 1}\right)^2. \tag{4.23}$$

This resonance condition can be rewritten for the transformed geometry. For instance, it yields for a dimer with nanowires of equal diameter ($D = D_1 = D_2$) [66, 74],

$$\left(\sqrt{\rho'} + \sqrt{1 + \rho'}\right)^{4n} = \mathrm{Re}\left\{\frac{\epsilon - 1}{\epsilon + 1}\right\}. \tag{4.24}$$

Note that this condition of resonance only depends on the ratio $\rho' = \delta/(2D)$ between the gap δ and the overall physical cross section $2D$. Each mode may give rise to a resonance which is red-shifted when the two nanowires approach each other, as it will be shown in Section 4.5. The transformation shown in Figure 4.8 tells us that these modes couple to each other in the narrow gap separating the two nanowires (Fig. 4.8f) or when the nanotube gets thinner (Fig. 4.8e): Their wavelength and velocity decrease, leading to an important field enhancement at this location. However, contrary to the kissing cylinders or the crescent-shaped cylinder (see Section 4.2.2), their velocity does not vanish, hence the LSPs turn infinitely around the nanoparticles before being absorbed. This accounts for their resonant behavior. This brief qualitative account will be completed by quantitative results in Section 4.5.

This section has presented the two main families of nanostructures that can be derived by conformal mapping from the classical MIM and IMI structures (Fig. 4.1). In the next three sections, we will present the main analytical results provided by this CT approach for each of these families.

4.3 BROADBAND LIGHT HARVESTING AND NANOFOCUSING

This section deals with the light harvesting and nanofocusing properties of touching dimers and crescent-shaped nanostructures (see Fig. 4.2). It reviews the main results presented in a series of three papers [45, 70, 71].

4.3.1 Broadband Light Absorption

The absorption cross section of nanostructures can be directly derived from the dipole power dissipated in the initial frame [70, 71]. The energy pumped into the SPs in the metal slab(s) (Fig. 4.2a) can be calculated from the scalar product of the induced electric field and the dipole [100]:

$$P = -\frac{\omega}{2} \operatorname{Im} \left\{ \Delta^{*}.\mathbf{E}(z = 0) \right\}. \tag{4.25}$$

This quantity can be calculated in the near-field approximation by picking out the poles due to the propagating SP modes [70, 71]. This dipole power dissipated maps directly onto the power absorbed from a plane wave incident on the nanostructure in the transformed geometry. Following this strategy, the absorption cross section σ_a of the crescent can be deduced for $\epsilon < -1$ [71]:

$$\sigma_a = \frac{\pi^2 \omega}{c} \left[\frac{\rho D_o}{1-\rho} \right]^2 \operatorname{Re} \left\{ \ln \left(\frac{\epsilon - 1}{\epsilon + 1} \right) \frac{4\epsilon}{1-\epsilon^2} \left(\frac{\epsilon - 1}{\epsilon + 1} \right)^{-\frac{2\rho}{1-\rho}} \right\}. \tag{4.26}$$

Note that all orientations of the incident electric field are equally effective at excitation. This isotropy is allowed by the contact between the two crescent tips. On the contrary, a highly anisotropic behavior is expected as soon as the tips are no longer touching [26–29, 65]. At the surface plasma frequency where $\epsilon = -1$, Equation 4.26 is singular if $\rho = D_i/D_o < 1/3$, that is, only for fat crescents. Figures 4.9a and 4.9b display the wave-length dependence absorption cross section σ_a as a fraction of the physical cross section, for $D_o = 20$ nm and considering silver [97]. For $\rho > 1/3$, the crescent exhibits a broadband spectrum that shifts toward red when the crescent gets thinner. The efficiency of the crescent in light harvesting is also significant since its absorption cross section is of the order of the physical cross section even for such a small particle size ($D_o = 20$ nm).

The same analytical calculation can be performed for two kissing nanowires of same diameter ($D = D_1 = D_2$). However, unlike the crescent, this plasmonic device is strongly dependent on the polarization of the incoming field. Indeed, the two

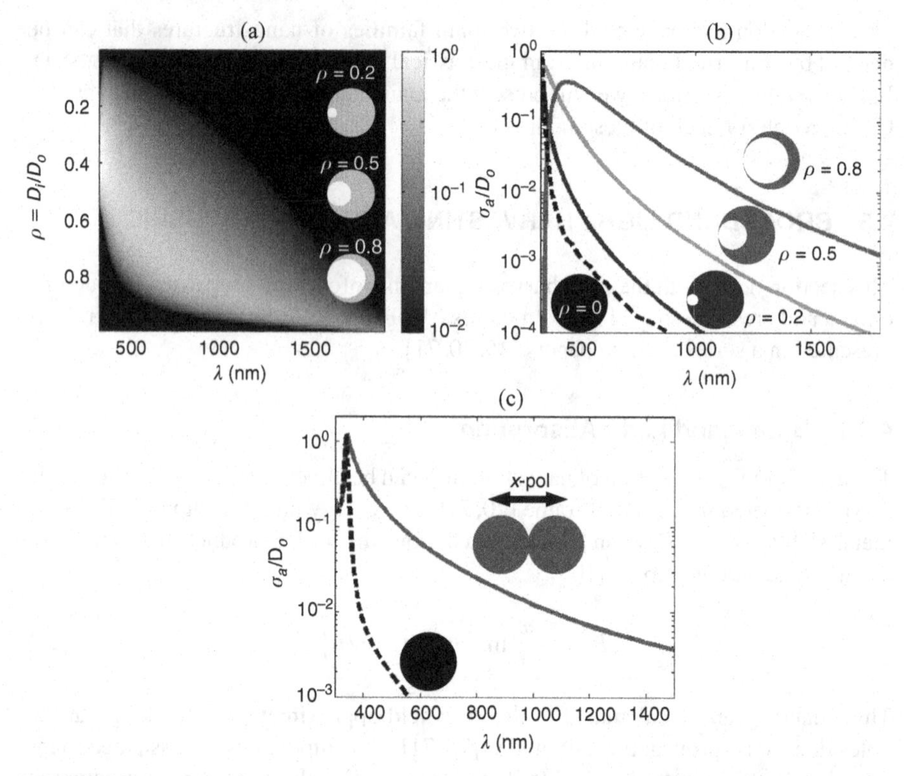

FIGURE 4.9 (a) Absorption cross section of the 2D crescent as a fraction of the physical cross section as a function of ρ and the wavelength. (b) The same result is shown for ρ =0.2, 0.5, 0.8 and compared to the case of a single nanowire [101]. (c) Absorption cross section of the kissing cylinders as a fraction of the physical cross section $D_o = 2D$ as a function of the wavelength λ for an incident electric field polarized along x'. The overall size D_o of each device is 20 nm. All the graphs are shown in log-scale.

kissing cylinders geometry is derived from an MIM plasmonic structure (Fig. 4.2d). This configuration only supports odd SP modes for $\epsilon < -1$ [70]. If the dipole is placed at the center of the two metal slabs, only its x-component can give rise to odd SP modes and its y-component is totally ineffective. In the transformed geometry, it means that only the x'-component of the incident electric field can excite SPs for two kissing cylinders of the same diameter. The corresponding absorption cross section σ_a can be derived for $\epsilon < -1$ [70]:

$$\sigma_a = \frac{\pi^2 \omega}{c} D^2 \operatorname{Re}\left\{\ln\left(\frac{\epsilon - 1}{\epsilon + 1}\right)\right\}. \tag{4.27}$$

Figure 4.9c displays σ_a as a fraction of the physical cross section ($D_o = 2D = 20\,\text{nm}$). Similar to the crescent, the two kissing cylinders are powerful light harvesting devices over a broadband spectrum for an incident wave polarized along x'. Note that, in both cases, σ_a/D scales linearly with D. Thus higher cross sections could be obtained for larger dimension but in this case our near-field analytic theory may not be valid [45,72] (see Section 4.6.2).

4.3.2 Balance between Energy Accumulation and Dissipation

The 2D crescent and kissing nanowires are capable of an efficient harvesting of light over the visible and near-infrared spectra. As we will now see, they are also strong far-field to near-field converters of energy, providing a considerable confinement and amplification of the electric field in the vicinity of their physical singularity. Figures 4.10a and 4.10b illustrate this fact by showing our analytical calculation of the field distribution in the two nanostructures. The metal is still silver [97]. In the slab frame, the SP modes transport the energy of the dipole out to infinity (see Fig. 4.2). In the

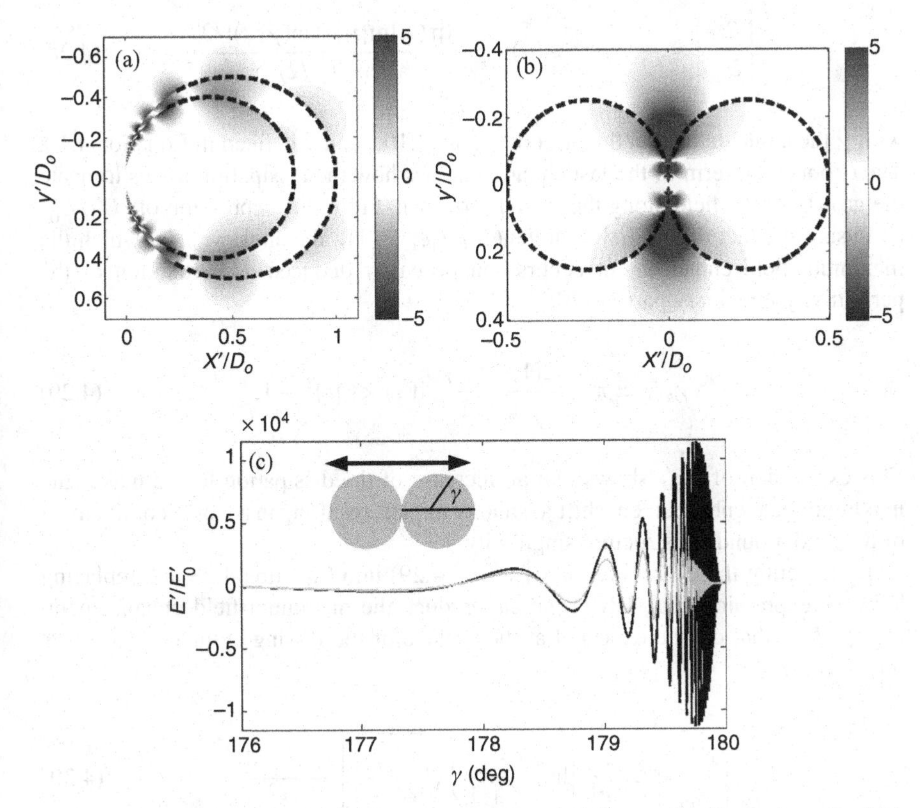

FIGURE 4.10 Amplitude of the x'-component of the electric field normalized by the incoming field (polarized along x'). The (a) and (b) panels display the field in the crescent (for $\rho = 0.8$) and in the two kissing cylinders, respectively, considering silver at $\omega = 0.9\omega_{sp}$. The black and white scale is restricted to $[-5\ 5]$ but note that the field magnitude is by far larger around the singularity of the structures. (c) The electric field is plotted along the surface of the kissing nanowires, plotted as a function of the angle γ, defined in the figure, for $\omega = 0.75\omega_{sp}$, $\epsilon = -7.06 + 0.21i$ [97] (black curve) and $\epsilon = -7.06 + 2 \times 0.21i$ (grey curve, losses $\times 2$). Both curves are normalized to the incoming field amplitude E'_0.

transformed frame, the same modes are excited in the fat part of the crescent and the diametrically opposite sides of the kissing cylinders. Then, they propagate around the claws of the crescent and the surface of nanowires in an adiabatic fashion.

Figure 4.10c shows our analytic calculation of the electric field induced at the surface of the kissing cylinders by a plane wave polarized along the x'-axis. As pointed out previously, the wavelength of SPPs shortens as they approach the touching point, leading to an enhancement of the electric field. The growth of the field is then truncated by absorption losses at a finite angle. A total field enhancement of 2×10^3 arises here at an angle $\theta = 179.75°$. Also shown is a second calculation in which losses are increased by a factor of two greatly reducing the enhancement and decreasing the angle at which maximum enhancement occurs.

For kissing nanowires, the field enhancement $|E'|/E'_{0x}$ along the nanowire can be expressed, for $\epsilon < -1$, as [70]

$$\left|\frac{E'}{E'_{0x}}\right| = \frac{\pi}{2}|\alpha|\,|\cosh(\alpha)|^{1/2}\,\frac{\exp\left(-\mathrm{Im}\{\gamma\}|\tan(\gamma/2)|/2\right)}{\cos^2(\gamma/2)}. \tag{4.28}$$

with γ the angle defined in the inset of Figure 4.10c, and α defined in Equation 4.10. The exponential term of the last equation shows how the dissipation losses truncate the growth of the field along the cylinder surface. From this expression of $|E'|/E'_{0x}$ and using the fact that $\mathrm{Im}\{\alpha\} = \arctan(2\epsilon_I/(|\epsilon|^2 - 1))$, the angle γ_{max} at which the maximum field enhancement occurs can be easily deduced as a function of the permittivity imaginary part ϵ_I [70]:

$$\gamma_{max} \simeq \pi - \frac{2(1-\rho)\epsilon_I}{|\epsilon|^2 - 1}, \text{ if } \epsilon_I \ll |\epsilon|^2 - 1. \tag{4.29}$$

This expression of γ_{max} shows that an increase of the dissipation losses makes the maximum field enhancement shift to smaller angles, resulting in a worse confinement of the field around the structure singularity.

By injecting the expression of γ_{max} (Eq. 4.29) into Equation 4.28 and replacing α by its expression (Eq. 4.10), one can deduce the maximum field enhancement, $|E'_{max}|/E'_{0x}$, that can be expected at the surface of the kissing cylinders, for $\epsilon_I \ll |\epsilon|^2 - 1$ [70]:

$$\left|\frac{E'_{max}}{E'_{0x}}\right| \simeq \frac{2\pi}{e^2}\left|\ln\left(\frac{\epsilon - 1}{\epsilon + 1}\right)\sqrt{\frac{\epsilon^2 + 1}{\epsilon^2 - 1}}\right|\frac{(|\epsilon|^2 - 1)^2}{\epsilon_I^2}. \tag{4.30}$$

The dissipation losses reduce the field enhancement as the inverse square of the permittivity imaginary part ϵ_I. This explains the ratio 4 observed between the black and grey curves in Figure 4.10c. The same analytical calculations can be performed for the crescent [71] or for a nanowire placed on top of a metal plate [70] (Fig. 4.2f).

Note that the field enhancement displayed by Figure 4.10c may be unrealistic in practice. There are indeed two limits to the CT approach:

- A micro-scale limit: When the size of the device becomes comparable with the wavelength, radiation losses are no longer negligible and will reduce the field enhancement induced by the nanostructure. This point has been addressed for kissing nanowires in Reference 72 and will be discussed in this review in Section 4.6.2 for the case of a dimer of nontouching nanowires.
- A nanoscale limit: At small length scales, continuum electrodynamics is no longer valid and nonlocal effects can result in an increase of the permittivity imaginary part [4, 102–106]. This issue will be addressed in Section 4.7. Furthermore, quantum mechanical effects, such as electron tunneling or screening, have to be taken into account in the vicinity of the structure singularity and may also reduce the field enhancement relative to classical predictions [107].

4.3.3 Extension to 3D

In Section 4.2.2, we have shown how the 2D conformal mapping of kissing nanowires can be extended to 3D leading to the study of the kissing spheres (Fig. 4.4). The problem cannot be solved exactly due to the spatial variations of the permittivity in the slab frame, but describes accurately the EM fields behavior at the touching point of the spheres. A quite simple expression can be found for the absorption cross section of two identical kissing spheres of radii R_1 with an incident polarization along x' [67]:

$$\sigma_a = \frac{64\pi^2}{3}\frac{\omega}{c}R_1^3\text{Re}\left\{\alpha^2 - \alpha\right\}. \tag{4.31}$$

Figure 4.11a plots the absorption cross section σ_a normalized by R_1^3 for kissing spheres of different dimensions calculated using COMSOL multiphysics (color or shaded dots). The metal is assumed to be silver with a permittivity taken from Palik [108]. The analytical spectrum obtained from Equation 4.31 is also shown as black solid line. The agreement between theory and simulations is remarkable for radii up to 35 nm, where the near-field approximation fails and radiation losses become significant. Note that the comparison for small R_1 worsens at high frequencies, as metal absorption prevents EM fields from reaching the touching point, where the theoretical result is the most accurate. Figure 4.11a indicates that σ_a for small dimers is of the order of the physical size even at frequencies well below the single sphere resonance (black dots). This demonstrates that touching dimers strongly interact with radiation over the whole optical spectrum and that their cross section presents a much smoother dependence on frequency than an isolated nanoparticle.

In order to gain a physical insight into the origin of the broadband response of touching spheres, the amplitude enhancement of the z-component of the electric field close to the contact point is plotted in Figure 4.11b. Fields are evaluated at 600 nm

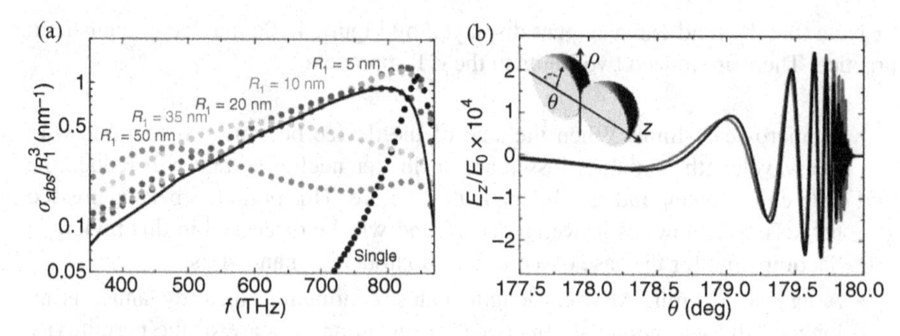

FIGURE 4.11 (a) Numerical absorption cross section σ_a versus incident frequency for twin Ag touching spheres of different radii R_1 (color or shaded dots) and a single 5 nm radius sphere (black dots). Black solid line plots the analytical result obtained from Equation 4.31. All the spectra are normalized to R_1^3. (b) Electric field amplitude enhancement versus angle θ (defined in the inset) at 500 THz ($R_1 = 20$ nm). Black (red or gray) line corresponds to analytical (numerical) calculations (Eq. 4.32). Reprinted figure with permission from [67]. Copyright (2010) by the American Physical Society.

(500 THz, $\epsilon = -16.0 + 2.1i$), where the dimer cross section is 50 times larger than the single sphere. For this magnitude, the TO approach yields [67]

$$\frac{E_z'}{E_0'} = \frac{\pi}{\sqrt{2}(1+\cos\theta)^{3/2}} \mathrm{Re}\left\{\alpha^2(3e^{\alpha/2} + e^{-\alpha/2}) J_0\left(\frac{\alpha}{2}\frac{\sin\theta}{1+\cos\theta}\right)\right\}, \quad (4.32)$$

where θ is defined in the inset of Figure 4.11b and J_0 is the Bessel function of rank 0. There is again a good agreement between theory (black) and simulations (red) for spheres of radius 20 nm. E_z'/E_0' is superior to 10^4 at the vicinity of the contact point of the spheres far from the single particle resonance. This is a clear demonstration of the broadband superfocusing capabilities of the structure. The physics is similar to the 2D case: Energy accumulates due to the drastic reduction of group velocity and effective wavelength that the surface waves experience while approaching the contact point of the dimer. Note that neither the TO model nor the numerical simulations include nonlocal effects (see Section 4.7).

4.3.4 Conclusion

This section has highlighted the power and elegance of the CT approach. On the one hand, this approach allows to design a path toward a broadband nanofocusing of light at the nanoscale. On the other hand, it provides novel insights into the physical behavior of SPs at the nanoscale. The nanostructures derived from conformal mapping should rely on perfectly shaped singularities on which drastic hot spots arise. In practice, such a geometry will be particularly difficult to implement. In the next section, this issue is addressed by considering the bluntness of plasmonic structures at the nanoscale, considering the conformal map described in Section 4.2.3.2.

4.4 SURFACE PLASMONS AND SINGULARITIES

This section deals with the light harvesting and nanofocusing properties provided by the family of nanostructures described in Figure 4.5 (V-groove, wedge, open crescent, rough surface, overlapping nanowires). These nanostructures have been studied in detail in a series of papers [65, 73, 75]. Here we will first focus on the effect of the vertex angle on the bandwidth of the light harvesting properties by considering the example of overlapping nanowires [73]. Second, we will show how conformal mapping can address the issue of not perfectly shaped singularities [68], inherent in any nanofabrication process.

4.4.1 Control of the Bandwidth with the Vertex Angle

The last section has addressed the issue of singular nanostructures (crescent, kissing nanowires) whose singularity is characterized by a zero vertex angle. This situation is quite unrealistic experimentally, hence one can wonder what happens when the vertex angle θ increases. Here, we will study the transition between kissing nanowires studied in Section 4.3 and overlapping nanowires (Fig. 4.5f) that has been studied in detail by Lei et al. [73].

Solving the problem in the slab frame allows to derive the absorption cross section σ_a following the same strategy as the one described in Section 4.3.1. A full-analytical expression of σ_a for two overlapping nanowires of same diameter D is not available but one can derive the following asymptotic limits for an incident wave polarized along x' [73]:

$$\sigma_a \sim 4\pi^2\sqrt{3}\frac{\omega}{c}\frac{\sin^2(\theta/2)}{\theta^2}D^2\mathrm{Re}\left\{\sqrt{\frac{\epsilon_c}{1-\epsilon_c^2}}\frac{1}{\sqrt{\epsilon-\epsilon_c}}\right\}, \text{ if } \mathrm{Re}\{\epsilon\}\to\epsilon_c, \quad (4.33)$$

$$\sigma_a \sim 4\pi^2\frac{\omega}{c}\frac{\sin^2(\theta/2)}{\theta^2}D^2\mathrm{Re}\left\{\ln\left(\frac{\epsilon-1}{\epsilon+1}\right)\right\}, \text{ if } \left|\ln\left(\frac{\epsilon-1}{\epsilon+1}\right)\right| \gg -\epsilon_c^{-1}. \quad (4.34)$$

with ϵ_c the permittivity threshold defined in Equation 4.18. The first asymptote of σ_a (Eq. 4.33) displays a square-root singularity at the cutoff frequency ω_c, for which $\mathrm{Re}\{\epsilon\} = \epsilon_c$. The second asymptote (Eq. 4.34) applies for larger frequencies than ω_c and is strictly identical to the expression derived for kissing cylinders (Eq. 4.27). Figure 4.12a displays the absorption cross section of overlapping nanowires normalized by the overall physical cross section D_o as a function of the wavelength λ and the vertex angle θ. Figure 4.12b displays the wavelength dependence of σ_a for different values of the vertex angle θ. The two asymptotes derived in Equations 4.33 and 4.34 are superimposed to σ_a. As shown in Figure 4.12, the absorption spectrum

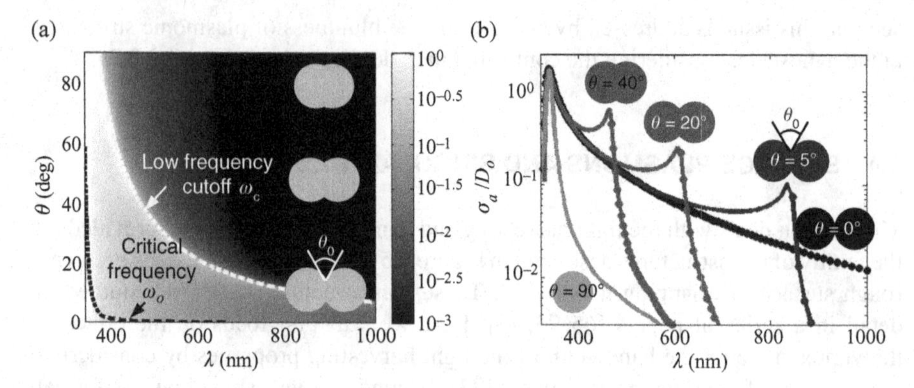

FIGURE 4.12 (a) Absorption cross section σ_a normalized by the physical cross section D_o as a function of θ and λ for an overlapping cylinders pair of overall size $D_o = 20$ nm. The white dashed line represents the low-frequency cutoff ω_c. The black dashed line represents the frequency ω_o (Eq. 4.35). The black and white bar is in log-scale. (b) The same quantity is displayed as a function of frequency for different values of $\theta = 5°, 20°, 40°$. The low-frequency asymptotes (shaded dots, Eq. 4.33) and the high-frequency asymptote (black dots, Eq. 4.34) are shown. At last, the single cylinder case is also shown for comparison [101]. For both panels, the metal is silver [97].

is strongly dependent on the vertex angle θ, which governs the cutoff frequency ω_c (Eq. 4.18). Three distinct regimes can be distinguished:

- Kissing cylinders regime ($\theta \to 0°$): The absorption spectrum of kissing nanowires corresponds to the high-frequency asymptote derived in Equation 4.34 (see Fig. 4.12b). In that case, the cutoff frequency ω_c is zero and the absorption cross section displays a continuous and broadband absorption spectrum over the whole visible and near-infrared spectra.
- Overlapping regime ($0° < \theta < 90°$): When the vertex angle θ increases, the cut-off frequency ω_c blue-shifts, which limits the bandwidth of the light harvesting process. An absorption peak is observed around ω_c and its line shape is well predicted by Equation 4.33 for $\omega < \omega_c$. When $\omega \to \omega_{sp}$, the device behaves like kissing cylinders (Eq. 4.34) in terms of light harvesting (see Fig. 4.12b). The comparison stops here since the nanofocusing behaviors differ in both situations.
- Single nanowire regime ($\theta \to 90°$): The two nanowires merge into a single one. The absorption spectrum exhibits one sharp resonance at the SP frequency ω_{sp} (see Fig. 4.12b).

In contrast to kissing nanowires, overlapping nanowires show a clear cutoff frequency ω_c in their absorption spectrum. This cutoff frequency can be adjusted by tailoring the overlap distance between the two nanowires. Note that the spectral properties shown by overlapping nanowires can be extended to all the structures displayed in Figure 4.5: A squeezed metallic wedge $\theta \to 0°$ can support SP modes over a broad bandwidth, whereas large angles $\theta \to 90°$ imply an extremely narrow line width.

The nanofocusing mechanism is on the contrary quite different compared to kissing nanowires (see Section 4.3). The field can be actually divergent at the singularity for the family of nanostructures shown in Figure 4.5. The field divergence condition can be rewritten as [65, 73]

$$\arctan\left[\frac{2\mathrm{Im}\{\epsilon\}}{|\epsilon|^2+1}\right] < \theta. \tag{4.35}$$

The term on the left is directly related to the dissipation losses in the metal with the imaginary part of the permittivity. The term on the right is the vertex angle which accounts for the field compression at the structure singularities. Let us introduce ω_0 the frequency for which the two terms of the last equation are equal. Below ω_0, Equation 4.35 is checked: The field compression dominates over dissipation losses and the field diverges at the structure singularities. Beyond ω_0, Equation 4.35 is no longer verified: The dissipation losses are large enough to make the field vanish at the structure singularity. The wavelengths corresponding to ω_0 and ω_c are shown as a function of θ and are superimposed on the absorption spectrum in Figure 4.12a. The comparison between the two curves shows that the frequency ω_0 is clearly larger than the cutoff frequency ω_c and that both frequencies blue-shift with the overlap distance. The divergence of the electric field occurs over most of the device bandwidth. This divergent behavior is surprising but not realistic in practice since it is strongly dependent on how sharp the singularity is.

4.4.2 Effect of the Bluntness

The effect of bluntness on singular nanostructures is now investigated considering the example of blunt-ended nanocrescents (see Fig. 4.7). However, note that this strategy can be applied to the whole set of singular plasmonic structures displayed in Figures 4.2 and 4.5. The absorption cross section σ_a of the crescent can be directly deduced from the power dissipated by the dipole in the original frame shown in Figure 4.7a. Analytical details are presented in Reference 68. Figure 4.13a shows the evolution of the absorption spectrum as a function of the bluntness δ of the tip crescent (Eq. 4.21).

Three distinct regimes can be distinguished:

- Single-resonance structure ($\delta/D > 10^{-1}$): The absorption spectrum displays only one sharp resonance at ω_{sp}. The system behaves like a single rounded nanoparticle.
- Multi-resonance regime ($0 < \delta/D < 10^{-1}$): When the tips of the nanocrescent become sharper, resonances associated to a small angular moment n start to arise at a smaller frequency than ω_{sp}. These resonances are red-shifted when δ decreases and the absorption spectrum displays several resonances in the visible spectrum in addition to the overall structure resonance at ω_{sp}.
- Broadband regime ($\delta/D \to 0$): When the tips become extremely sharp, the absorption spectrum becomes continuous and broadband with a lower cut-off frequency ω_c depending only on the vertex angle θ.

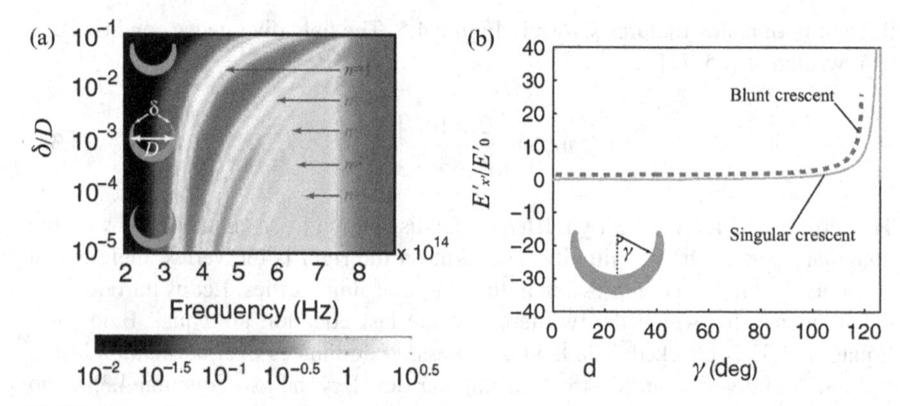

FIGURE 4.13 (a) Normalized absorption cross sections σ/D as a function of frequency and the tip bluntness δ for a nanocrescent with $\theta = 9^\circ$ and $t = 0.16D$. (b) Normalized electric fields $E'_{x'}/E'_0$ along the inner boundaries of the blunt and the corresponding singular crescents at $f = 368$ THz (the resonance of the dipole mode ω_c). Here, the geometry parameters are set as $\theta = 9^\circ$, $t = 0.16D$, and $\delta = 0.01D$ (for the blunt crescent). The metal is silver considering Palik data [108]. Reprinted figure with permission from [68]. Copyright (2012) by the American Physical Society.

Figure 4.13a indicates the degree of sharpness required to obtain a broadband harvesting of light. In the conditions of Figure 4.13a ($\theta = 9^\circ, t = 0.16D$), a bluntness $\delta/D = 10^{-2}$ is sufficient to get an efficient light absorption from $f = 350$ THz ($\lambda = 850$ nm) to $f = 900$ THz ($\lambda = 333$ nm), that is, over the whole visible spectrum.

Figure 4.13b shows the effect of the bluntness on the field enhancement that arises in the vicinity of the singularity. The normalized electric field $E'_{x'}/E'_0$ is plotted along the inner boundaries as a function of the angle γ defined in the inset. The case of the blunt (red dashed line) and the corresponding singular (blue solid line) crescents are compared. As expected, the maximum field enhancement is distinctly decreased around 25 as the claw tips are rounded. However, away from the tip, the field only undergoes minor changes due to the bluntness. It should be pointed out that quantum mechanical effects are not considered in our calculation. For subnanometer bluntness dimensions, a more general quantum description and nonlocal constitutive relation of the metal may be necessary to further improve the analytical model [15, 102].

4.5 PLASMONIC HYBRIDIZATION REVISITED WITH TRANSFORMATION OPTICS

This section deals with the light harvesting and nanofocusing properties provided by the family of resonant nanostructures described in Figure 4.8 (metallic nanotubes, nanowire dimer, etc.). Such nanostructures have been widely studied both theoretically, numerically, and experimentally in the literature [10, 11, 13, 15, 17, 18, 20, 33–39, 88–90]. An elegant physical picture to describe these resonant structures is the plasmon hybridization model [15, 36]. In analogy with molecular orbital theory, the dimer plasmons can be viewed as bonding and antibonding combinations of the individual nanoparticle plasmons. However, albeit elegant, the plasmon hybridization picture is a limited tool: Numerical simulations are still needed to calculate the optical response of the nanostructures. On the contrary, the CT approach provides an analytical description of the plasmonic hybridization and a unique physical insight into the propagation of SPs in that kind of nanostructures [66,74].

4.5.1 A Resonant Behavior

In this section, we will consider the example of a dimer of nanowires [66, 74] to illustrate the plasmonic hybridization concept. However, note that the same strategy can be applied to solve the problem of off-axis nanotubes [81] (Fig. 4.8e).

The absorption cross section σ_a of a cylinder pair (Fig. 4.8f) can be deduced in the quasi-static limit from the power absorbed by each dipole in the original frame (Fig. 4.8b). The following expression of σ_a is found for an incident wave polarized along x' [66]:

$$\sigma_a = \text{Im}\left\{16\pi\frac{\omega}{c}\rho'(1+\rho')D^2\chi\right\}, \tag{4.36}$$

$$\text{with } \chi = \frac{\epsilon-1}{\epsilon+1}\sum_{n=1}^{+\infty}\frac{n}{\left[\sqrt{\rho'}+\sqrt{1+\rho'}\right]^{4n}-\frac{\epsilon-1}{\epsilon+1}}, \tag{4.37}$$

with ρ' the ratio between the gap δ between the two nanowires and $2D$ their added diameters. The parameter χ in Equation 4.37 displays the sum of each contribution due to the LSP modes supported by the cylinder pair and denoted by their angular moment n. Each mode may give rise to a resonance at a frequency satisfying the condition previously defined in Equation 4.24.

Figure 4.14 illustrates this resonant feature by displaying $\sigma_a/(2D)$ as a function of frequency and $\rho' = \delta/2D$, for an overall physical cross section $2D = 20$ nm. The metal is silver with a permittivity taken from Johnson and Christy [97]. As shown by Figure 4.14, the absorption spectrum is strongly dependent on the ratio $\rho' = \delta/(2D)$ and shows three distinct regimes:

- Weak coupling regime ($\rho' > 0.5$, i.e, for a gap larger than the cylinder diameter): All the modes resonate at the vicinity of the surface plasma frequency ω_{sp}. The coupling between the two nanoparticles is weak and the system exhibits the same absorption spectrum as an individual cylinder.
- Strong coupling regime ($\rho' < 0.5$): When the two nanoparticles are approached by less than one diameter, resonances for small n start to arise at a smaller frequency than ω_{sp}. These resonances are red-shifted when the gap decreases and the absorption spectrum displays several resonances in the visible spectrum in addition to the individual LSPs resonance at ω_{sp}.
- Kissing cylinders regime ($\rho' \to 0$): The number of excited modes becomes infinite, leading to a continuous and broadband absorption spectrum (Fig. 4.9c).

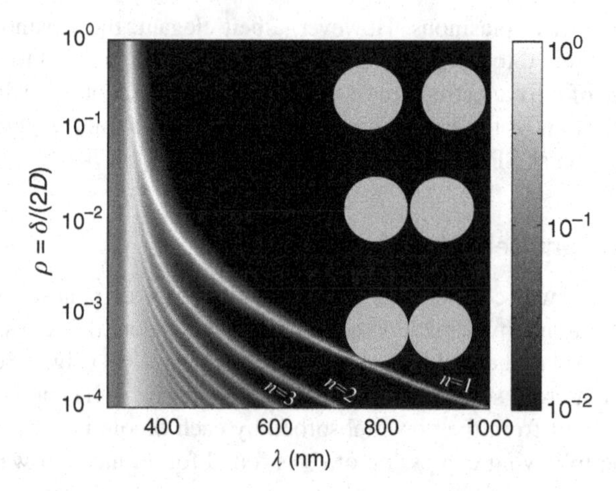

FIGURE 4.14 Absorption cross section σ_{ext} normalized by the physical cross section $2D$ as a function of ρ' and wavelength for a cylinders pair with $D_o = 10$ nm. The incident field is polarized along x'. The black and white bar is in log-scale.

4.5.2 Nanofocusing Properties

The electric field $\mathbf{E}'$ induced by the dimer can be deduced from the electrostatic potential solved in the original frame(Fig. 4.8b). It can be decomposed as an infinite sum of modes $\Psi^{(\mathbf{n})}$: $\mathbf{E}' = \sum_{n=1}^{\infty} \Psi^{(\mathbf{n})}$. The analytical expression of the modes $\Psi^{(\mathbf{n})}$ is derived in Reference 74. Note that in the near-field approximation, the enhancement of electric field is independent of the size of the system.

Figures 4.15a and 4.15b represent the imaginary part of the field distribution along x' and y', respectively. These figures can be easily interpreted with CT. In the slab frame, the SP modes transport the energy of the dipoles along the surface of the metal slabs (see Fig. 4.8b). The same modes are excited in the transformed frame and propagate along the cylinder surface. As they approach the gap separating the two nanoparticles, the LSPs supported by each nanoparticle couple to each other, their wavelength shortens, and group velocity decreases in proportion. This leads to an enhancement of the field in the narrow gap. However, contrary to the kissing cylinders (Section 4.3), the velocity of LSPs does not vanish and energy cannot accumulate infinitely in the narrow gap. Instead, LSPs propagate indefinitely around the cylinders, leading to the resonant behavior pointed out previously.

Figure 4.15c represents the imaginary part of the field along the surface of the cylinders. The comparison between each mode allows to confirm our previous qualitative description: The angular momentum n associated to each mode corresponds to the number of spatial periods covered by the SP when propagating around one nanowire. Figure 4.15c also highlights the drastic field enhancement that can be induced within the gap between the two nanoparticles. Typically, for $\delta = D/50$, the field enhancement $|E'|/E_0$ can reach a value of 600. Note that the field enhancement

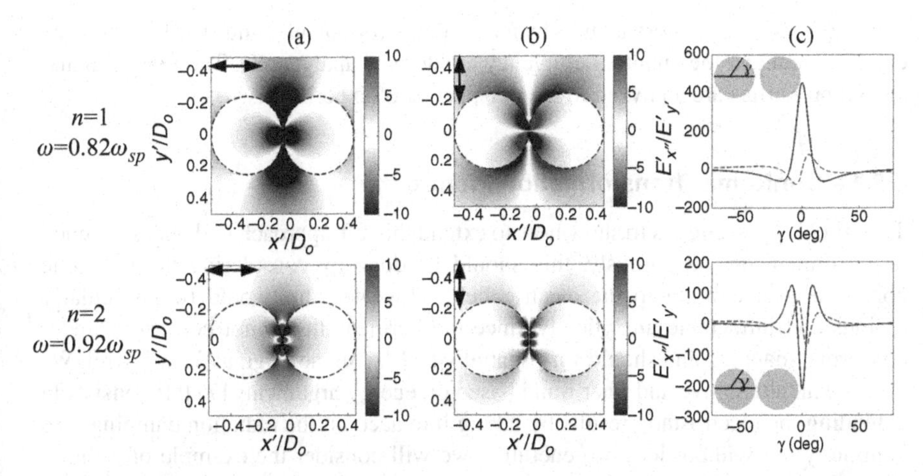

FIGURE 4.15 Electric field for $\rho' = 10^{-2}$ associated with the two first modes $n = 1, 2$ (from top to bottom) at their corresponding resonant frequencies (Eq. 4.24) for an incident wave polarized along x'. (a) Amplitude of the imaginary part of $\psi_{x'}^{(n)}$ normalized by the incoming field E_0' (polarized along x'). (b) Amplitude of the imaginary part of $\psi_{y'}^{(n)}$ normalized by the incoming field E_0' (polarized along y'). For (a) and (b) plots, the color scale is restricted to $[-10\ 10]$ but note that the field magnitude can by far be larger in the narrow gap between the structures. (c) Amplitude of the imaginary part of $\psi_{x'}^{(n)}/E_0'$ continuous curve and $\psi_{y'}^{(n)}/E_0'$ dashed curve along the cylinder surface as a function of the angle γ defined in the figure.

is less than one order of magnitude of the value obtained for kissing cylinders ($\sim 10^4$) (Fig. 4.10c).

Note that the results displayed by Figures 4.14 and 4.15 are valid in the quasi-static limit, that is, for $D < 20$ nm. In the next section, we show how to take into account the radiative losses in the CT picture and extend the range of validity of conformal mapping. Note also that the problem of a nanowire separated from a metal slab by an insulator thin layer (Fig. 4.8f) has also been solved explicitly and the analytical details are presented in Reference 74.

4.6 BEYOND THE QUASI-STATIC APPROXIMATION

Until now, the CT approach has been quite restrictive since it can only apply to nanostructures of a few tens of nanometers (typically 20 nm). In this section, the range of validity of this approach is extended by taking into account the radiation damping [72]. The radiative losses are shown to map directly onto the power dissipated by a fictive absorbing particle in the original frame. We apply this approach to the case of the nanowire dimer studied in the previous section. Radiative losses are shown to limit the light harvesting process but improve its broadband feature. The field enhancement induced by the nanostructure decreases with the structure dimension but remains significant. In a second part, we take advantage of this strategy to study the interaction of a dipole emitter (e.g., molecule or quantum dot) with complex plasmonic nanostructures by means of conformal mapping. The fluorescence enhancement and the quantum efficiency are derived analytically. Their spectral and spatial properties are analyzed in the perspective of experiments.

4.6.1 Conformal Transformation Picture

The aim of this section is to show how to extend this CT approach to devices of much larger dimensions. Rigorously, this should be done by considering the magnetic component in addition to the electric field. However, this would be particularly tedious and intractable analytically since the permeability is not conserved under conformal mapping and should vary spatially as $1/r^2$ in the original frame [109]. We propose an alternative and finer route based on energy arguments [72]. It consists in extending the electrostatic results by taking into account the radiation damping. For simplicity but without loss of generality, we will consider the example of a dimer of nanowires illuminated by an incident plane wave $\mathbf{E}'_0$. A dipole $\mathbf{p}'$ is placed in the vicinity of the plasmonic system and accounts for the transition dipole moment of a fluorophore (Fig. 4.16a). This complex problem can be mapped to a much simpler geometry through the transformation described in Figure 4.16 and already studied in the quasi-static limit in Section 4.5.

In the dimer frame, the incoming beam is both absorbed and scattered by the nanostructure. The scattered field can be taken as uniform in the near field of the nanoparticles. Similar to the incident field that is transformed into an array of dipoles in the slab frame, the counterpart of the scattered field in the slab geometry is an array of fictive absorbing particles, Δ_s, of polarizability $\gamma_s = -i\frac{\pi^2}{2}\epsilon_o g^2\omega^2/c^2$ [74], superimposed on the emitting dipoles Δ (Fig. 4.16b). These absorbers account for the radiative losses in the transformed geometry.

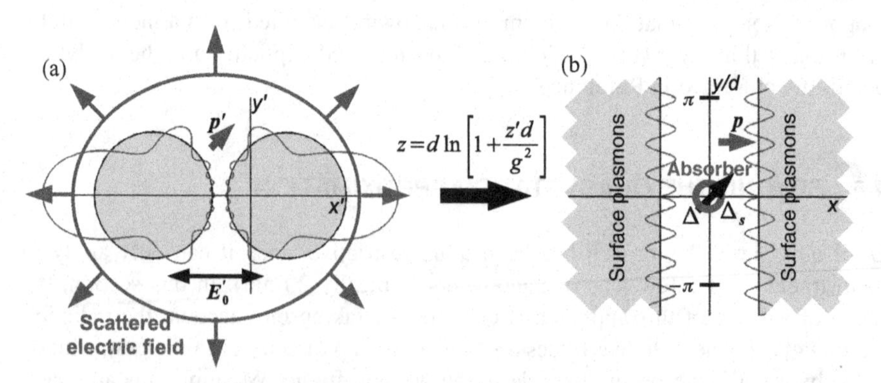

FIGURE 4.16 (a) A pair of metallic nanowires of diameter D and separated by a distance δ is illuminated by an incident plane wave $\mathbf{E}'_0$. The dipole $\mathbf{p}'$ accounts for a fluorescent molecule placed in the vicinity of the nanostructure. (b) Using the transformation already described in Section 4.2.4, the latter system is transformed into two semi-infinite metallic slabs separated by a thin dielectric film. $\mathbf{E}'_0$ and $\mathbf{p}'$ are transformed into two dipoles Δs and $\mathbf{p}$, respectively. An absorbing particle Δ_s accounts for the radiative losses in the initial frame [72].

Radiative losses in the transformed geometry correspond to the power dissipated by this small absorber in the original frame. The radiative reaction is then considered to go beyond the electrostatic approximation. Taking into account the radiative reaction is necessary in order to satisfy the optical theorem (i.e., energy conservation) [110]. In the transformed frame, it corresponds to the self-interaction between the particle and its own scattered field [110–112]. In the slab frame, the radiative reaction corresponds to the field scattered by the fictive absorbing particle and back-emitted toward the metal slabs. The analytical results taking into account radiation damping will be presented in Section 4.6.2 and will be confronted to numerical simulations.

The fluorescent molecule $\mathbf{p}'$ located at z_0' in the dimer frame is transformed into an array of dipoles, whose location z_0 and dipole moment $\mathbf{p}$ are given by

$$z_0 = d \ln\left(1 + \frac{g^2}{z_0' d}\right), \quad \overline{p} = -\frac{g^2/z_0'^2}{1 + \frac{g^2}{z_0' d}} \overline{p'}. \tag{4.38}$$

On the one hand, we will be able to study the total decay rate of the fluorescent molecule $\mathbf{p}'$ by investigating the total power of the dipole $\mathbf{p}$ dissipated in the slab frame. On the other hand, the radiative decay channel of $\mathbf{p}'$ will correspond to the power dissipated by each small absorber Δ_s in the slab geometry. This problem will be treated in Section 4.6.3.

4.6.2 Radiative Losses

The introduction of a fictive absorbing particle at the origin of the slab frame allows to take into account radiation damping in the transformed frame. It yields the following expression for the extinction cross section σ_{ext} of a dimer illuminated with a x'-polarized incident field [66]:

$$\sigma_{ext} = \operatorname{Im}\left\{\frac{16\pi \frac{\omega}{c} \rho'(1+\rho') D^2 \chi}{1 - i2\pi \rho'(1+\rho') D^2 \frac{\omega^2}{c^2} \chi}\right\}. \tag{4.39}$$

with the parameter χ defined in Equation 4.37. Equation 4.39 should be compared with the previous expression of the absorption cross section σ_a derived under the quasi-static limit (Eq. 4.36). Radiation losses lead to a renormalization of σ_{ext} by the denominator of Equation 4.39. Figure 4.17a shows the effect of radiative damping on the extinction spectrum for different size of dimers at a fixed ratio $\rho' = \delta/(2D) = 0.01$. The theoretical predictions (Eq. 4.39) are compared to the results of numerical simulations with the software COMSOL. An excellent agreement is found in the quasi-static limit ($D = 10$ nm). For larger structure dimensions, the numerical results are slightly red-shifted compared to our theoretical predictions. This is due to the retardation effects which are not considered by the CT approach [72]. However, the magnitude and line shape of resonances are nicely predicted for structure dimension up to 200 nm. Figure 4.17a shows that radiative damping broadens the line width of

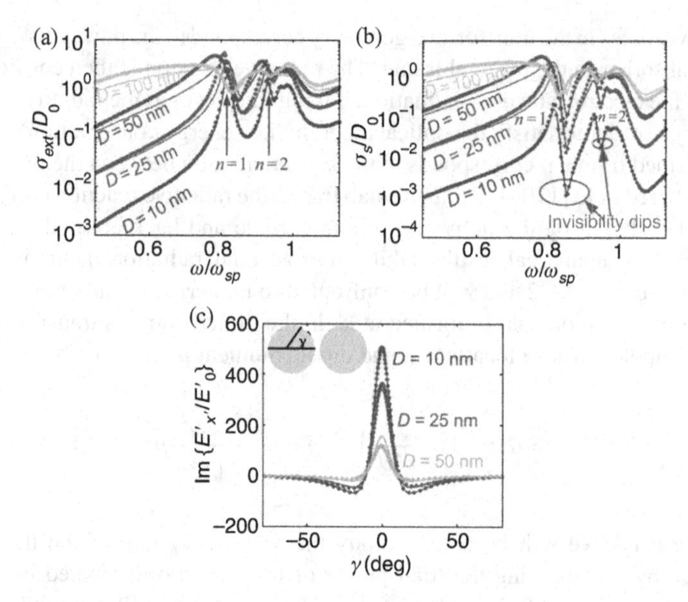

FIGURE 4.17 Extinction cross section σ_{ext} (a) and scattering cross section σ_s (b), normalized by the physical cross section $2D$, plotted as a function of frequency for nanowire dimers of different size with $\rho' = 0.01$. The theoretical predictions (continuous lines, Eqs. 4.39 and 4.40) are compared to the results of numerical simulations (dots). (c) Imaginary part of $E'_{x'}/E'_0$ along the surface of the nanowire at the first resonant frequency $\omega = 0.8\omega_{sp}$, for $\rho' = 0.01$. The theoretical predictions (continuous lines, Eq. 4.41) and numerical results (dots) are compared for different nanowire diameters D: 10 nm, 25 nm, 50 nm. For all panels, the metal is assumed to be silver [97].

each resonance and leads to the saturation of the extinction cross section at the level of the physical cross section.

An expression of the scattering cross section σ_s can also be derived by computing the power dissipated by the fictive absorber in the slab frame:

$$\sigma_s = \frac{32\pi^2 \frac{\omega^3}{c^3} \rho'^2 (1+\rho')^2 D^4 \, |\chi|^2}{\left|1 - i2\pi\rho'(1+\rho')D^2 \frac{\omega^2}{c^2}\chi\right|^2}. \tag{4.40}$$

The radiative spectrum depends on $|\chi|^2$. Hence, the resonances defined by Equation 4.24 also occur in the scattering spectrum. This is confirmed by Figure 4.17b which displays $\sigma_s/(2D)$ as a function of frequency for different size of dimers at a fixed ratio $\rho' = 0.01$. There is a perfect agreement between our theoretical prediction and the numerical result in the quasi-static limit ($D = 10$ nm). The resonances displayed in Figure 4.17b clearly display an asymmetric line shape. We stress on the fact that these are not Fano resonances which appear usually in the extinction spectrum and correspond to the coupling between bright and dark modes [113–116]. The sharp dips observed in Figure 4.17b originate from the destructive interference between each successive bright mode. Typically, the sharp dip observed at $\omega = 0.85\omega_{sp}$ results from the destructive interference between the modes $n = 1$ and $n = 2$ which resonate on each side of the dip [74]. This feature can be promising in perspective of sensing applications, since the ratio between the absorption and scattering cross sections can reach for instance a value of 150 in the conditions considered in Figure 4.17b. At these invisibility frequencies, the nanowire dimer can harvest light efficiently from the far-field and focus its energy at the nanoscale, without any scattering in the surrounding area. The dimer acts then as an invisible/noninvasive sensor. This idea is in relation with the concept of sensor cloaking proposed by Alú and Engheta [117]. Note that the dimer is invisible in all the directions except in the forward scattering one where the absorption of light will induce an attenuation of the incident beam.

As for the extinction cross section, radiation damping leads to a renormalization of the scattering cross section for large structure dimensions (see the denominator in Eq. 4.40) which makes σ_s saturate at the level of the physical cross section (see Fig. 4.17b). A satisfying agreement is found between numerical and analytical results in Figure 4.17b, except for the slight red shift explained by retardation effects. The Q-factor of the invisibility dips decreases for large structure dimensions and the nanowire dimer may keep its invisible feature only for nanowire diameter inferior to 50 nm.

In the quasi-static limit, the near-field enhancement does not depend on the size of the device. However, radiative damping breaks this property and limits the nanofocusing properties of the dimer. Radiative reaction requires the renormalization of the electric field by a factor ζ [74]:

$$\zeta = 1 - 2i\pi\rho'(1+\rho')D^2\frac{\omega^2}{c^2}\chi. \tag{4.41}$$

The effect of the radiative losses on the field enhancement is shown in Figure 4.17c. The imaginary part of $E'_{x'}/E'_0$ along the surface of the nanowire is displayed at the first resonant frequency $\omega = 0.8\omega_{sp}$, for $\rho' = 0.01$. The theoretical and numerical results are compared for different dimensions $D = 10$, 25, 50 nm and a good agreement is found. The radiation losses lead to a renormalization of the electric field by the factor ζ which increases with the structure dimension (Eq. 4.41). For $D = 50$ nm, a slight disagreement starts to appear between theory and numerical simulations, due to the retardation effects which are not taken into account by our model [72].

Note that the same strategy can be applied to determine the radiative losses in all the structures that can be derived via conformal mapping. For instance, the effect of radiative damping has been studied analytically for kissing nanowires as well [72].

4.6.3 Fluorescence Enhancement

We now apply the CT strategy beyond the quasi-static limit to predict analytically the fluorescence enhancement for molecules placed in the vicinity of metallic nanostructures. We first briefly recall the main point of the spontaneous emission by a molecule in the near-field of a nanoantenna. Then we will show how the CT approach can be used to predict analytically the fluorescence enhancement of molecules placed in the vicinity of metallic nanostructures. As the absorption and emission processes occur at different frequencies, a perfect knowledge of the spectral and spatial properties of the field enhancement might be decisive for the implementation of new types of biosensors.

4.6.3.1 Fluorescence Enhancement in the Near-Field of Nanoantenna

A plasmonic device can strongly influence the fluorescence of a molecule placed in its near field [44, 118–122]. On the one hand, LSPs can induce intense fields at the nanoscale that are likely to increase the number of photons absorbed by a molecule at its excitation frequency. On the other hand, at the emission frequency, the radiative and nonradiative properties of the emitter strongly depend on its environment. This is the so-called Purcell effect [123]. The interplay between local-field enhancements, and radiative and nonradiative decay channels can lead to an enhancement of the fluorescence signal or to quenching.

The fluorescence emission by a molecule is governed by an important parameter, the decay rate γ, which describes the amplitude decay suffered by the transition dipole of the molecule in time. Two contributions to this overall decay can be identified: γ^R is the radiative decay rate that gives rise to the emission of a photon in the far field; γ_{int}^{NR} is an intrinsic nonradiative decay rate that accounts for internal losses. The emission efficiency of a fluorophore is then quantified by the intrinsic quantum yield:

$$\eta_0 = \frac{\gamma_R^\circ}{\gamma^\circ} = \frac{\gamma_R^\circ}{\gamma_R^\circ + \gamma_{NR}^\circ}. \tag{4.42}$$

The quantum yield represents the probability that an excited fluorescent molecule emits one photon. The presence of a nanoparticle in the vicinity of the emitter can modify the quantum yield of the molecule, such that

$$\eta = \frac{\gamma_R}{\gamma} = \frac{\gamma_R}{\gamma_R + \gamma_{NR} + \gamma_{NR}^\circ} \tag{4.43}$$

We can identify three different decay channels in the system now: the radiative rate, γ_R, the external nonradiative decay rate induced by dissipation in the metal, γ_{NR}, and the nonradiative decay rate intrinsic to the emitter, γ_{NR}°. A metallic nanostructure sustains highly confined modes and induces strong radiative losses. Hence, it may give rise to a large enhancement of the radiative decay rate experienced by any emitter placed in its vicinity. The ratio γ_R/γ_R° is usually termed as the Purcell factor.

The fluorescence of a molecule is given by the product of two different factors: the emitter quantum yield, η, which measures its ability to re-radiate once excited, and the EM energy that it is able to absorb from the incident electric field in the excitation process, which is proportional to $|\mathbf{p}'.\mathbf{E}'|^2$ (where $\mathbf{p}'$ is the transition dipole moment and $\mathbf{E}'$ is the electric field evaluated at the emitter position). The fluorescence enhancement then reads

$$S/S_0 = \left(\eta \left|\mathbf{p}'.\mathbf{E}'\right|^2\right) / \left(\eta_0 \left|\mathbf{p}'.\mathbf{E}'_0\right|^2\right). \quad (4.44)$$

The radiative decay rate γ_R is proportional to the power radiated in the far field, while the nonradiative decay rate γ_{NR} is proportional to the power absorbed by the metallic nanostructures.

4.6.3.2 The CT Approach

As an example, we consider the interaction of a fluorophore with a dimer of nanowires as already described in Figure 4.16. To solve this problem, this configuration is mapped onto a much simpler one consisting of a dipole sandwiched between two semi-infinite metal slabs (Fig. 4.16b). The solution in the latter geometry allows to derive analytically different quantities such that the fluorescence enhancement as well as the quantum efficiency in presence of the metallic nanostructure. Note that previous works have investigated theoretically similar problems [44, 124, 125] but none of them have led to a full-analytical solution.

The radiative decay rate is proportional to the power radiated by the nanostructure. In the slab frame (Fig. 4.16a), it consists in calculating the power dissipated by the small absorber accounting for radiative losses. Normalizing it by the power radiated by the molecule in free space leads to the Purcell factor γ_R/γ_R°. The wavelength dependence of the Purcell factor is displayed in Figure 4.18a for a transition dipole moment $\mathbf{p}'$ aligned along x' and located at the center of the gap. The Purcell factor displays dramatic peaks ($> 10^7$) at the LSPs resonances (Eq. 4.24). Figures 4.18b and 4.18c show the spatial dependence of the Purcell factor for a transition dipole moment $\mathbf{p}'$ aligned along x' and y', respectively. These graphs are shown at the first LSP resonance($n = 1$) for a gap $\delta = D/50$ between the two nanowires. The Purcell factor displays strong spatial variations in the near field of the nanowire with the highest values within the gap. These results show the extreme tunability of the Purcell factor according to the dipole position and the wavelength.

The quantum yield is an important parameter for future experiments. It can be computed through the CT approach by calculating the ratio between the power dissipated by the small absorber accounting for radiative damping (proportional to the radiative decay rate) and the total power dissipated by the array of dipoles $\mathbf{p}$ in the slab frame (proportional to the total decay rate). This is valid under the assumption that the nonradiative decay rate intrinsic to the emitter, γ_{NR}°, is negligible compared to the nonradiative decay rate due to the dissipation losses in the metal, γ_{NR}°. Figures 4.19a and 4.19b show the spatial dependence of the quantum yield η (Eq. 4.43) for a transition dipole moment $\mathbf{p}'$ aligned along x' and y', respectively. These graphs are shown at the first LSP resonance($n = 1$) for a gap $\delta = D/50$ between the two nanowires. Similar to the Purcell factor, the quantum yield displays strong variations in the near field of the dimer. For a transition dipole moment oriented along x', the quantum yield is maximum at the exact center of the gap but strongly decreases when

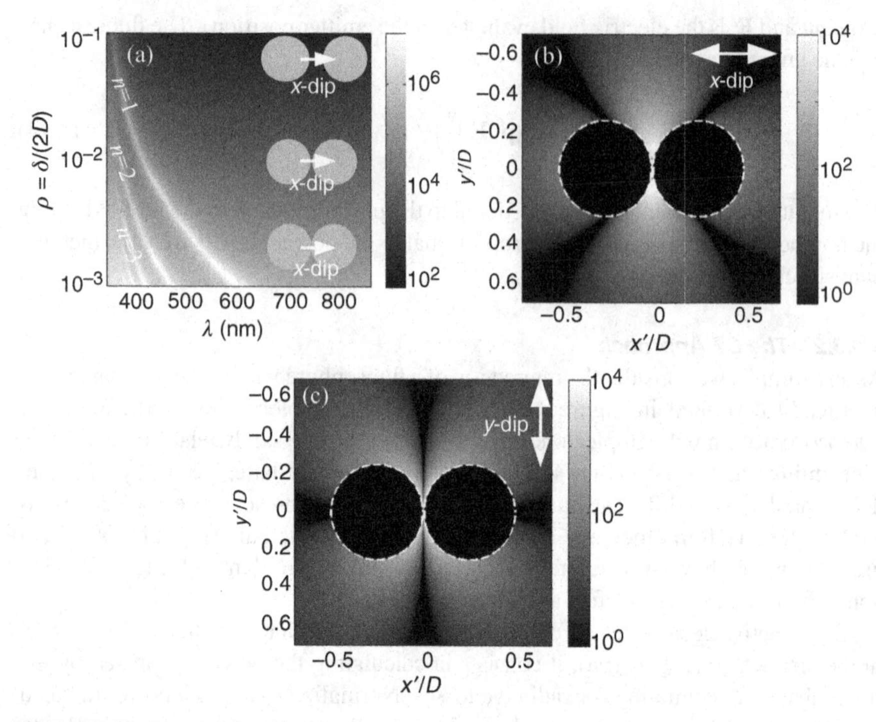

FIGURE 4.18 Purcell factor for a dimer of silver nanowires (with $D = 20$ nm, permittivity taken from Reference 97). (a) Wavelength dependence of the Purcell factor as a function of the distance between the two nanowires for a fluorophore placed exactly at the center of the gap and a transition dipole moment aligned along x'. (b and c) Spatial dependence of the Purcell factor at the first resonance frequency ($n = 1$) for a transition dipole moment aligned along x' and y', respectively. The dimer size is 20 nm and the gap δ is $D/50$.

the molecule moves along the axis of symmetry of the dimer. The lines defined by $y' = \pm x'$ correspond to the area outside of the gap where a molecule, whose dipole moment is oriented along the x'-axis, is quenched. For a dipole moment oriented along the y'-axis, the molecule is quenched if it is placed along the axis of the dimer, hence in the gap as well.

Figures 4.19c and 4.19d show the fluorescence enhancement (Eq. 4.44) in the same conditions as for the quantum yield (Figs. 4.19a and 4.19b), assuming that the absorption and emission processes occur at the first resonance frequency of the dimer. Of course, this is a strong hypothesis but, note that, in practice, different absorption and emission frequencies can be incorporated in the model according to the fluorophore used in the experiment. The intrinsic quantum yield η_0 has been arbitrarily fixed to 0.66. Again, the fluorescence enhancement highly depends on the position of the fluorophore and on the orientation of its dipole moment. Maximal values for the fluorescence enhancement (up to 10^5) occur in the gap for a dipole oriented along x'. Hence, plasmon hybridization offers spectral and spatial degrees

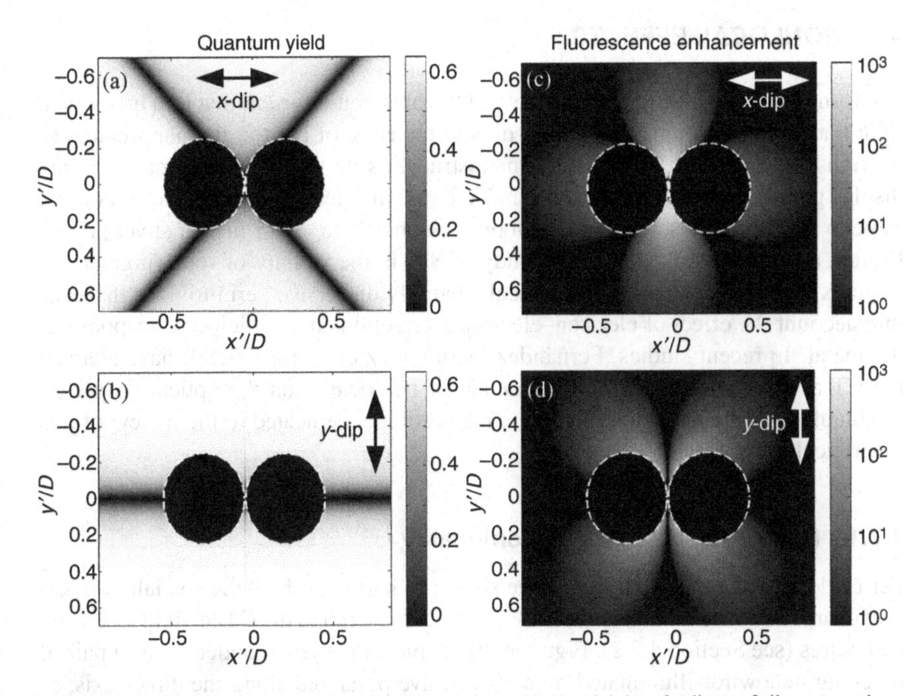

FIGURE 4.19 Fluorescent properties of a molecule placed in the vicinity of a dimer of silver nanowires at the first resonance $n = 1$ (with $D = 20$ nm, $\delta = D/50$, permittivity taken from Reference 97). (a and c) Spatial variations of the quantum yield for a dipole oriented along x' and y', respectively. (b and d) Spatial variations of the fluorescence enhancement for a dipole oriented along x' and y', respectively. The intrinsic quantum yield η_0 has been fixed to 0.66.

of freedom that can be used to tune the fluorescence and the apparent quantum yield with high sensitivity.

To conclude, note that the CT strategy is really powerful since it provides a fully analytical solution to describe the physics of the interaction between a fluorophore and a plasmonic nanostructure. Unlike a numerical approach, an efficient and rapid optimization of the nanostructures can be performed. Our analytical model predicts quantitatively the quantum yield and the fluorescence enhancement induced by these metallic structures for nanowire diameters below 100 nm. As the absorption and emission processes occur at different frequencies, a perfect knowledge of the spectral and spatial properties of the field enhancement might be decisive for the implementation of an experimental setup aiming at single molecule detection. The main perspective of this work is to extend this study to a 3D configuration and in particular to consider a 3D dipole rather than a 2D dipole, which has no physical existence. We believe our 2D model provides a qualitative knowledge for the spatial and wavelength dependence of the Purcell factor, the quantum yield, and the fluorescence enhancement. However, note that the values provided here are not quantitative predictions of what would happen for a 3D transition dipole moment like in a real experiment.

4.7 NONLOCAL EFFECTS

Conformal mapping designs a path toward the plasmonics *graal*, that is, a broadband light harvesting accompanied with a drastic focusing of light at the nanoscale (see Sections 1.3 and 1.4). However, the nanostructures derived by conformal mapping display geometrical singularities in which the spatial extent of the EM fields is comparable to the Coulomb screening length in noble metals (δ_c =0.1 nm for silver [126]). From a theoretical perspective, the study of SPs in the vicinity of such singularities requires the implementation of nonlocal, spatially dispersive permittivities that take into account the effect of electron–electron interactions in the dielectric response of the metal. In recent studies, Fernández-Domínguez et al. [69, 79, 80] have adapted the CT approach to investigate the impact of nonlocality on the optical properties of singular plasmonic nanostructures. This section is dedicated to the review of this theoretical breakthrough.

4.7.1 Conformal Mapping of Nonlocality

Fernández-Domínguez et al. [69] have shown in particular how the spatially dispersive character of the metal permittivity can be considered in the CT leading to kissing nanowires (see Section 4.2.2). Figure 4.20b depicts the system under study, a pair of touching nanowires illuminated by a plane wave polarized along the dimer axis, as already investigated under the local approximation in Section 4.3. Now, the metal permittivity is modeled using the so-called hydrodynamical Drude model [127], which yields the following dielectric functions ϵ'_T and ϵ'_M for transverse and longitudinal EM fields, respectively:

$$\epsilon'_T(\omega) = \epsilon_\infty \left[1 - \frac{\omega_P^2}{\omega(\omega + i\gamma)}\right], \quad \epsilon'_L(\omega, \mathbf{k}) = \epsilon_\infty \left[1 - \frac{\omega_P^2}{\omega(\omega + i\gamma) - \xi'^2|\mathbf{k}|^2}\right] \tag{4.45}$$

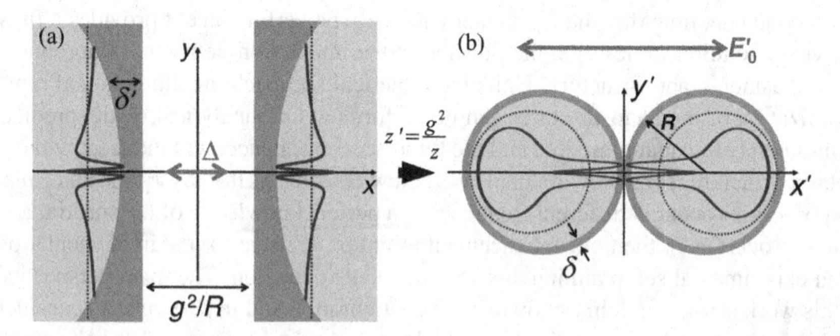

FIGURE 4.20 Conformal map transforming an MIM structure (a) into kissing nanowires (b) with a nonlocal description of the interaction of light with the nanostructure. The shaded areas in both panels represent the decay length of the longitudinal plasmons excited in the system. Solid lines show artistic plots of the electrostatic potential describing propagating SPPs in both frames. Reprinted figure with permission from [69]. Copyright (2012) by the American Physical Society.

The usual Drude constants are obtained from the fitting to experimental data for silver [108], whereas ξ', which measures the degree of nonlocality, is set as a free parameter for the moment. The dark ring in Figure 4.20b accounts for the decay length of the longitudinal plasmons into the metal, $\delta' \sim \frac{\xi'}{\omega_p}$ $(\omega \ll \omega_p)$. δ' represents the extension of the surface charges induced in the nanowires by the incident light.

Through conformal inversion (Eq. 4.5), the problem consists in an incident field illuminating the touching nanowires (Fig. 4.20b) mapping onto a dipole sandwiched between two semi-infinite metallic slabs (Fig. 4.20a). Whereas $\epsilon'_T(\omega)$ is preserved under the transformation, the $\mathbf{k}'$-dependence of $\epsilon'_L(\omega, \mathbf{k}')$ makes it sensitive to the inversion. Assuming that the nonlocal length scale is much smaller than the spatial range in which dz/dz' varies significantly, the general relation between the original longitudinal permittivity and its transformed counterpart reads [69]

$$\epsilon_l(\mathbf{k}, \omega) = \epsilon'_l\left(\left|\frac{dz}{dz'}\right|\mathbf{k}, \omega\right). \tag{4.46}$$

Hence, conformal mapping leads to a wavevector stretching weighted by $|dz/dz'|$ and to a longitudinal permittivity characterized by a degree of nonlocality $\xi(z) = \xi'|z|^2/g^2$. The surface charge thickness δ in the slab frame now exhibits a y'-dependence (Fig. 4.20b). The problem in the slab frame is solved under the so-called eikonal (WKB) approximation [69]. This assumption relies on the fact that $\delta(y)$ varies in space much more slowly than the oscillating EM fields. Note that this approach omits electron tunneling effects, which could have some influence in the plasmonic response of nanoparticle dimers as well [107].

4.7.2 Toward the Physics of Local Dimers

Solving the problem in the slab frame leads to an analytical solution for the nonlocal kissing nanowires problem. The analytical expressions of the absorption cross section and the field enhancement can be found in the original article [69]. Here we will present the main results and insist on the physics hidden behind them. The plasmonic behavior of nonlocal touching nanowires is qualitatively different from the local prediction. In the local approximation there is a continuous spectrum of modes that propagate toward the touching point but can never reach it. In the nonlocal description, the modes can sneak past the touching point and circulate round and round the structure. Hence the spectrum is now discrete and the physics is closely related to the case of nontouching local nanowires (see Section 4.5).

Figure 4.21a displays the absorption spectra for an $R = 10$ nm nanowire dimer. Analytical results are compared to numerical ones obtained with COMSOL. For comparison, σ_a is also displayed for a local dimer (gray), and local (solid black) and nonlocal ($\delta = 10^{-2}\lambda_p$, dashed line) single nanowires. Whereas nonlocality only blue-shifts the dipolar resonance of isolated nanowires, it gives rise to a set of absorption peaks for kissing nanowires. These nonlocal resonances shift to higher frequencies when ξ increases, in a similar manner as the dipolar peak of a single nanowire. Importantly, Figure 4.21a shows that the broadband response of touching dimers is

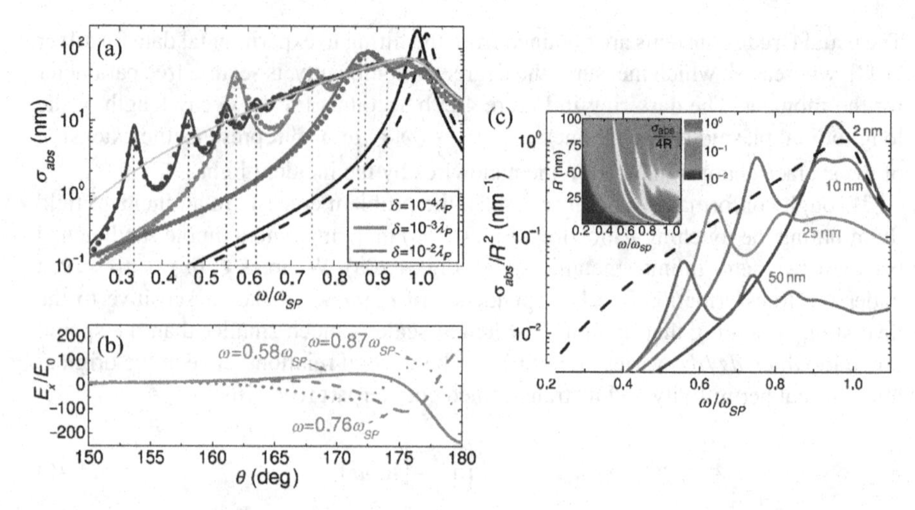

FIGURE 4.21 (a) Theoretical (solid lines) and numerical (dots) absorption spectra for 10 nm radii nonlocal nanowires compared with the cross section obtained within the local approximation (gray dashed line). σ_a for a local (nonlocal, $\delta = 10^{-2}\lambda_p$) single nanowire is also plotted in solid (dashed) black line. (b) Field enhancement close to the touching point at the three lowest nonlocal resonances for $\delta = 10^{-3}\lambda_p$ (value appropriate for silver). (c) Absorption cross section normalized to the structure area for Ag ($\beta = 3.6 \times 10^{-3}c$) for nanowires of different radii. Color solid lines plot the nonlocal results and dashed black line corresponds to the electrostatic local result for all R. The inset shows the absorption efficiency versus the incident frequency and nanowires radius. Reprinted figure with permission from [69]. Copyright (2012) by the American Physical Society.

lost at frequencies below the lowest nonlocal resonance. Figure 4.21b plots the field enhancement along the nanowire surface evaluated at the three first resonances for a value of $\delta' = 10^{-3}\lambda_p$. In analogy with nontouching local nanowires (Section 4.5), the rank n of the modes can be related to the number of nodes that the SPP fields present near the singularity. Note that the nonlocal effects drastically reduce the maximum field enhancement that can be reached (typically 2×10^2 instead of 2×10^4 under the local approximation). The electric field now spreads over a larger range of angles than under the local approximation. The good point is that the area of field amplification will be much more accessible to molecules in surface-enhanced Raman scattering and fluorescence enhancement experiments.

Figure 4.21c shows the absorption spectra normalized to R^2 for various nanowire sizes (solid lines). Within the electrostatic local approximation (Eq. 4.27) [45], σ_a scales as R^2, hence plotting σ_a/R^2 allows to investigate simultaneously nonlocal and radiative effects on the optical properties of the system. Radiation damping is taken into account incorporating a fictive absorber in the slab frame (see Section 4.6). The comparison between the nonlocal results and the electrostatic local prediction clarifies how nonlocality and radiation losses modify the optical properties of the system. For small sizes ($R = 2$ nm), radiation is negligible, but the weight of spatial dispersion, δ'/R, is high. The absorption spectrum shows a single peak with a narrow line width like a single nanowire (see Fig. 4.21a). For larger sizes, the absorption cross section decreases compared to the electrostatic prediction due to radiative reaction, while the impact of nonlocality is diminished. Thus, the dimers interact less efficiently with the incoming light, but over a larger bandwidth. In the inset, the absorption efficiency σ_a/R is shown as a function of ω/ω_{sp} and the nanowire radius R. The light collection performance of the structure is optimized for nanowire radii between 5 and 50 nm, where the balance between radiative and nonlocal effects is the most convenient.

Note that the study of nonlocality with conformal mapping is not restricted to the case of kissing nanowires but can be extended to all the singular structures mentioned earlier in the review and displayed in Figures 4.2 and 4.5.

4.8 SUMMARY AND OUTLOOK

To briefly conclude, this review shows how CT provides an elegant tool to design plasmonic structures capable of an efficient harvesting of light over the whole visible spectrum. SP modes are shown to propagate toward the structure singularity where their velocity vanishes and energy accumulates. Strong field enhancement ($> 10^3$) and confinement at the nano-scale have been predicted within the classical approach. Subsequent refinements of the CT theory have allowed to take into account the bluntness of the nanostructure that will be inevitable in practice, as well as radiative losses and nonlocality. The consideration of these limits makes the CT approach really powerful, capable of predicting analytically the optical properties of a large variety of nanostructures and designing a path toward a broadband nanofocusing of light. Experiments are currently performed to confirm the theoretical predictions [42, 128, 129]. The proposed plasmonic nanostructures may find great potential applications in solar cells [130–132], surface-enhanced Raman scattering [96, 133], single molecule detection [134, 135], plasmon-controlled fluorescence [44, 118–122], high-harmonic generation [26], and SPASERS [90, 136].

ACKNOWLEDGMENTS

The authors wish to thank Dang Yuan Lei, Yu Luo, Antonio Fernández-Domínguez, Yannick Sonnefraud, and Stefan A. Maier for all the collaborative work and the fruitful discussions within the framework of the *transformation optics for plasmonics* project at Imperial College. This work was supported by the European Community project PHOME (Contract No. 213390) and by the U.K. Engineering and Physical Sciences Research Council (EPSRC).

REFERENCES

1. Mie G (1908). Beiträge zur optik trüber medien, speziell kolloidaler metallösungen. *Ann. Phys.* 25: 377–445.

2. Ritchie RH (1957) Plasma losses by fast electrons in thin films. *Phys. Rev.* 106: 874–881.
3. Raether H (1988) *Surface Plasmons on Smooth and Rough Surfaces and on Gratings.* Berlin: Springer.
4. Kreibig U, Vollmer M (1995) *Optical Properties of Metal Clusters.* Berlin: Springer.
5. Maier SA (2007) *Plasmonics: Fundamentals and Applications.* New York: Springer.
6. Maier SA, Atwater HA (2005) Plasmonics: localization and guiding of electromagnetic energy in metal/dielectric structures. *J. Appl. Phys.* 98: 011101.
7. Pendry JB (2000) Negative refraction makes a perfect lens. *Phys. Rev. Lett.* 85: 3966–3969.
8. Fang N, Lee H, Sun C, Zhang X (2005) Subdiffraction-limited optical imaging with a silver superlens. *Science* 308: 534–537.
9. Kelly KL, Coronado E, Zhao LL, Schatz GC (2003) The optical properties of metal nanoparticles: the influence of size, shape, and dielectric environment. *J. Phys. Chem. B* 107: 668–677.
10. Hao E, Schatz GC (2003) Electromagnetic fields around silver nanoparticles and dimers. *J. Chem. Phys.* 120: 357–366.
11. Kottmann JP, Martin OJF (2001) Plasmon resonant coupling in metallic nanowires. *Opt. Express* 8: 655–663.
12. Su K-H, Wei Q-H, Zhang X, Mock JJ, Smith DR, Schultz S (2003) Interparticle coupling effects on plasmon resonances of nanogold particles. *Nano Lett.* 3: 1087–1090.
13. Li K, Stockman MI, Bergman DJ (2003) Self-similar chain of metal nanospheres as an efficient nanolens. *Phys. Rev. Lett.* 91: 227402.
14. Atay T, Song J-H, Nurmikko AV (2004) Strongly interacting plasmon nanoparticle pairs: from dipole-dipole interaction to conductively coupled regime. *Nano Lett.* 4: 1627–1631.
15. Nordlander P, Oubre C, Prodan E, Li K, Stockman MI (2004) Plasmon hybridization in nanoparticle dimers. *Nano Lett.* 4: 899–903.
16. Talley CE, Jackson JB, Oubre C, Grady NK, Hollars CW, Lane SM, Huser TR, Nordlander P (2005) Surface-enhanced Raman scattering from individual Au nanoparticles and nanoparticle dimer substrates. *Nano Lett.* 5: 1569–1574.
17. Sweatlock LA, Maier SA, Atwater HA, Penninkhof JJ, Polman A (2005) Highly confined electromagnetic fields in arrays of strongly coupled Ag nanoparticles. *Phys. Rev. B* 71: 235408.
18. Romero I, Aizpurua J, Bryant GW, García de Abajo FJ (2006) Plasmons in nearly touching metallic nanoparticles. *Opt. Express* 14: 9988–9999.
19. Romero I, Teperik TV, García de Abajo FJ (2008) Plasmon molecules in overlapping nanovoids. *Phys. Rev. B* 77: 125403.
20. Britt Lassiter J, Aizpurua J, Hernandez LI, Brandl DW, Romero I, Lal S, Hafner JH, Nordlander P, Halas NJ (2008) Close encounters between two nanoshells. *Nano Lett.* 8: 1212–1218.

21. Härtling T, Alaverdyan Y, Hille A, Wenzel MT, Käll M, Eng LM (2008) Optically controlled interparticle distance tuning and welding of single gold nanoparticle pairs by photochemical metal deposition. *Opt. Express*, 16: 12362.

22. Yang L, Yan B, Reinhard BM (2008) Correlated optical spectroscopy and transmission electron microscopy of individual hollow nanoparticles and their dimers. *J. Phys. Chem. C* 112: 15989–15996.

23. Busson MP, Rolly B, Stout B, Bonod N, Larquet E, Polman A, Bidault S (2011) Optical and topological characterization of gold nanoparticle dimers linked by a single DNA double strand. *Nano Lett.* 11: 5060–5065.

24. Halas NJ, Lal S, Chang W-S, Link S, Nordlander P (2011) Plasmons in strongly coupled metallic nanostructures. *Chem. Rev.* 111: 3913–3961.

25. Lu Y, Liu GL, Kim J, Mejia YX, Lee LP (2005) Nanophotonic crescent moon structures with sharp edge for ultrasensitive biomolecular detection by local electromagnetic field enhancement effect. *Nano Lett.* 5: 119–124.

26. Kim J, Liu GL, Lu Y, Lee LP (2005) Intra-particle plasmonic coupling of tip and cavity resonance modes in metallic apertured nanocavities. *Opt. Express* 13: 8332–8338.

27. Rochholz H, Bocchio N, Kreiter M (2007) Tuning resonances on crescent-shaped noble-metal nanoparticles. *New J. Phys.* 9: 53.

28. Bukasov R, Shumaker-Parry JS (2007) Highly tuneable infrared extinction properties of gold nanocrescents. *Nano Lett.* 7: 1113–1118.

29. Ross BM, Lee LP (2008) Plasmon tuning and local field enhancement maximization of the nanocrescent. *Nanotechnology* 19: 275201.

30. Feng L, Van Orden D, Abashin M, Wang Q-J, Chen Y-F, Lomakin V, Fainman Y (2009) Nanoscale optical field localization by resonantly focused plasmons. *Opt. Express* 17: 4824–4832.

31. Wu LY, Ross BM, Lee LP (2009) Optical properties of the crescent-shaped nanohole antenna. *Nano Lett.* 9: 1956–1961.

32. Bukasov R, Ali TA, Nordlander P, Shumaker-Perry JS (2010) Probing the plasmonic near-field of gold nanocrescent antennas. *ACS Nano* 4: 6639–6650.

33. Oldenburg SJ, Averitt RD, Westcott SL, Halas NJ (1998) Nanoengineering of optical resonances. *Chem. Phys. Lett.* 288: 243–247.

34. Averitt RD, Westcott SL, Halas NJ (1999) Linear optical properties of gold nanoshells. *J. Opt. Soc. Am. B* 16: 1824–1832.

35. Graf C, van Blaaderen A (2001) Metallodielectric colloidal core-shell particles for photonic applications. *Langmuir* 18: 524–534.

36. Prodan E, Radloff C, Halas NJ, Nordlander P (2003) A hybridization model for the plasmon response of complex nanostructures. *Science* 302: 419–422.

37. Sun Y, Wiley B, Li Z-Y, Xia Y (2004) Synthesis and optical properties of nanorattles and multiple-walled nanoshells/nanotubes made of metal alloys. *J. Am. Chem. Soc.* 126: 9399–9406.

38. Hirsch LR, Gobin AM, Lowery AR, Tam F, Drezek RA, Halas NJ, West JL (2006) Metal nanoshells. *Ann. Biomed. Eng.* 34: 15–22.

39. García-Vidal FJ, Pendry JB (1996) Collective theory of surface enhanced Raman scattering. *Phys. Rev. Lett.* 77: 1163–1166.

40. Li K, Clime L, Cui B, Veres T (2008) Surface enhanced Raman scattering on long-range ordered noble-metal nanocrescent arrays. *Nanotechnology* 19: 145305.

41. Liu GL, Lu Y, Kim J, Lee LP, Doll JC (2005) Magnetic nanocrescents as controllable surface-enhanced Raman scattering nanoprobes for biomolecular imaging. *Adv. Mater.* 17: 2683–2688.

42. Hill RT, Mock JJ, Urzhumov Y, Sebba DS, Oldenburg SJ, Chen S-Y, Lazarides AA, Chilkoti A, Smith DR (2010) Highly confined electromagnetic fields in arrays of strongly coupled Ag nanoparticles. *Nano Lett.* 10: 4150–4154.

43. Zhang J, Fu Y, Chowdhury MH, Lakowicz JR (2007) Metal-enhanced single-molecule fluorescence on silver particle monomer and dimer: coupling effect between metal particles. *Nano Lett.* 7: 2101–2107.

44. Vandenbem C, Brayer D, Frouze-Pérez LS, Carminati R (2010) Controlling the quantum yield of a dipole emitter with coupled plasmonic modes. *Phys. Rev. B* 81: 085444.

45. Aubry A, Lei DY, Fernandez-Dominguez A, Sonnefraud Y, Maier SA, Pendry JB (2010) Plasmonic light-harvesting devices over the whole visible spectrum. *Nano Lett.* 10: 2574–2579.

46. Ward AJ, Pendry JB (1996) Refraction and geometry in Maxwell's equations. *J. Mod. Opt.* 43: 773–793.

47. Shyroki DM (2007) Note on transformation to general curvilinear coordinates for Maxwell's curl equations. ArXiv: physics/0307029v1.

48. Kundtz NB, Smith DR, Pendry JB (2011) Electromagnetic design with transformation optics. *Proc. IEEE* 99: 1622–1633.

49. Leonhardt U (2006) Optical conformal mapping. *Science* 312: 1777–1780.

50. Pendry JB, Schurig D, Smith DR (2006) Controlling electromagnetic fields. *Science* 312: 1780–1782.

51. Schurig D, Mock JJ, Justice BJ, Cummer SA, Pendry JB, Starr AF, Smith DR (2006) Metamaterial electromagnetic cloak at microwave frequencies. *Science* 314: 977–980.

52. Li J, Pendry JB (2008) Hiding under the carpet: a new strategy for cloaking. *Phys. Rev. Lett.* 101: 203901.

53. Liu R, Ji C, Mock JJ, Chin JY, Cui TJ, Smith DR (2009) Broadband ground-plane cloak. *Science* 323: 366–369.

54. Valentine J, Li J, Zentgraf T, Bartal G, Zhang X (2009) An optical cloak made of dielectrics. *Nat. Mater.* 8: 568–571.

55. Gabrielli LH, Cardenas J, Pointras CB, Lipson M (2009) Silicon nanostructure cloak operating at optical frequencies. *Nat. Photonics* 3: 461–463.

56. Ergin T, Stenger N, Brenner P, Pendry JB, Wegener M (2010) Three-dimensional invisibility cloak at optical wavelengths. *Science* 328: 337–339.

57. Huidobro PA, Nesterov ML, Martín-Moreno L, García-Vidal FJ (2010) Transformation optics for plasmonics. *Nano Lett.* 10: 1985–1990.

58. Liu Y, Zentgraf T, Bartal G, Zhang X (2010) Transformational plasmon optics. *Nano Lett.* 10: 1991–1997.

59. Renger J, Kadic M, Dupont G, Aćimović SS, Guenneau S, Quidant R, Enoch S (2010) Hidden progress: broadband plasmonic invisibility. *Opt. Express* 18: 15757–15768.

60. Zentgraf T, Liu Y, Mikkelsen MH, Valentine J, Zhang X (2011) Plasmonic Luneburg and Eaton lenses. *Nat. Nanotech.* 6: 151–155.

61. Huidobro PA, Nesterov ML, Martín-Moreno L, García-Vidall FJ (2011) Moulding the flow of surface plasmons using conformal and quasiconformal mappings. *New J. Phys.* 13: 033011.

62. Zhang J, Xiao S, Wubs M, Mortensen NA (2011) Surface plasmon wave adapter designed with transformation optics. *ACS Nano* 5: 4359–4364.

63. Kadica M, Dupont G, Guenneau S, Enoch S (2011) Controlling surface plasmon polaritons in transformed coordinates. *J. Mod. Opt.* 58: 994–1003.

64. Gharghi M, Gladden C, Zentgraf T, LiuY, Yin X, Valentine J, Zhang X (2011) A carpet cloak for visible light. *Nano Lett.* 11: 2825–2828.

65. Luo Y, Pendry JB, Aubry A (2010) Surface plasmons and singularities. *Nano Lett.* 10: 4186–4191.

66. Aubry A, Lei DY, Maier SA, Pendry JB (2010) Interaction between plasmonic nanoparticles revisited with transformation optics. *Phys. Rev. Lett.* 105: 233901.

67. Fernández-Domínguez A, Maier SA, Pendry JB (2010) Collection and concentration of light by touching spheres: a transformation optics approach. *Phys. Rev. Lett.* 105: 266807.

68. Luo Y, Lei DY, Maier SA, Pendry JB (2012) Broadband light harvesting nanostructures robust to edge bluntness. *Phys. Rev. Lett.* 108: 023901.

69. Fernández-Domínguez A, Maier SA, Pendry JB (2012) Transformation optics description of nonlocal effects in plasmonic nanostructures. *Phys. Rev. Lett.* 108: 106802

70. Lei DY, Aubry A, Maier SA, Pendry JB (2010) Broadband nano-focusing of light using kissing nanowires. *New J. Phys.* 12: 093030.

71. Aubry A, Lei DY, Maier SA, Pendry JB (2010) Broadband plasmonic device concentrating the energy at the nanoscale: the crescent-shaped cylinder. *Phys. Rev. B* 82: 125430.

72. Aubry A, Lei DY, Maier SA, Pendry JB (2010) Conformal transformation applied to plasmonics beyond the quasistatic limit. *Phys. Rev. B* 82: 205109.

73. Lei DY, Aubry A, Luo Y, Maier SA, Pendry JB (2011) Plasmonic interaction between overlapping nanowires. *ACS Nano* 5: 597–607.

74. Aubry A, Lei DY, Maier SA, Pendry JB (2011) Plasmonic hybridization between nanowires and a metallic surface: a transformation optics approach. *ACS Nano* 5: 3293–3308.

75. Luo Y, Aubry A, Pendry JB (2011) Electromagnetic contribution to surface-enhanced Raman scattering from rough metal surfaces: a transformation optics approach. *Phys. Rev. B* 83: 155422.

76. Zeng Y, Liu J, Werner DH (2011) General properties of two-dimensional conformal transformations in electrostatics. *Opt. Express* 19: 20035–20047.

77. Pendry JB, Aubry A, Smith DR, and Maier SA (2012) Transformation Optics and Subwavelength Control of Light. *Science* 337: 549–552.

78. Fernández-Domnguez AI, Maier SA, Pendry JB (2012) Transformation optics description of touching metal nanospheres, *Phys. Rev. B* 85: 165148.

79. Fernández-Dominguez AI, Luo Y, Wiener A, Pendry JB, Maier SA (2012) Theory of Three-Dimensional Nanocrescent Light Harvesters, *Nano Lett.* 12: 5946–5953.

80. Fernández-Domnguez AI, Zhang P, Luo Y, Maier SA, Garca-Vidal FJ, Pendry JB (2012) Transformation-optics insight into nonlocal effects in separated nanowires, *Phys. Rev. B* 86: 241110.

81. Zhang J and Zayats A (2013) Multiple Fano resonances in single-layer nonconcentric core-shell nanostructures. *Opt. Exp.* 21: 8426–8436.

82. Schinzinger R, Laura PAA (2003) *Conformal Mapping Methods and Applications*. New York: Dover.

83. Smith GS, Barakat R (1975) Electrostatics of two conducting spheres in contact. *Appl. Sci. Res.* 30: 418–432.

84. McPhedran RC, McKenzie DR (1980) Electrostatic and optical resonances of arrays of cylinders. *Appl. Phys. A* 23: 223–235.

85. McPhedran RC, Perrins WT (1981) Electrostatic and optical resonances of cylinder pairs. *Appl. Phys. A* 24: 311–318.

86. McPhedran RC, Milton GW (1987) Transport properties of touching cylinder pairs and of the square array of touching cylinders. *Proc. R. Soc. A* 411: 313–326.

87. Paley AV, Radchik AV, Smith GB (1993) Quasistatic optical response of pairs of touching spheres with arbitrary dielectric permeability. *J. Appl. Phys.* 73: 3446–3453.

88. Nordlander P, Le F (2006) Plasmonic structure and electromagnetic field enhancements in the metallic nanoparticle-film system. *Appl. Phys. B Lasers Opt.* 84: 35–41.

89. Oulton RF, Sorger VJ, Genov DA, Pile DFP, Zhang X (2008) A hybrid plasmonic waveguide for subwavelength confinement and long-range propagation. *Nat. Photonics* 2: 496–500.

90. Oulton R, Sorger VJ, Zentgraf T, Ma R-M, Gladden C, Dai L, Bartal G, Zhang X (2009) Plasmon lasers at deep subwavelength scale. *Nature* 461: 629–632.

91. Stockman MI. Nanofocusing of optical energy in tapered plasmonic waveguides. *Phys. Rev. Lett.* 93: 137404, 2004.

92. Nerkararyan KhV (1997) Superfocusing of a surface polariton in a wedge-like structure. *Phys. Lett. A* 237: 103–105.

93. Pile DFP, Ogawa T, Gramotnev DK, Okamoto T, Haraguchi M, Fukui M, Matsuo S (2005) Theoretical and experimental investigation of strongly localized plasmons on triangular metal wedges for subwavelength waveguiding. *Appl. Phys. Lett.* 87: 061106.

94. Fleischmann M, Hendra PJ, McQuillan AJ (1974) Raman spectra of pyridine adsorbed at a silver electrode. *Chem. Phys. Lett.* 26: 163–166.

95. Albrecht MG, Creighton JA (1977) Anomalously intense Raman spectra of pyridine at a silver electrode. *J. Am. Chem. Soc.* 99: 5215–5217.

96. Moskovits M (1985) Surface-enhanced spectroscopy. *Rev. Mod. Phys.* 57: 783–826.

97. Johnson PB, Christy RW (1972) Optical constants of the noble metals. *Phys. Rev. B* 6: 4370–4379.

98. Wang H, Brandl DW, Nordlander P, Halas NJ (2007) Plasmonic nanostructures: artificial molecules. *Acc. Chem. Res.* 40: 53–62.

99. Bardhan R, Grady NK, Cole JR, Joshi A, Halas NJ (2009) Fluorescence enhancement by Au nanostructures: nanoshells and nanorods. *ACS Nano* 3: 744–752.

100. Ford GW, Weber WH (1984) Electromagnetic interactions of molecules with metal surfaces. *Phys. Rep.* 113: 195–287.

101. Bohren CH, Huffman DR (1983) *Absorption and Scattering of Light by Small Particles.* New York: John Wiley & Sons Ltd.

102. García de Abajo FJ (2008) Nonlocal effects in the plasmons of strongly interacting nanoparticles, dimers, and waveguides. *J. Phys. Chem. C* 112: 17983–17987.

103. McMahon JM, Gray SK, Schatz GC (2010) Optical properties of nanowire dimers with a spatially nonlocal dielectric function. *Nano Lett.* 10: 3473–3481.

104. Raza S, Toscano G, Jauho A-P, Wubs M, Mortensen NA (2011) Unusual resonances in nanoplasmonic structures due to nonlocal response. *Phys. Rev. B* 84: 121412.

105. David C, García de Abajo FJ (2011) Spatial nonlocality in the optical response of metal nanoparticles. *J. Phys. Chem. C* 115: 19470–19475.

106. Cirac C, Hill RT, Mock JJ, Urzhumov Y, Fernández-Domínguez AI, Maier SA, Pendry JB, Chilkoti A, and Smith DR (2012) *Science* 337: 1072–1074.

107. Zuloaga J, Prodan E, Nordlander P (2009) Quantum description of the plasmon resonances of a nanoparticle dimer. *Nano Lett.* 9: 887–891.

108. Palik ED (1985) *Handbook of Optical Constants of Solids.* New York: Academic Press.

109. Pendry JB (2003) Perfect cylindrical lenses. *Opt. Express* 11: 755–760.

110. Draine BT (1988) The discrete-dipole approximation and its application to interstellar graphite grains. *Astrophys. J.* 333: 848–872.

111. Wokaun A, Gordon JP, Liao PF (1982) Radiation damping in surface-enhanced Raman scattering. *Phys. Rev. Lett.* 48: 957960.

112. Carminati R, Greffet J-J, Henkel C, Vigoureux JM (2006) Radiative and non-radiative decay of a single molecule close to a metallic nanoparticle. *Opt. Commun.* 261: 368–375.

113. Hao F, Sonnefraud Y, Van Dorpe P, Maier SA, Halas NJ, Nordlander P (2008) Symmetry breaking in plasmonic nanocavities: subradiant LSPR sensing and a tuneable fano resonance. *Nano Lett.* 8: 3983–3988.

114. Zhang S, Genov DA, Wang Y, Liu M, Zhang X (2008) Plasmon-induced transparency in metamaterials. *Phys. Rev. Lett.* 101: 047401.

115. Verellen N, Sonnefraud Y, Sobhani H, Hao F, Moshchalkov VV, Van Dorpe P, Nordlander P, Maier SA (2009) Fano resonances in individual coherent plasmonic nanocavities. *Nano Lett.* 9: 1663–1667.

116. Liu N, Langguth L, Weiss T, Kastel J, Fleischhauer M, Pfau T, Nordlander P, Giessen H (2009) Plasmonic analogue of electromagnetically induced transparency at the Drude damping limit. *Nat. Mat.* 8: 758–762.

117. Alú A, Engheta N (2009) Cloaking a sensor. *Phys. Rev. Lett.* 102: 233901.

118. Dulkeith E, Morteani AC, Niedereichholz T, Klar TA, Feldmann J, Levi SA, van Veggel FCJM, Reinhoudt DN, Müller M, Gittins DI (2002) Fluorescence quenching of dye molecules near gold nanoparticles: radiative and nonradiative effects. *Phys. Rev. Lett.* 89: 203002.

119. Kuhn S, Hakanson U, Rogobete L, Sandoghdar V (2006) Enhancement of single-molecule fluorescence using a gold nanoparticle. *Phys. Rev. Lett.* 97: 017402.

120. Anger P, Bharadwaj P, Novotny L (2006) Enhancement and quenching of single-molecule fluorescence. *Phys. Rev. Lett.* 96: 113002.

121. Giannini V, Fernández-Domínguez AI, Heck SC, Maier SA (2011) Plasmonic nanoantennas: fundamentals and their use in controlling the radiative properties of nanoemitters. *Chem. Rev.* 111: 3888–3912.

122. Ming T, Chen H, Jiang R, Li Q, Wang J (2012) Plasmon-controlled fluorescence: beyond the intensity enhancement. *J. Phys. Chem. Lett.* 3: 191–202.

123. Purcell EM (1946) Spontaneous emission probabilities at radio frequencies. *Phys. Rev.* 69: 681.

124. Klimov VV, Guzatov DV (2007) Strongly localized plasmon oscillations in a cluster of two metallic nanospheres and their influence on spontaneous emission of an atom. *Phys. Rev. B* 75: 024403.

125. Klimov VV, Guzatov DV (2007) Optical properties of an atom in the presence of a two-nanosphere cluster. *Quantum Electron.* 37: 209–230.

126. Kittel C (1966) *Introduction to Solid State Physics*. New York: Wiley.

127. Boardman A (1982) *Electromagnetic Surface Modes*. New York: Wiley.

128. Lei DY, Fernández-Domínguez AI, Sonnefraud Y, Appavoo K, Haglund RF, Pendry JB, Maier SA (2012) Revealing plasmonic gap modes in particle-on-film systems using dark-field spectroscopy. *ACS Nano*. 6: 1380–1386.

129. Ward DR, Hüser F, Pauly F, Cuevas JC, Natelson D (2010) Optical rectification and field enhancement in a plasmonic nanogap. *Nat. Nanotech.* 5: 732–736.

130. Atwater HA, Polman A (2010) Plasmonics for improved photovoltaic devices. *Nat. Mater.* 9: 205–213.

131. Aydin K, Ferry VE, Briggs RM, Atwater HA (2011) Broadband polarization-independent resonant light absorption using ultrathin plasmonic super absorbers. *Nat. Commun.* 2: 517.

132. Atre AC, García-Etxarri A, Alaeian H, Dionne JA (2012) Toward high-efficiency solar upconversion with plasmonic nanostructures. *J. Opt.* 14: 024008.

133. Campion A, Kambhampati P (1998) Surface-enhanced Raman scattering. *Chem. Soc. Rev.* 27: 241–250.

134. Nie S, Emory SR (1997) Probing single molecules and single nanoparticles by surface-enhanced Raman scattering. *Science* 275: 1102–1106.

135. Kneipp K, Wang Y, Kneipp H, Perelman LT, Itzkan I, Dasari RR, Feld MS (1997) Single molecule detection using surface-enhanced Raman scattering (SERS). *Phys. Rev. Lett.* 78: 1667–1670.

136. Bergman DJ, Stockman MI (2003) Surface plasmon amplification by stimulated emission of radiation: quantum generation of coherent surface plasmons in nanosystems. *Phys. Rev. Lett.* 90: 027402.

5

Loss Compensation and Amplification of Surface Plasmon Polaritons

PIERRE BERINI
School of Electrical Engineering and Computer Science, University of Ottawa, Ottawa, Canada

5.1 INTRODUCTION

Surface plasmon polaritons (SPPs) exhibit interesting and useful properties such as energy asymptote in dispersion curves, resonant modes, field localization and enhancement, high sensitivities (surface and bulk), and subwavelength confinement [1]. Because of these attributes, SPPs have found uses in several areas including nanophotonics [2, 3], biosensing [4], and integrated optical circuits [5].

Attenuation, however, limits the SPP propagation length, which in turn limits the scope for applications. SPPs dissipate their energy mostly by interacting with the metal. For a metal bounded by an ideal dielectric, loss is caused by free-electron scattering and absorption via interband transitions at short enough wavelengths. These loss mechanisms are fundamental: The latter may be avoided by selecting a long operating wavelength, but the former cannot be eliminated although its effects can be reduced via improved fabrication. Roughness along the metal interface causes additional loss by scattering SPPs into bulk waves.

Loss compensation and amplification can be achieved by adding optical gain to the dielectric(s) bounding the metal [6]. Loss compensation means that the SPP attenuation is reduced compared to the passive case whereas amplification means that it is overcompensated. The amplification of SPPs potentially leads to several interesting, useful, and compelling applications, motivating much of the research on this topic. SPP amplifiers are envisaged as stand-alone components, or "gain blocks" integrated

Active Plasmonics and Tuneable Plasmonic Metamaterials, First Edition. Edited by Anatoly V. Zayats and Stefan A. Maier.

with plasmonic elements, biosensors, or integrated circuits to compensate losses and improve performance. Combining feedback with amplification leads to SPP oscillation, which is of interest as an SPP source. Such applications can be achieved with nanoscale structures because SPPs can be confined to deep subwavelength dimensions. This chapter reviews work conducted on SPP loss compensation and amplification in SPP waveguides; the chapter is organized in terms of SPP waveguide types.

5.2 SURFACE PLASMON WAVEGUIDES

An SPP is a transverse magnetic (TM)-polarized optical surface wave that propagates along a metallo-dielectric structure, typically at visible or infrared wavelengths [1]. It is a coupled excitation, which comprised a charge density wave in the metal and electromagnetic fields that peak at metal–dielectric interfaces and decay exponentially away.

5.2.1 Unidimensional Structures

The simplest SPP waveguide is the single metal–dielectric interface sketched in Figure 5.1a. The wavenumber of the single-interface SPP is given by [1] $\alpha + j\beta = j(2\pi/\lambda_0)[\varepsilon_{r,m}\varepsilon_{r,d}/(\varepsilon_{r,m} + \varepsilon_{r,d})]^{1/2}$, where $\varepsilon_{r,m}$ and $\varepsilon_{r,d}$ are the relative permittivity of the

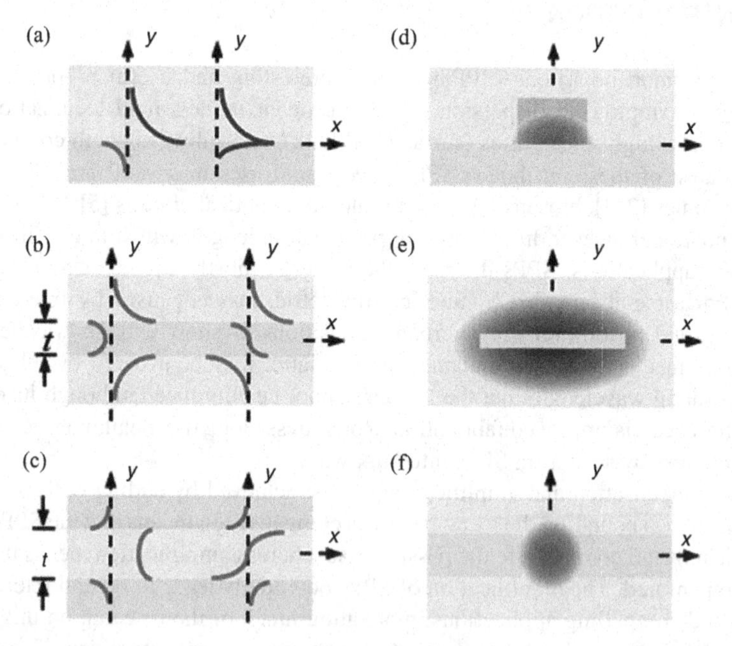

FIGURE 5.1 (a–c) 1D and (d–f) 2D SPP waveguides. (a) Metal (dark gray)–dielectric (light gray) interface with the SPP mode fields (Re{*Ey*}—left, Re{*Hx*}—right) transverse to the direction of propagation (*z*). (b) Symmetric metal film and (c) symmetric metal clads; Re{*Ey*} of the symmetric (left) and asymmetric (right) SPP supermodes also shown. (d) Dielectric-loaded (dark gray), (e) metal stripe, and (f) gap waveguides; the distribution of the transverse electric field magnitude of the main symmetric SPP mode is also shown. (Adapted from Reference 6.)

TABLE 5.1 Characteristics of the SPP on an Ag–SiO_2 Interface [7]

λ_0 (nm)	n_{eff}	δ_w (nm)	2α (cm^{-1})	2α (dB/μm)	L(μm)
360	2.537	44	5×10^5	218	0.02
633	1.565	176	1.6×10^3	0.71	6
1550	1.457	1269	1×10^2	0.044	100

metal and dielectric, respectively, and λ_0 is the free-space wavelength ($e^{+j\omega t}$ time-harmonic form implied; the SPP fields evolve as $e^{-(\alpha+j\beta)z}$ for propagation in the $+z$-direction).

Metals have $\mathrm{Re}\{\varepsilon_{r,m}\} < 0$ and are dispersive, which causes the SPP wavenumber to diverge at a specific wavelength—the photon energy corresponding to this wavelength is termed the energy asymptote. Table 5.1 gives the modal characteristics of the SPP on an Ag–SiO_2 interface at three wavelengths, including near the asymptote ($\lambda_0 = 360$ nm) [7]; $n_{eff} = \beta\lambda_0/2\pi$, δ_w is the 1/e mode field width, and $L = 1/(2\alpha)$ is the 1/e propagation length. As noted, the properties of the SPP reach extrema at the asymptote. However, losses limit its depth (and the extrema) by bending back the dispersion curve to the left of the light line where the SPP becomes radiative, as shown in Figure 5.2.

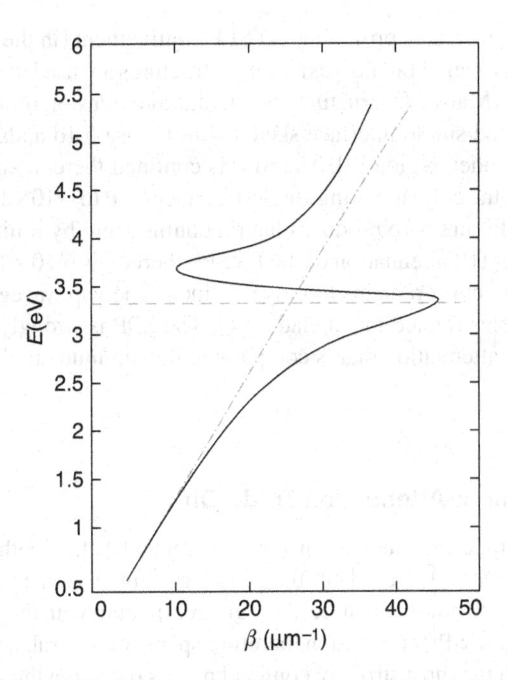

FIGURE 5.2 Dispersion of the single-interface SPP on an Ag–SiO_2 interface ($E = \hbar\omega$ vs. β); the dashed-dot curve is the light line in SiO_2 and the dashed line marks the energy asymptote. (Adapted from Reference 7.)

The thin metal film bounded by symmetric dielectrics, shown in Figure 5.1b, supports two SPP supermodes created from symmetric or asymmetric couplings of single-interface SPPs [8]. The attenuation of the symmetric mode decreases as the thickness of the metal film decreases ($t \rightarrow 0$), to several orders of magnitude lower than the attenuation of the single-interface SPP; the symmetric mode is therefore termed a long-range SPP (LRSPP). The reduced attenuation of the LRSPP, however, comes with reduced confinement [7]. The asymmetric mode behaves oppositely as t decreases and is termed the short-range SPP (SRSPP) [8]. Neither mode has a cutoff thickness. In the Ag–SiO_2 system, the power attenuation (2α) of the LRSPP and SRSPP is 2.8 and 1×10^3 cm^{-1} (0.0012 and 0.45 dB/μm), respectively, for $t = 20$ nm at $\lambda_0 = 1550$ nm, increasing as λ_0 decreases [7].

Figure 5.1c shows the complementary structure, a thin dielectric film bounded by metal claddings, which also supports SPPs [7]. If the dielectric is thick then several TE and TM parallel-plate-like modes are supported. As the thickness of the dielectric decreases ($t \rightarrow 0$) only the symmetric SPP is supported. This mode has no cutoff thickness, so it can be confined to an arbitrarily small width. However, as it shrinks, its attenuation increases, so it is an SRSPP. In the Ag–SiO_2 system, its attenuation is 2×10^3 cm^{-1} (0.85 dB/μm) for $t = 50$ nm at $\lambda_0 = 1550$ nm, increasing as λ_0 decreases [7].

5.2.2 Bidimensional Structures

Metallo-dielectric structures providing 2D SPP confinement (in the plane transverse to the direction of propagation) also exist; such structures are used to enable integrated optical circuits [5]. Many 2D structures are a natural extension of a 1D counterpart. The dielectric-loaded single interface sketched in Figure 5.1d adds confinement by integrating a high-index region [9]. The SPP is confined therein, and its attenuation ranges from that of the corresponding single-interface SPP to $\sim 10\times$ higher. The metal stripe sketched in Figure 5.1e produces lateral confinement by limiting the width of the metal film [10]. The attenuation of the LRSPP thereon is $\sim 10\times$ lower than on the infinitely wide film, but it remains loosely confined. The gap waveguide sketched in Figure 5.1f resembles rotated metal clads [11]. The SPP is strongly confined within the gap and has an attenuation that is comparable in magnitude to the corresponding metal clads.

5.2.3 Confinement-Attenuation Trade-Off

SPP waveguides trade off attenuation for confinement [7]. Whether it appears by varying the wavelength (Table 5.1) or the geometry (Fig. 5.1), the trade-off is fundamental: Increased SPP confinement leads to greater overlap with the metal(s), leading to greater attenuation. SPP attenuation therefore spans, in general, a very broad range (1–10^5 cm^{-1}), with the most strongly confined modes requiring the greatest gains for loss compensation.

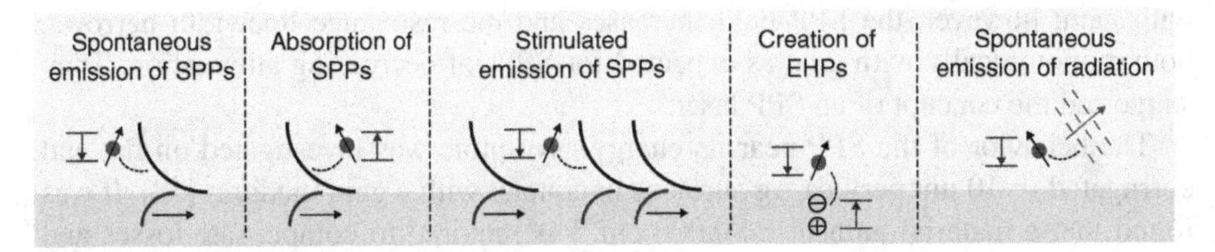

FIGURE 5.3 Optical processes occurring for dipoles (circles) near a single metal–dielectric interface. The magnitude of the transverse electric field of the SPP involved in the processes is shown along with the associated dipole energy transitions. The dotted curves indicate energy transfer. (Adapted from Reference 6.)

5.2.4 Optical Processes Involving SPPs

SPPs are bosons, energy quantized in the usual way by analogy with a quantum harmonic oscillator (including material dispersion [12, 13] and absorption [14]). An SPP can gain or lose a quantum of energy in processes involving atoms, molecules, or continuous media, such as spontaneous and stimulated emission, or absorption. Figure 5.3 sketches optical processes that may occur for dipoles near a single metal–dielectric interface. In addition to processes involving SPPs, dipoles may interact with electron–hole pairs in the metal via dipole–dipole transitions if they are close enough or they may emit radiative modes. SPP densities of states can also be defined and used in the assessment of spontaneous emission rates [15].

5.3 SINGLE INTERFACE

Prism and grating coupling configurations are commonly used to excite single-interface SPPs [1]. The Kretschmann configuration consists of a metal film on the base of a high-index prism bounded by a lower-index medium on the other side. The grating configuration involves a periodic perturbation in the metal such as corrugations, bumps, or slits. The role of the prism or of the grating is to increase the in-plane momentum of (p-) TM-polarized incident light to match that of the SPP. Several studies involving the single-interface SPP propagating though a gain medium were conducted in such configurations.

5.3.1 Theoretical

The first study of SPP amplification is that of Plotz et al. [16] who considered an Ag film on a prism with the other Ag surface in contact with a gain medium. They computed the reflectance of the glass–metal interface versus the angle of incidence as a function of the material gain, showing that the usual reflectance drop which occurs at the SPP excitation angle [1] becomes monotonically shallower as the gain increases, eventually becoming larger than one, corresponding to *amplified* total reflection. Sudarkin and Demkovich [17] investigated a similar structure over a broader range of parameters. They found that the reflectance does not always increase monotonically with gain; however, the SPP field increases and the resonance linewidth narrows, both monotonically with gain as expected for SPPs of decreasing attenuation. They suggested the concept of an SPP laser.

The behavior of the SPP near its energy asymptote was investigated on flat and corrugated (~10 nm period) Ag surfaces in contact with a gain medium [18]. It was found that a material gain of ~80,000 cm^{-1} is required to compensate losses and restore the properties of the SPP near the energy asymptote, spoiled in the passive case by bend-back. A low group velocity (~1 km/s), a large SPP effective index (~29), and strong SPP localization (few nanometers) are predicted on gain-compensated gratings. The large material gain required is challenging, but it is interesting to note the potential for extreme performance.

Nezhad et al. [19] investigated SPP amplification along the interface of Ag and an InGaAsP gain medium, predicting lossless propagation at $\lambda_0 = 1550$ nm for a material gain of ~1260 cm^{-1}. Kumar et al. [20] also investigated the interaction of SPPs with a semiconductor gain medium (GaAs), modeling the diode structure and carrier injection levels required for amplification.

Lu et al. [21] modeled the optical parametric amplification of SPPs in an Ag–$LiNbO_3$ structure. They find seed and pump wavelengths for which phase matching occurs and then predict gains of ~30 dB over 3 mm of interaction length for a pump intensity of 50 MW/cm.

5.3.2 Experimental

Oscillation (lasing) involving SPPs was first demonstrated in quantum cascade lasers operating at far-infrared wavelengths [22,23]. In one device [23] a bimetallic (Ti/Au) grating was used to form a distributed-feedback SPP laser emitting at $\lambda_0 \sim 17$ μm. This device owes its success to the low loss of metals at long wavelengths.

The first observations of stimulated emission of SPPs at optical wavelengths were reported in Kretschmann structures consisting of 39 and 67 nm thick Ag films on BK7 prisms with the other Ag surface in contact with the gain medium [24]. Rhodamine 101 and cresyl violet in ethanol (concentrations of $N = 7\times10^{17}$ cm^{-3}) were used as gain media. TM-polarized light at $\lambda_0 = 633$ nm was used to excite probe SPPs at the Ag/dye interface, while the dye was excited at $\lambda_0 = 580$ nm (~10 mW) by pump SPPs counter-propagating along the same interface. The differential reflectance, measured as a function of the probe incidence angle with and without pumping, was attributed to stimulated emission of probe SPPs. Increased reflectance was noted over the SPP excitation angle for the 39 nm Ag films, in qualitative agreement with theory [16], whereas a narrowing and deepening of the reflectance dip was noted for the 67 nm Ag films, also in qualitative agreement with theory [17].

A similar structure was investigated by Noginov et al. [25] but with the gain medium formed as a 10 μm thick layer of PMMA highly doped with Rhodamine-6G (R6G, $N = 2.2\times10^{22}$ cm^{-3}). They strongly pumped the structure from the back (~18 mJ pulses, ~10 ns duration, $\lambda_0 = 532$ nm) and excited probe SPPs via the prism at $\lambda_0 = 594$ nm. They measured an increase in reflectance at the SPP excitation angle during pump pulses due to stimulated emission of SPPs and estimated via modeling an SPP loss reduction of 35% and a material gain of ~420 cm^{-1}. In other experiments incorporating a thinner gain layer (~3 μm), they measured the fluorescence decoupled from the prism during pumping in the absence of the probe, observing characteristics of amplified spontaneous emission (ASE) which they attributed to the stimulated emission of SPP [26].

The amplified spontaneous emission of SPPs (ASE-SPP) was also observed by Bolger et al. [27] on a 100 nm thick Au film coated with a ~1 μm thick film of PMMA doped with PbS quantum dots (QDs). A grating was created in the metal film to out-couple SPPs at wavelengths near the QD emission peak, as shown in Figure 5.4a. Output spectra narrowed as the pump power increased, as shown in Figure 5.4b, indicating ASE-SPP. They estimate a ~30% increase in SPP propagation length due to stimulated emission and point out that significant gain depletion may occur due to ASE-SPP.

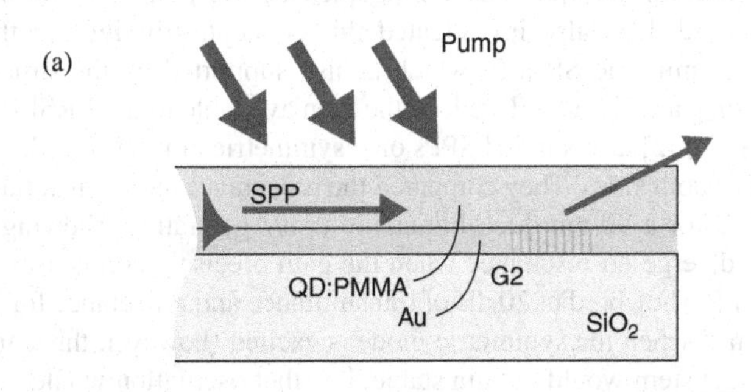

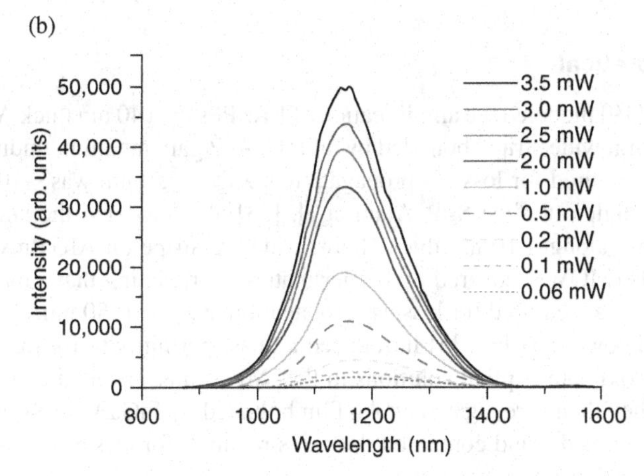

FIGURE 5.4 (a) SPPs propagating through a QD-doped PMMA layer; grating G2 out-couples SPPs into bulk waves. (b) Measured ASE-SPP spectra for increasing pump irradiance. (Adapted from Reference 27.)

5.4 SYMMETRIC METAL FILMS

As discussed in Section 5.2, a thin metal film (Fig. 5.1b) or stripe (Fig. 5.1e) bounded by nearly symmetric dielectrics supports LRSPPs [8]. Less gain is required to amplify LRSPPs compared to single-interface SPPs due to the significantly lower attenuation of the former.

5.4.1 Gratings

It was suggested that a 2D corrugated grating formed from a thin metal film supporting LRSPPs would operate as an SPP bandgap laser if a gain medium having its peak emission within the bandgap was incorporated into the structure [28–30]. An ~20 nm thick corrugated Ag film, bounded symmetrically by DCM-doped Alq_3 as the gain medium, was proposed and it was argued that sufficient gain could be provided to overcome LRSPP losses. In this concept, standing LRSPP waves would be amplified by stimulated emission and partially out-coupled by the grating to form the laser output. Winter et al. [31] also investigated this concept, pointing out that dipole decay into the asymmetric SRSPP (which is also supported by the structure—see Section 5.2) is substantial and will reduce the gain available to the LRSPP.

Kovyakov et al. [32] also studied SPPs on a symmetric grating bounded by a gain layer(s) on one or both sides. They computed the reflectance and transmittance under normal incidence for a 90 nm thick bimetallic (Au/Ag) grating, showing that they simultaneously diverge on resonance when the gain precisely compensates the loss of the excited SPP. They predict 20 dB of transmittance and reflectance for a material gain of 4670 cm^{-1} when the symmetric mode is excited (however, this computation assumes that the system would remain stable, i.e., that oscillation would not occur).

5.4.2 Theoretical

Nezhad et al. [19] modeled the amplification of LRSPPs on a 40 nm thick Ag film and a similar 400 nm wide stripe, bounded by an InGaAsP gain medium, finding that the material gain required for lossless propagation at $\lambda_0 = 1550$ nm was ~10× smaller than for the single-interface SPP. Alam et al. [33] also modeled the amplification of LRSPPs, but along a 10 nm thick, 1 μm wide Ag stripe on AlGaInAs multiple quantum wells (MQWs) covered by barrier material, predicting that a material gain of ~400 cm^{-1} was required for lossless propagation at $\lambda_0 = 1550$ nm.

Genov and coworkers [34, 35] introduced a quasi-permittivity for the metal film such that approximate explicit solutions to the IMI's transcendental equation could be derived. They then investigated an Ag film bounded by InGaAsP [34] and InGaN [35] MWQ gain media and computed the gains required for lossless propagation of the symmetric and asymmetric SPPs.

Chen and Guo [36] computed the Q and threshold gain of LRSPP whispering gallery modes in GaAs-based microdisk cavities incorporating a thin metal film, finding high Qs (~4000) and a low threshold gain (~200 cm^{-1}) at $\lambda_0 = 1400$ nm.

De Leon and Berini [37,38] constructed a theoretical model for SPP amplification, considering a four-level gain medium, an inhomogeneous pump intensity distribution, and a position-dependent dipole lifetime, finding that if both distributions are neglected the material gain required for lossless propagation can be underestimated by ~10×. They considered R6G in solvent as the gain medium, and predicted lossless LRSPP propagation at λ_0 = 560 nm under modest pumping (λ_0 = 532 nm, ~200 kW/cm^2) for a modest dye concentration ($N = 1.8\times10^{18}$ cm^{-3}) assuming a 20 nm thick Ag film covered by the gain medium. They also predict lossless propagation of the single-interface SPP in a similar system but under significantly stronger pumping (~3.5 MW/cm^2) and a higher dye concentration ($N = 2.4\times10^{19}$ cm^{-3}).

5.4.3 Experimental

Ambati et al. [39] observed stimulated emission of LRSPPs at λ_0 = 1532 nm on a ~20 nm thick, 8 μm wide Au stripe embedded in Er-doped glass. They used a co-propagating LRSPP pump/probe arrangement.

De Leon and Berini [40] reported measurements of LRSPP amplification in an insertion arrangement, as sketched in Figure 5.5a. The structure investigated consisted of an Au stripe (20 nm thick, 1 μm wide, 2.7 mm long) on 15 μm of SiO_2 on Si, covered by ~100 μm of dye solution (IR-140, $N = 6\times10^{17}$ cm^{-3}) index-matched to SiO_2. The structure was pumped from the top (λ_0 = 808 nm, 20 mJ/cm^2, 8 ns pulses) and probed at λ_0 = 882 nm via end-fire-aligned polarization-maintaining single-mode optical fibers having a 92% coupling efficiency per facet to the LRSPP. The insertion gain was measured versus amplifier (pump) length l_a, as shown in Figure 5.5b, from which the slope of the best-fitting linear model yields the LRSPP small-signal mode power gain coefficient, in this case, of γ = 8.55 dB/mm (~20 cm^{-1}).

In a subsequent study [41] they measured ASE into the LRSPP, revealing spectrum narrowing with increasing pump length. The output power at the peak ASE-LRSPP wavelength was measured as a function of amplifier length, as shown in Figure 5.6a. The images in the inset of Figure 5.6a show the output of a structure with the pump on but the probe off (left), and the pump off but the probe on (right). An exponential model for the output power was fitted to the measurements (dashed curve), from which an effective input noise power per unit bandwidth of ~3.3 photons per (LRSPP) mode was deduced along with an LRSPP gain of ~17 cm^{-1}. They relate this low noise power to the low spontaneous emission rate of dipoles into the LRSPP. The computed normalized dipole (isotropic) decay rates into the various channels supported by a symmetric metal slab ($w = \infty$) reveal that LRSPPs receive less spontaneous emission compared to other channels, as shown in Figure 5.6b. Using these decay rates, a theoretical model for ASE-LRSPP was constructed and compared to the experimental results (blue curve, Fig. 5.6a), revealing good agreement. Details on their measurement setup and characterization techniques were also reported [42].

Gather et al. [43] reported LRSPP gain at visible wavelengths in a symmetric structure comprising a 4 nm thick Au film and a 1 μm thick polymeric gain layer. The structure was pumped from the top (λ_0 = 532 nm, 5 ns pulses), and ASE-LRSPP was measured as a function of pump intensity and pump length. They report

(a)

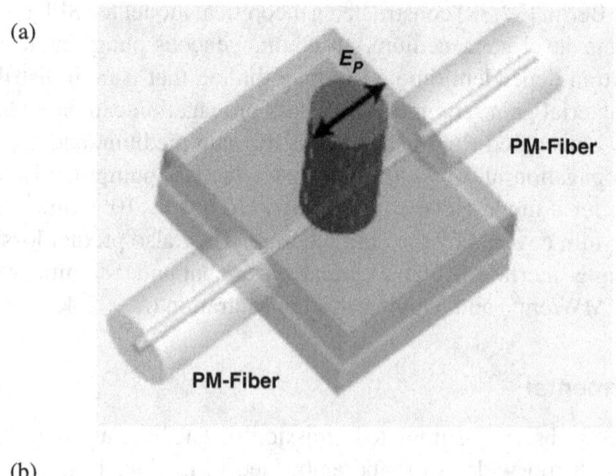

(b)

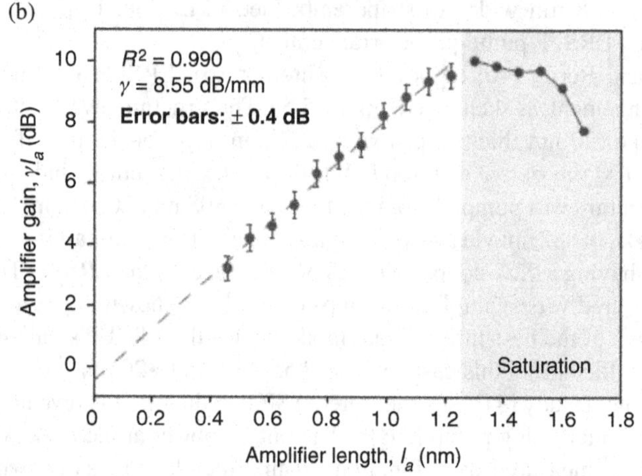

FIGURE 5.5 (a) LRSPP amplifier formed as an Au stripe on SiO_2 (on Si) covered by a dye gain medium (IR-140). The pump is polarized along the stripe length (E_p) and applied to the top of the structure. End-fire-coupled input/output polarization-maintaining fibers are also shown. (b) Measured amplifier gain at a probe wavelength of $\lambda_0 = 882$ nm versus amplifier (pump) length l_a; the slope of the curve gives the LRSPP mode power gain ($\gamma = 8.55$ dB/mm). (Adapted from Reference 40.)

spectrum narrowing, a threshold behavior in the emitted intensity, and an LRSPP gain coefficient of $\sim$8 cm^{-1} at $\lambda_0 \sim 600$ nm.

Flynn et al. [44] demonstrated LRSPP lasing at $\lambda_0 \sim 1.46$ μm in a symmetric InP-based structure comprised of a 15 nm thick Au film placed between MQW stacks providing TM gain. Each stack was formed from eight tensile InGaAs QWs separated by compressive InAlAs barriers. LRSPP lasing from a 1 mm long, 100 μm wide Fabry–Perot cavity was observed as TM-polarized light emitted from an end facet while pumping from the top ($\lambda_0 = 1.06$ μm, 140 ns pulses). The emitted power was linearly dependent on the pump intensity (beyond a clear threshold), and the emitted spectrum narrowed with increasing pump intensity.

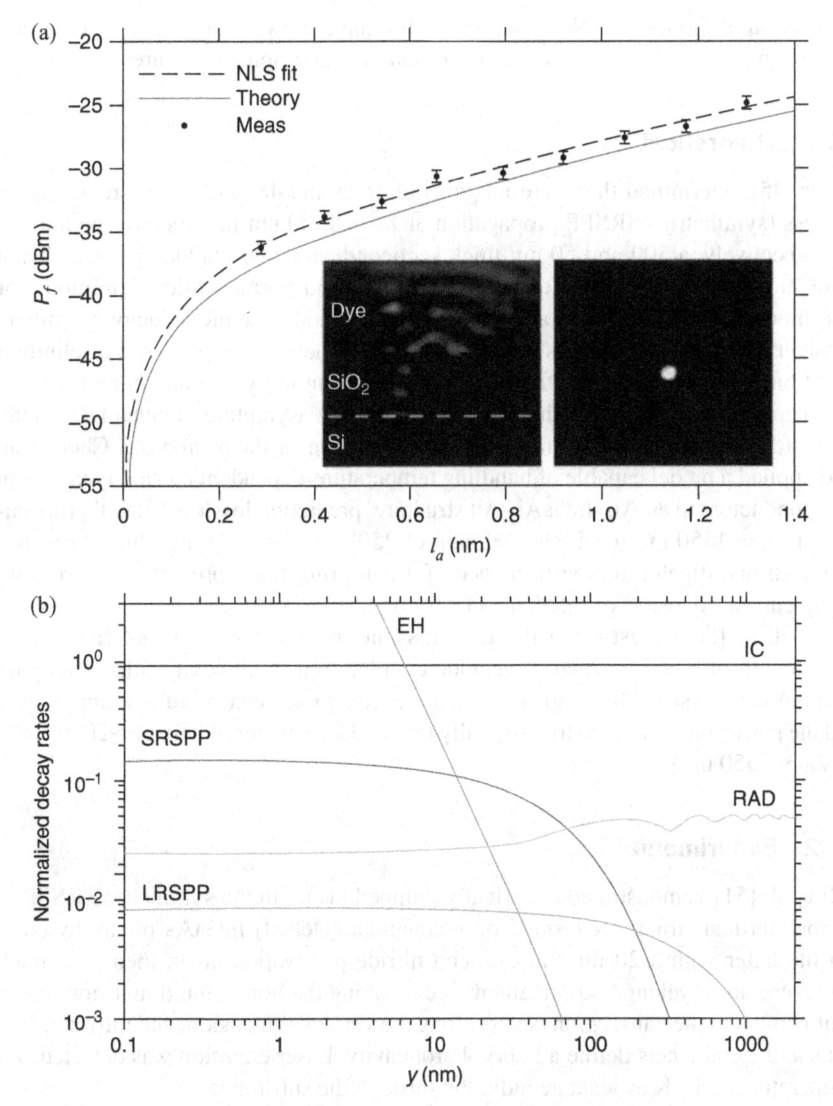

FIGURE 5.6 (a) Measured ASE-LRSPP (dots), corresponding to nonlinear least squares (NLS) best-fitting curve (dashed) and theoretical curve (solid). The images in inset show the output of a structure with the pump on but probe off (left) and the pump off but probe on (right). (b) Normalized dipole (isotropic) decay rates into the various channels supported by a symmetric metal slab ($w = \infty$). (Adapted from Reference 41.)

5.5 METAL CLADS

As discussed in Section 5.2, a thin dielectric or semiconducting film bounded by metal clads (Fig. 5.1c) supports strongly confined symmetric SRSPPs. More gain is required to amplify SRSPPs compared to single-interface SPPs due to the significantly higher attenuation of the former. Nonetheless, studies have shown that semiconductor gain media can provide enough gain to compensate losses in some structures.

5.5.1 Theoretical

Maier [45] determined that material gains of 1625 and 4830 cm^{-1} are required for lossless (symmetric) SRSPP propagation at $\lambda_0 = 1500$ nm in structures comprised of, respectively, a 500 and 50 nm thick semiconductor film cladded by Au. Chang and Chuang [46, 47] derived quasi-orthogonality and normalization conditions for SPP modes, then formulated a model for metal-cladded semiconductor amplifiers considering various definitions for the confinement factor (they proposed a definition based on energy and showed that it remains less than unity). Li and Ning [48] find that the mode power gain of the SPP near its energy asymptote is much larger than the material gain, due to low mode energy velocity near the asymptote. Chen et al. [49] applied a model capable of handling temperature-dependent carrier dynamics in semiconductors to an Au–InGaAs–Au structure, predicting lossless SRSPP propagation at $\lambda_0 = 1550$ nm for a material gain of 2500 cm^{-1} in a 75 nm thick structure; they also investigated the performance of a nano-ring laser, predicting a threshold lasing current (density) of ~550 nA (1 kA/cm^2).

Yu et al. [50] investigated the transmission and reflection characteristics of a short length of metal-cladded waveguide coupled to a small cavity filled with gain medium (InGaAsP). They showed that the cavity losses can be fully compensated and the performance of the structure fully restored for a material gain of ~2000 cm^{-1} (at $\lambda_0 \sim 1550$ nm).

5.5.2 Experimental

Hill et al. [51] demonstrated electrically pumped lasing in the symmetric SRSPP in narrow vertical structures formed on rectangular (etched) InGaAs pillars by coating the latter with a 20 nm thick silicon nitride passivation layer, then by a thick Ag layer. Subwavelength confinement occurs along the horizontal dimension, index confinement to the gain region occurs along the vertical dimension, and mirror reflections at the end facets define a Fabry–Perot cavity. Laser emission was detected at a temperature of 10 K as leakage radiation through the substrate.

5.6 OTHER STRUCTURES

5.6.1 Dielectric-Loaded SPP Waveguides

As discussed in Section 5.2 (Fig. 5.1d), a dielectric region having a refractive index higher than that of the surrounding environment (usually air) provides additional confinement by index contrast. Grandidier et al. [52] investigated waveguides comprised of PMMA stripes (~600×400 nm^2) doped with PbS QDs defined on a ~40 nm thick Au film, as sketched in Figure 5.7a. Measurements of spontaneous and stimulated emission into SPPs were obtained via Fourier-plane leakage radiation microscopy

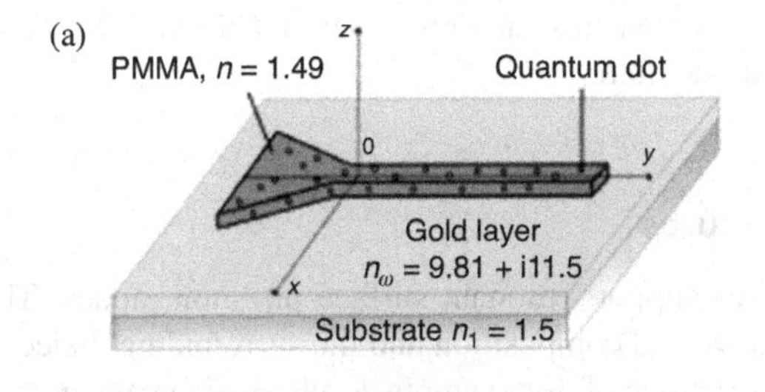

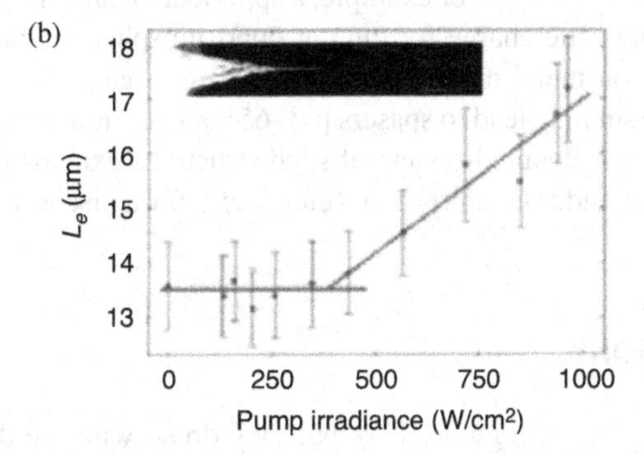

FIGURE 5.7 (a) Sketch of a dielectric-loaded SPP waveguide comprising a QD-doped PMMA ridge. (b) Measured SPP propagation length (L_e) for increasing pump irradiance; the inset shows a near-field scan captured above the waveguide (without pumping). (Adapted from Reference 52.)

(FP-LRM) at $\lambda_0 \sim 1550$ nm (near the QD emission peak) while pumping from the top at $\lambda_0 = 532$ nm. The SPP propagation length was measured as a function of pump intensity while probing at $\lambda_0 = 1550$ nm, as shown in Figure 5.7b. A distinct threshold is observed beyond which the propagation length increases linearly with the pump; a maximum increase of ~27% was noted. In a subsequent paper [53] they investigated CdSe/ZnSe QD-doped PMMA stripes on an Ag film, measuring a comparable increase in propagation length at $\lambda_0 = 632.8$ nm. Radko et al. [54] reported a 32% increase in SPP propagation length in PbS QD-doped PMMA stripes on Au at $\lambda_0 \sim 860$ nm. Colas Des Francs et al. [55] modeled structures including gain, reproducing measurements reported earlier [52], while extracting the cross section for stimulated emission of PbS QDs. Rao and Tang [56] modeled structures at wavelengths close to the SPP energy asymptote, finding that nanoscale dielectric cross sections ($\sim 22 \times 20$ nm^2) with $\sim 5 \times 10^4$ cm^{-1} of gain in the dielectric would provide lossless subwavelength propagation at $\lambda_0 = 450$ nm.

5.6.2 Hybrid SPP Waveguide

A metal surface covered by a low-index dielectric then by a high-index dielectric supports a TM-polarized hybrid SPP exhibiting field localization in the low-index dielectric and a high attenuation (compared to LRSPPs) [57, 58]. The high-index dielectric can be a gain medium, leading to loss compensation and lasing in the hybrid SPP, when the latter overlaps sufficiently with the gain medium. Laser-like behavior was observed in such structures, where a CdS nanowire [59] or patch [60] was used as the gain medium.

5.6.3 Nanostructures

Small metal particles support resonant surface plasmon modes. The latter depend on particle shape, size, and composition and the surrounding dielectric [61, 62]. The fundamental resonant mode of, for example, a spherical metal nanoparticle is dipolar, with densities of opposite charge forming at opposite spherical caps. The behavior of resonant modes on metal nanoparticles coupled to a gain medium is of interest because such a system may lead to spasers [63–65] or exhibit near-singular scattering properties [66–68]. (A detailed review of such structures extends beyond the scope of our chapter—the reader is referred to Reference 6 for a discussion.)

5.7 CONCLUSIONS

Metallo-dielectric structures guide SPPs but they do so with the drawback of high attenuation. Incorporating optical gain for loss compensation allows SPPs to propagate longer distances and alters the confinement–attenuation trade-off.

The *approximate* small-signal material gains required for lossless SPP propagation are (in order of increasing confinement) 1–200 cm^{-1} for LRSPPs along thin ($\sim$20 nm) metal films; 1000–2000 cm^{-1} for single-interface SPPs; 2000–5000 cm^{-1} for SPPs along thin ($\sim$50 nm) metal-cladded waveguides; and 80,000 cm^{-1} for SPPs near their energy asymptote. Except perhaps for the latter, such gain levels are available from dyes and semiconductor quantum structures. Optical dipoles (e.g., dyes) can be incorporated into dielectrics (polymers, glasses) and integrated with metal structures. Semiconductors (epitaxial) can be pumped electrically; however, they can be difficult to integrate with a target metallic structure and the gain may be polarization-dependent so alignment of the medium with the SPP fields is necessary.

Significant progress on SPP amplification and lasing has been reported, but much work remains, particularly on the development of applications where performance and usability are paramount. Aspects requiring improvements and further work include *pumping* (electrical is more convenient), *resonators* (Q worsens as confinement increases), *power dissipation* (severely limited in several applications), *energy efficiency* (more is better), *signal-to-noise* (noise constrains communication), *operating temperature* (room temperature is more practical), and *stability* (large material gains can lead to undesired oscillation).

Although challenges remain, the prospects for SPP amplifiers and lasers are bright. Demonstrations of amplification (overcompensation of loss) and lasing (oscillation) have already been reported involving several kinds of SPPs on many metallic structures, including single-interface SPPs [23], LRSPPs on stripes [40] and slabs [43, 44], and SPPs in metal-cladded [51] and hybrid [60] waveguides. Building on such demonstrations should lead to useful applications.

REFERENCES

1. Maier SA (2007) *Plasmonics: Fundamentals and Applications*. Springer.
2. Barnes WL, Dereux A, Ebbesen TW (2003) Surface plasmon subwavelength optics. *Nature* 424: 824–830.
3. Gramotnev DK, Bozhevolnyi SI (2010) Plasmonics beyond the diffraction limit. *Nat. Photonics* 4: 83–91.
4. Homola J (2008) Surface plasmon resonance sensors for detection of chemical and biological species. *Chem. Rev.* 108: 462–493.
5. Ebbesen TW, Genet C, Bozhevolnyi SI (2008) Surface plasmon circuitry. *Phys. Today* 61: 44–50.
6. Berini P, De Leon I (2012) Surface plasmon-polariton amplifiers and lasers. *Nat. Photonics.* doi: 10.1038/nphoton.2011.285.
7. Berini P (2006) Figures of merit for surface plasmon waveguides. *Opt. Express* 14: 13030–13042.
8. Berini P (2009) Long-range surface plasmon polaritons. *Adv. Opt. Photonics* 1: 484–588.
9. Hohenau A, et al. (2005) Dielectric optical elements for surface plasmons. *Opt. Lett.* 30: 892–895.
10. Berini P (2000) Plasmon-polariton waves guided by thin lossy metal films of finite width: bound modes of symmetric structures. *Phys. Rev. B* 61: 10484–10503.
11. Pile DFP, et al. (2005) Two-dimensionally localized modes of a nanoscale gap plasmon waveguide. *Appl. Phys. Lett.* 87: 261114.
12. Nkoma J, Loudon R, Tilley DR (1974) Elementary properties of surface plasmons. *J. Phys. C* 7: 3547–3559.
13. Archambault A, Marquier F, Greffet J-J (2010) Quantum theory of spontaneous and stimulated emission of surface plasmons. *Phys. Rev. B* 82: 035411.
14. Matloob R, Loudon R, Barnett SM, Jeffers J (1995) Electromagnetic field quantization in absorbing dielectrics. *Phys. Rev. A* 52: 4823–4838.
15. Barnes WL (1998) Fluorescence near interfaces: the role of photonic mode density. *J. Mod. Opt.* 45: 661–669.
16. Plotz G, Simon H, Tucciarone J (1979) Enhanced total reflection with surface plasmons. *J. Opt. Soc. Am.* 69: 419–422.
17. Sudarkin AN, Demkovich PA (1988) Excitation of surface electromagnetic waves on the boundary of a metal with an amplifying medium. *Sov. Phys. Tech. Phys.* 34: 764–766.
18. Avrutsky I (2004) Surface plasmons at nanoscale relief gratings between a metal and a dielectric medium with optical gain. *Phys. Rev. B.* 70: 155416.
19. Nezhad MP, Tetz K, Fainman Y (2004) Gain assisted propagation of surface plasmon polaritons on planar metallic waveguides. *Opt. Express* 12: 4072–4079.
20. Kumar P, Tripathi VK, Liu CS (2008) A surface plasmon laser. *J. Appl. Phys* 104: 033306.
21. Lu FF, et al. (2011) Surface plasmon polariton enhanced by optical parametric amplification in nonlinear hybrid waveguide. *Opt. Express* 19: 2858–2865.

22. Sirtori C, et al. (1998) Long-wavelength ($\lambda \approx 11.5\mu m$) semiconductor lasers with waveguides based on surface plasmons. *Opt. Lett.* 23: 1366–1368.

23. Tredicucci A, et al. (2000) Single-mode surface-plasmon laser. *Appl. Phys. Lett.* 76: 2164–2166.

24. Seidel J, Grafstrom S, Eng L (2005) Stimulated emission of surface plasmons at the interface between a silver film and an optically pumped dye solution. *Phys. Rev. Lett.* 94: 177401.

25. Noginov MA, et al. (2008) Compensation of loss in propagating surface plasmon polariton by gain in adjacent dielectric medium. *Opt. Express* 16: 1385–1392.

26. Noginov MA, et al. (2008) Stimulated emission of surface plasmon polaritons. *Phys. Rev. Lett.* 101: 226806.

27. Bolger PM, et al. (2010) Amplified spontaneous emission of surface plasmon polaritons and limitations on the increase of their propagation length. *Opt. Lett.* 35: 1197–1199.

28. Okamoto T, H'Dhili F, Kawata S (2004) Towards plasmonic band gap laser. *Appl. Phys. Lett.* 85: 3968.

29. Okamoto T, Simonen J, Kawata S (2008) Plasmonic band gaps of structured metallic thin films evaluated for a surface plasmon laser using the coupled-wave approach. *Phys. Rev. B* 77: 115425.

30. H'Dhili F, Okamoto T, Simonen J, Kawata S (2011) Improving the emission efficiency of periodic plasmonic structures for lasing applications. *Opt. Commun.* 284: 561–566.

31. Winter G, Wedge S, Barnes WL (2006) Can lasing at visible wavelength be achieved using the low-loss long-range surface plasmon-polariton mode? *New J. Phys.* 8: 125.

32. Kovyakov A, Zakharian AR, Gundu KM, Darmanyan SA (2009) Giant optical resonances due to gain-assisted Bloch surface plasmon. *Appl. Phys. Lett.* 94: 151111.

33. Alam MZ, Meier J, Aitchison JS, Mojahedi M (2007) Gain assisted surface plasmon polariton in quantum well structures. *Opt. Express* 15: 176–182.

34. Genov DA, Ambati M, Zhang X (2007) Surface plasmon amplification in planar metal films. *IEEE J. Quant. Electron.* 43: 1104–1108.

35. Ambati M, Genov DA, Oulton RF, Zhang X (2008) Active plasmonics: surface plasmon interaction with optical emitters. *IEEE J. Sel. Top. Quant. Electron.* 14: 1395–1403.

36. Chen Y-H, Guo LJ (2011) High Q long-range surface plasmon polariton modes in subwavelength metallic microdisk cavity. *Plasmonics* 6: 183–188.

37. De Leon I, Berini P (2008) Theory of surface plasmon-polariton amplification in planar structures incorporating dipolar gain media. *Phys. Rev. B* 78: 161401(R).

38. De Leon I, Berini P (2009) Modeling surface plasmon-polariton gain in planar metallic structures. *Opt. Express* 17: 20191–20202.

39. Ambati M, et al. (2008) Observation of stimulated emission of surface plasmon polaritons. *Nano Lett.* 8: 3998–4001.

40. De Leon I, Berini P (2010) Amplification of long-range surface plasmons by a dipolar gain medium. *Nat. Photonics* 4: 382–387.

41. De Leon I, Berini P (2011) Spontaneous emission in long-range surface plasmon-polariton amplifiers. *Phys. Rev. B* 83: 081414(R).

42. De Leon I, Berini P (2011) Measuring gain and noise in active long-range surface plasmon-polariton waveguides. *Rev. Sci. Instrum.* 82: 033107.

43. Gather MC, Meerholz K, Danz N, Leosson K (2010) Net optical gain in a plasmonic waveguide embedded in a fluorescent polymer. *Nat. Photonics* 4: 457–461.

44. Flynn RA, et al. (2011) A room-temperature semiconductor spaser operating near 1.5 μm. *Opt. Express* 19: 8954–8961.

45. Maier SA (2006) Gain-assisted propagation of electromagnetic energy in sub-wavelength surface plasmon polariton gap waveguides. *Opt. Commun.* 258: 295–299.

46. Chang S-W, Chuang SL (2009) Normal modes for plasmonic nanolasers with dispersive and inhomogeneous media. *Opt. Lett.* 34: 91–93.

47. Chang S-W, Chuang SL (2009) Fundamental formulation for plasmonic nanolasers. *IEEE J. Quant. Electron.* 45: 1014–1023.

48. Li DB, Ning CZ (2009) Giant modal gain, amplified surface plasmon-polariton propagation, and slowing down of energy velocity in a metal-semiconductor-metal structure. *Phys. Rev. B* 80: 153304.

49. Chen X, Bhola B, Huang Y, Ho ST (2010) Multi-level multi-thermal-electron FDTD simulation of plasmonic interaction with semiconducting gain media: applications to plasmonic amplifiers and nano-lasers. *Opt. Express* 18: 17220–17238.

50. Yu Z, Veronis G, Fan S, Bongersma ML (2008) Gain-induced switching in metal-dielectric-metal plasmonic waveguides. *Appl. Phys. Lett.* 92: 041117.

51. Hill MT, et al. (2009) Lasing in metal-insulator-metal sub-wavelength plasmonic waveguides. *Opt. Express* 17: 11107–11112.

52. Grandidier J, et al. (2009) Gain-assisted propagation in a plasmonic waveguide at telecom wavelength. *Nano Lett.* 9: 2935–2939.

53. Grandidier J, et al. (2010) Leakage radiation microscopy of surface plasmon coupled emission: investigation of gain-assisted propagation in an integrated plasmonic waveguide. *J. Microsc.* 239: 167–172.

54. Radko IP, Nielsen MG, Albrektsen O, Bozhevolnyi SI (2010) Stimulated emission of surface plasmon polaritons by lead-sulfide quantum dots at near infra-red wavelengths. *Opt. Express* 18: 18633–18641.

55. Colas Des Francs G, et al. (2010) Optical gain, spontaneous and stimulated emission of surface plasmon polaritons in confined plasmonic waveguide. *Opt. Express* 18: 16327–16334.

56. Rao R, Tang T (2011) Study on active surface plasmon waveguides and design of a nanoscale lossless surface plasmon waveguide. *J. Opt. Soc. Am. B* 28: 1258–1265.

57. Alam MZ, Meier J, Aitchison JS, Mojahedi M (2007) Super mode propagation in low index medium. *Proc. CLEO*, paper JThD112.

58. Oulton RF, Sorger VJ, Genov DA, Pile DFP, Zhang X (2008) A hybrid plasmonic waveguide for subwavelength confinement and long-range propagation. *Nat. Photonics* 2: 496–500.

59. Oulton RF, et al. (2009) Plasmon lasers at deep subwavelength scale. *Nature* 461: 629–632.

60. Ma R-M, Oulton RF, Sorger VJ, Bartal G, Zhang X (2011) Room-temperature sub-diffraction-limited plasmon laser by total internal reflection. *Nat. Mater.* 10: 110–113.
61. Kelly KL, Coronado E, Zhao LL, Schatz GC (2003) The optical properties of metal nanoparticles: the influence of size, shape, and dielectric environment. *J. Phys. Chem. B* 107: 668–677.
62. Pelton M, Aizpurua J, Bryant G (2008) Metal-nanoparticle plasmonics. *Laser Photonics Rev.* 2: 136–159.
63. Bergman DJ, Stockman MI (2003) Surface plasmon amplification by stimulated emission of radiation: quantum generation of coherent surface plasmons in nanosystems. *Phys. Rev. Lett.* 90: 027402.
64. Stockman MI (2008) Spasers explained. *Nat. Photonics* 2: 327–329.
65. Noginov MA, et al. (2009) Demonstration of a spaser-based nanolaser. *Nature* 460: 1110–1113.
66. Lawandy NM (2004) Localized surface plasmon singularities in amplifying media. *Appl. Phys. Lett.* 85: 5040–5042.
67. Gordon JA, Ziolkowski RW (2007) The design and simulated performance of a coated nano-particle laser. *Opt. Express* 15: 2622–2653.
68. Li Z-Y, Xia Y (2010) Metal nanoparticles with gain toward single-molecule detection by surface-enhanced Raman scattering. *Nano Lett.* 10: 243–249.

6

Controlling Light Propagation with Interfacial Phase Discontinuities

NANFANG YU, MIKHAIL A. KATS, PATRICE GENEVET, FRANCESCO AIETA, ROMAIN BLANCHARD, GUILLAUME AOUST, ZENO GABURRO, AND FEDERICO CAPASSO
School of Engineering and Applied Sciences, Harvard University, Cambridge, Massachusetts, USA

This chapter is devoted to the phase response of plasmonic antennas and their applications. Plasmonic antennas are optical resonators that convert light propagating in free space into spatially localized energy. These antennas have a wide range of potential applications [1–6]. One of the main themes of previous research efforts has been the capability of optical antennas in efficiently capturing and concentrating light power; however, the antenna phase response and its implications in controlling the propagation of light have not been systematically studied and will be the focus of this chapter. This chapter is organized as follows.

Part I of the chapter discusses a few models with varying complexity that illustrate the physics behind the phase response of optical antennas. In Section 6.1, we provide an intuitive physical picture in analogy to the impedance of electric circuit elements, where the evolution of charge, current, and near-field distributions of an optical antenna across a resonance is explained. Sections 6.2 and 6.3 discuss 1D and 2D oscillator models considering explicitly the radiation damping effects besides the ohmic absorption losses for plasmonic antennas supporting one plasmonic mode and two independent, orthogonal plasmonic modes, respectively. Section 6.4 presents the results of two analytical models based on solving rigorous integral equations of antenna currents that consider retardation effects and near-field interactions. We adapt methods used in radio frequency antennas for optical frequencies and verify the validity of our models by comparing the results with full-wave simulations. Section 6.5 discusses the properties of V-shaped optical antennas by mapping out their two

Active Plasmonics and Tuneable Plasmonic Metamaterials, First Edition. Edited by Anatoly V. Zayats and Stefan A. Maier.

plasmonic eigenmodes and polarization conversion properties. These antennas are ideal examples of 2D plasmonic structures and are the building blocks of antenna arrays to be discussed later in the chapter.

Part II of the chapter discusses several applications of phased optical antenna arrays, dubbed "meta-interfaces," which are able to mold the incident wavefronts into desired shapes over a propagation distance comparable to the optical wavelength. These include demonstration of generalized laws of reflection and refraction in the presence of linear interfacial phase distributions, demonstration of giant and tuneable birefringence, and generation of optical vortices that carry orbital angular momentum.

6.1 PHASE RESPONSE OF OPTICAL ANTENNAS

6.1.1 Introduction

In the field of physical optics, which explores the propagation of electromagnetic waves, the general function of most optical devices can be described as the modification of the wavefront of light by altering its phase and amplitude in a desired manner. The class of optical components with a spatially varying phase response includes lenses, prisms, spiral phase plates [7], axicons [8], phase retarders, and more generally spatial light modulators (SLMs), which are able to imitate many of these components by means of a dynamically tuneable spatial phase response [9]. Another class of optical components such as gratings and holograms is based on diffractive optics [10], where diffracted waves from different parts of the components interfere in the far field to produce the desired optical pattern. All of these optical components shape optical wavefronts by relying on gradual phase shifts accumulated during different optical lengths. This approach is generalized in transformation optics [11,12] which utilizes metamaterials to engineer the spatial distribution of refractive indices and therefore bend light in unusual ways, achieving such phenomena as negative refraction, subwavelength-focusing, and optical cloaking [13,14].

It is possible to break away from our reliance on the propagation effect and attain new degrees of freedom in optical design by introducing abrupt phase changes into the optical path [15–20]. This can be achieved by using the large and controllable phase shift between the excitation and radiated light of optical resonators. This approach enables the design of a new class of optical devices with pixelated phase elements, which are thin compared to the wavelength of light. The choice of optical resonators is potentially wide-ranging, from nanocavities [21,22], to nanoparticles clusters [23,24] and optical antennas [1,2]. We concentrated on the latter in this chapter, due to their widely tailorable optical properties and the ease of fabricating planar antennas of nanoscale thickness. In general, resonant behavior can be found for any type of vibration, including mechanical, electrical, optical, and acoustic, among others, and can be utilized in the manipulation of these various kinds of waves (e.g., [25,26]).

The phase shift between the scattered and incident light of an optical resonator sweeps a range of $\sim\pi$ across a resonance. This π phase shift across resonance

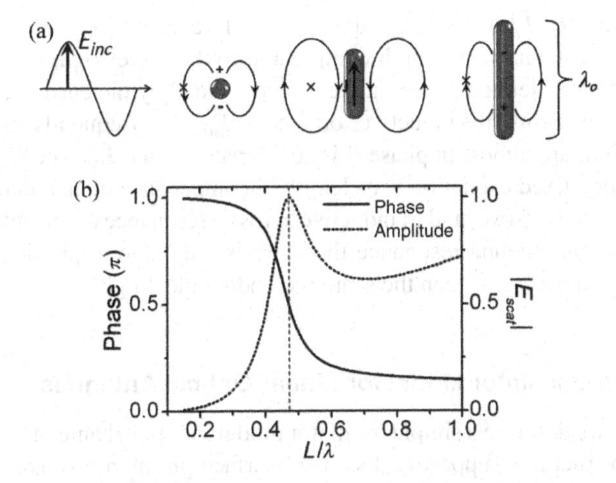

FIGURE 6.1 (a) Schematics showing the instantaneous distributions of charge, current, and near field (represented by electric field lines) of rod antennas of different sizes when the incident electric field (first panel) is upward and has the largest amplitude. A perfectly conducting metal is assumed and antennas are in the free space. The second panel shows that maximum amount of charges accumulates at the ends of an optically small antenna with length $L << \lambda_o$ and the current on the antenna is zero. The third panel shows that the instantaneous current on a resonant antenna with $L \approx \lambda_o/2$ is maximum and is in phase with the incident field, while there are no charge accumulations. In the last panel, for an antenna with $L \approx \lambda_o$, again the charge accumulation is maximum (but located at a distance of about $L/4$ from the ends [28]) and there is no antenna current. Let us take a monitor point $\lambda_o/4$ away from the antennas (the crosses in the second, third and fourth panels of Fig. 6.1a) and count the phase difference $\Delta\Phi$ between the antenna radiation at that point and the incident field. One can easily tell that $\Delta\Phi$ is π, $\pi/2$, and 0 for the three antennas with $L << \lambda_o$, $L \approx \lambda_o/2$, and $L \approx \lambda_o$, respectively, by comparing the electric field lines around the antennas and the instantaneous incident field. (b) Amplitude and phase response of a rod antenna as a function of antenna length calculated by analytically solving Hallén's integral equations for linear antennas [28]. The antenna is located in vacuum and has a circular cross-section with radius $r = L/50$. Incident monochromatic light impinges normal to the rod and is polarized along it. The scattered light is monitored in the far field along the direction of the incident light.

is observed in many other resonator systems between the response of a resonator and its driving force (e.g., when the frequency of an applied force tunes over the resonant frequency of a mechanical harmonic oscillator). Without resorting to any rigorous theoretical models or full-wave simulations, let us first gain a physical understanding of this phenomenon qualitatively by analyzing the phase response of a straight optical antenna of length L. If the antenna is optically small ($L/\lambda_o << 1$), its charge distribution instantaneously follows the incident field (Fig. 6.1a, panel two), that is, $\tilde{\sigma} \propto \tilde{E}_{inc} = E_{inc}\exp(i\omega t)$, where $\tilde{\sigma}$ is the charge density at one end of the antenna. Therefore, the emitted electric field, which is proportional to the acceleration of the charges (Larmor formula [27]), is $\tilde{E}_{scat} \propto \partial^2\tilde{\sigma}/\partial t^2 \propto -\omega^2\tilde{E}_{inc}$. That is, the incident and scattered fields are π out of phase. At antenna resonance ($L/\lambda_o \sim 1/2$), the incident field is in phase with the current at the center of the antenna (Fig. 6.1a, panel three), that is, $\tilde{I} \propto \tilde{E}_{inc}$, and therefore drives the current most efficiently. As a result,

$\tilde{E}_{scat} \propto \partial^2\tilde{\sigma}/\partial t^2 \propto \partial\tilde{I}/\partial t \propto i\omega\tilde{E}_{inc}$; the phase difference between $\tilde{E}_{scat}$ and $\tilde{E}_{inc}$ is $\pi/2$. For a long antenna with length comparable to the wavelength ($L/\lambda_o \sim 1$), the antenna impedance (defined as the incident field divided by the current at the center of the antenna) is primarily inductive, or $\tilde{I} \propto -i\tilde{E}_{inc}$. Consequently, the scattered and incident light are almost in phase (Fig. 6.1a, panel four), $\tilde{E}_{scat} \propto \partial\tilde{I}/\partial t \propto \tilde{E}_{inc}$. In summary, for a fixed excitation wavelength, the impedance of an antenna changes from capacitive, to resistive, and to inductive across a resonance as the antenna length increases. A single antenna resonance therefore is only able to provide a range of phase change at most π between the scattered and incident light.

6.1.2 Single Oscillator Model for Linear Optical Antennas

In this section, we describe a simple oscillator model for optical antennas and, in general, any nanostructures supporting localized surface plasmon resonances (LSPRs) [29, 30]. The model treats the resonant, collective oscillations of electrons in the nanostructure as a damped, driven harmonic oscillator consisting of a charge on a spring. Unlike previously proposed models in which all damping mechanisms were combined into a single loss term proportional to the charge velocity [31–33], we explicitly account for two decaying channels for LSPR modes: free carrier absorption (internal damping) and emission of light into free space (radiation damping) [16].

We begin by analyzing a system in which a charge q located at $x(t)$ with mass m on a spring with spring constant κ (Fig. 6.2a) is driven by an incident electric field with frequency ω and experiences internal damping with damping coefficient Γ_a:

$$m\frac{d^2x}{dt^2} + \Gamma_a\frac{dx}{dt} + \kappa x = qE_0e^{i\omega t} + \Gamma_s\frac{d^3x}{dt^3}. \tag{6.1}$$

In addition to the internal damping force $F_a(\omega) = \Gamma_a dx/dt$, the charge experiences an additional force $F_s(\omega) = \Gamma_s d^3x/dt^3$ due to radiation reaction, where $\Gamma_s = q^2/6\pi\varepsilon_0 c^3$. This term describes the recoil that the accelerating charge feels when it emits radiation that carries away momentum and is referred to as the Abraham–Lorentz force or the radiation reaction force [34]. The recoil can also be seen as the force that the field produced by the charge exerts on the charge itself [35]. For our charge-on-a-spring model, the radiation reaction term has to be included for physical consistency and cannot be absorbed into the internal damping coefficient Γ_a.

By assuming harmonic motion $x(\omega, t) = x(\omega)e^{i\omega t}$ we can immediately write down the steady-state solution to Equation 6.1 as

$$x(\omega, t) = \frac{(q/m)E_0}{(\omega_0^2 - \omega^2) + i\frac{\omega}{m}(\Gamma_a + \omega^2\Gamma_s)}e^{i\omega t} = x(\omega)e^{i\omega t}, \tag{6.2}$$

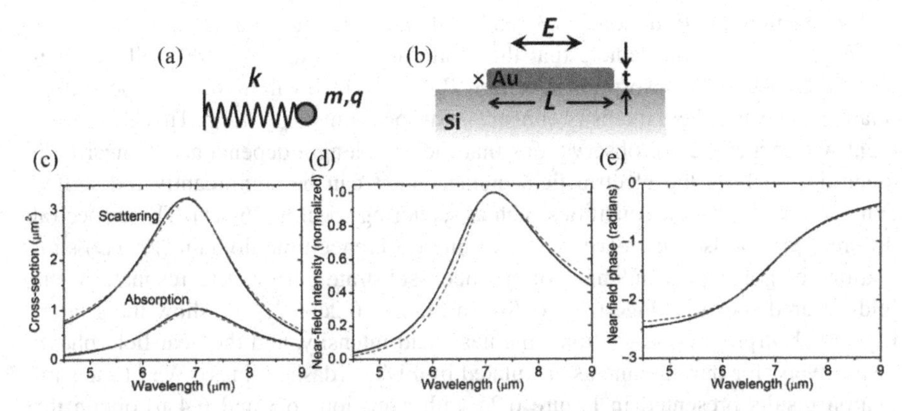

FIGURE 6.2 (a) Representation of an optical antenna in oscillator form, where q is the charge, m is the inertial mass, and $x(t)$ is the displacement from the equilibrium position. (b) Schematics for FDTD simulations. A gold optical antenna (length $L = 1$ μm, thickness $t = 50$ nm, width $w = 130$ nm) sits on a silicon substrate and is illuminated by a normally incident plane wave polarized along the antenna axis. The cross represents the point ~4 nm away from the antenna edge where the near field is calculated. Complex permittivity of gold is taken from Palik [42]. (c) Scattering and absorption cross-sections as calculated via FDTD (dashed curves) and the model (solid curves). (d) Near-field intensity calculated by the oscillator model (solid curve) and via FDTD (dashed curve) at the location identified by the cross, with the incident field subtracted off. (e) Oscillator phase (solid curve) and the phase of the near field calculated via FDTD (dashed curve).

where $x(\omega)$ contains the amplitude and phase response of the oscillator and $\omega_0 = \sqrt{k/m}$. The time-averaged absorbed power by the oscillator can be written as $P_{abs}(\omega) = F_a(\omega)^*(i\omega x(\omega))$, where $F_a(\omega)^*$ is the complex conjugate of the internal damping force. Similarly, the time-averaged scattered power by the oscillator is $P_{scatt}(\omega) = F_s(\omega)^*(i\omega x(\omega))$. Therefore, we have

$$P_{abs}(\omega) = \omega^2 \Gamma_a \left|x(\omega)\right|^2, \tag{6.3}$$

$$P_{scat}(\omega) = \omega^4 \Gamma_s \left|x(\omega)\right|^2, \tag{6.4}$$

where

$$|x(\omega)|^2 = \frac{(q/m)^2 E_0^2}{(\omega_0^2 - \omega^2)^2 + \frac{\omega^2}{m^2}(\Gamma_a + \omega^2 \Gamma_s)^2}. \tag{6.5}$$

Our oscillator model can shed light on the relationship between the near-field, absorption, and scattering spectra in optical antennas. If we interpret the optical antenna as an oscillator that obeys Equations 6.1, 6.2, 6.3, 6.4, and 6.5, we can associate P_{scat} and P_{abs} in Equations 6.3 and 6.4 with the scattering and absorption spectra of the antenna. Furthermore, we can calculate the near-field intensity enhancement at the tip of the antenna as $|E_{near}(\omega)|^2 \propto |x(\omega)|^2$ [30].

By examining Equations 6.3 and 6.4 and noting that $P_{scat} \propto \omega^2 P_{abs} \propto \omega^4 |E_{near}(\omega)|^2$ we can deduce that the scattering spectrum $P_{scat}(\omega)$ will be blue-shifted relative to the absorption spectrum $P_{abs}(\omega)$, which will in turn be blue-shifted relative to the near-field intensity enhancement spectrum $|E_{near}(\omega)|^2$. This is in agreement with experimental observations that the wavelength dependence of near-field quantities such as the electric field enhancement can be significantly red-shifted compared with far-field quantities such as scattering spectra [36–41]. These spectral differences can also be clearly seen in finite difference time domain (FDTD) simulations of gold linear antennas on a silicon substrate designed to resonate in the mid-infrared spectral range (Fig. 6.2b). In Figures 6.2c–6.2e, we show the scattering and absorption cross-sections, the near-field intensity, and the near-field phase, respectively, for our antenna as calculated by FDTD (dashed lines). We fit the simulation results presented in Figure 6.2c with Equations 6.3 and 6.4 to obtain the parameters q, m, ω_0, and Γ_a. The resulting model is able to explain the peak spectral position and general shape of the near-field intensity (Fig. 6.2d), as well as the phase response of the antenna (Fig. 6.2e). Note that no additional fitting was done to obtain the near-field curves in Figures 6.2d and 6.2e. This result suggests that this model can predict the near-field amplitude and phase response from experimental far-field spectra of antennas, which are much easier to obtain than near-field measurements.

The parameters obtained from the fit are consistent with the interpretation that the driving and radiative damping of the antenna mode are due to conduction electrons in the antenna. Assuming a carrier density of 5.9×10^{22} cm^{-3} in gold [43] and volume $V = 6.5 \times 10^{-15}$ cm^3 for the antenna in Figure 6.2b, we expect $\sim 3.8 \times 10^8$ free electrons to be present in the antenna. A Drude model fit to the mid-IR data for gold in Palik [42] yields a plasma frequency of $\omega_p \sim 1.2 \times 10^{16}$/s, which suggests an electron optical effective mass $m_e^* \sim 1.35 m_e$ (as defined in Reference 44) where m_e is the free electron mass. On the other hand, the parameters q and m from our fit correspond to $\sim 2.3 \times 10^8$ electrons of effective mass $m_e^* \sim 1.5 m_e$, close to the previous values. The total charge participating appears to be ~60% of the combined charge of the conduction electrons in the antenna, which is consistent with the fact that not all of the electrons interact equally with the driving and the scattered fields due to the skin depth effect.

Our model shows that in LSPR systems the near-field, absorption, and scattering spectra are all expected to peak at different frequencies and have distinct profiles, which agree very well with electromagnetic simulations of plasmonic antennas in the mid-infrared, and are consistent with experiments.

6.1.3 Two-Oscillator Model for 2D Structures Supporting Two Orthogonal Plasmonic Modes

To gain full control over an optical wavefront, we need a subwavelength optical element able to tailor the phase of the radiated light relative to that of the incident light over a range of 2π. Single oscillators such as the linear optical antennas shown in the last section cannot be engineered in a way to provide an arbitrary phase response over the entire 2π range while maintaining a large scattering cross-section (Figs. 6.2c and 6.2e), so a more elaborate oscillator element is required. In this section, we show that an element consisting of two independent and orthogonally oriented oscillator modes is sufficient to provide arbitrary amplitude and phase response and is therefore suitable for the creation of designer optical interfaces. We derive the phase and amplitude properties of this two-oscillator system and illustrate how the phase coverage is extended due to the dual oscillator modes and a phase contribution from a coordinate transformation [17].

We focus on lithographically defined nanoscale V- and Y-shaped plasmonic antennas as examples of two-oscillator systems. These antennas exhibit two noninteracting plasmonic modes, each of which can be treated as an independent oscillator. We choose our axes such that the two oscillators are oriented along x- and y-axes, respectively, with the incident light propagating along z and its electric field oriented along an axis w, which lies in the x–y plane at an angle θ from the y-axis (Fig. 6.3b). According to Equation 6.4 the fields scattered by the oscillators oriented along the x- and y-axes can be written respectively as

$$E_{s,x}(\omega) = -D_x(\vec{r})\sqrt{\Gamma_{s,x}}\omega^2 x(\omega), \tag{6.6}$$

$$E_{s,y}(\omega) = -D_y(\vec{r})\sqrt{\Gamma_{s,y}}\omega^2 y(\omega), \tag{6.7}$$

where $D_{x(y)}(\vec{r})$ contains the angular and radial dependence of the emitted field. In general, light is scattered by our two-oscillator element into some elliptical polarization state. We focus on light scattered only into the polarization state along the v-axis in Figure 6.3b, which is the cross-polarized direction relative to the incident polarization. The reason for this particular choice is twofold. The first is a matter of experimental convenience as it allows us to fully decouple the scattered light from the incident light by simply filtering out the former with a linear polarizer. The second is more subtle; as we will show, this configuration provides an additional phase shift which extends the potential phase coverage of our elements to the full 2π range.

Given an incident field polarized along the w-axis (Fig. 6.3b), we wish to study the component of the emitted field polarized along the v-direction $E_{s,v}(\omega)$. We can break up this polarization conversion process into two steps: the in-coupling of incident light into the two oscillator modes and the out-coupling of cross-polarized light from the oscillators. The in-coupling process depends on θ because it involves the projection of the incident field along the two oscillator modes, that is, $E_{0,x} = E_0\hat{w}\cdot\hat{x} = E_0\sin(\theta)$ and $E_{0,y} = E_0\hat{w}\cdot\hat{y} = E_0\cos(\theta)$. For the out-coupling process, we project the field scattered by each oscillator onto the v-axis (Fig. 6.3b):

$$E_{s,x}\hat{v}\cdot\hat{x} = -D_x(\vec{r})\sqrt{\Gamma_{s,x}}\omega^2 x(\omega)E_{0,x}\cos(\theta), \tag{6.8}$$

$$E_{s,y}\hat{v}\cdot\hat{y} = D_y(\vec{r})\sqrt{\Gamma_{s,y}}\omega^2 y(\omega)E_{0,y}\sin(\theta). \tag{6.9}$$

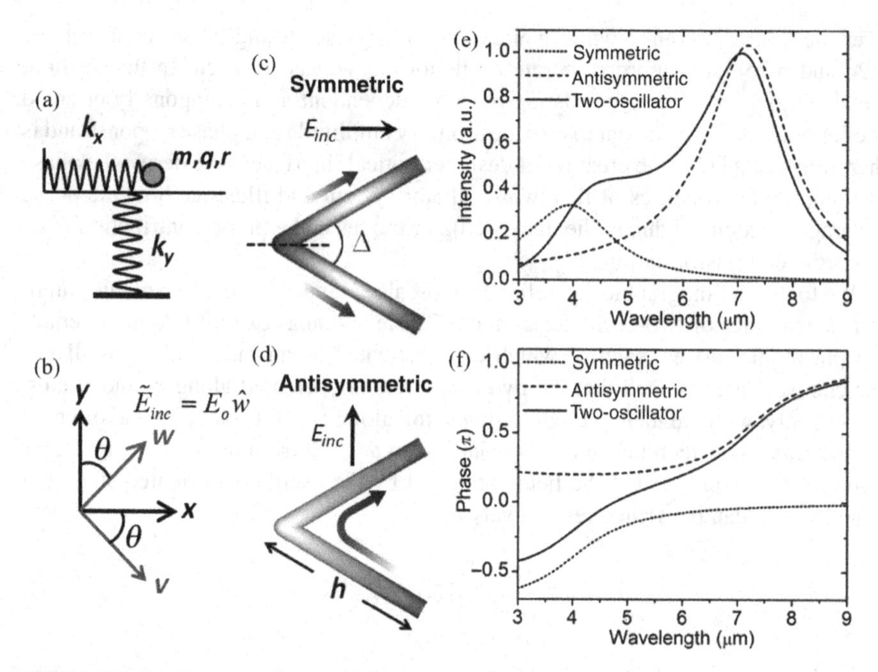

FIGURE 6.3 (a) Charge-oscillator model for a two-oscillator element, where q is the charge and m is the inertial mass. (b) Two coordinate systems related by a rotation by angle θ. The x–y axes are along the two fundamental oscillator modes, the w-axis is along the polarization of the incident field, and the v-axis is along the cross-polarization direction with respect to the incident polarization. A V-shaped optical antenna is a simple example of a plasmonic two-oscillator element. Its two orthogonal modes, that is, symmetric and antisymmetric modes, are shown, respectively, in (c) and (d). The schematic current distribution on the antenna is represented in gray scale with lighter tones indicating larger current density. The instantaneous direction of current flow is indicated by arrows with gradient. (e) Calculated intensity ($|E|^2$) of the field scattered into the cross-polarization by individual oscillators representing the two modes of a V-shaped antenna (dashed and dotted curves) with $\Delta = 90°$ and $h = 650$ nm and by the two-oscillator system representing the V-antenna (solid curve) for $\theta = 45°$. Note that the solid curve is not simply the sum of the dashed and dotted curves because of the coherent addition of fields. (f) Phase of the field scattered into the cross-polarization by the individual oscillators (dashed and dotted curves) and by the two-oscillator system (solid curve).

After summing these projections, the total cross-polarized field emitted by the structure $E_{s,v}$ can be written as

$$E_{s,v}(\omega) = D(\vec{r})\frac{E_0}{2}\sin(2\theta)\omega^2\left[\sqrt{\Gamma_{s,x}}x(\omega)e^{i\pi} + \sqrt{\Gamma_{s,y}}y(\omega)\right], \qquad (6.10)$$

where we assumed that $D_x(\vec{r}) \approx D_y(\vec{r}) = D(\vec{r})$, which is true for light emitted approximately normal to the orientation of the two oscillators. Equation 6.10 provides a complete description of the generation of cross-polarized light by our two-oscillator system.

A large class of plasmonic elements can support two orthogonally orientated modes. We choose V-shaped antennas consisting of two arms of equal length h

connected at one end at an angle Δ. They support "symmetric" and "antisymmetric" modes (Figs. 6.3c and 6.3d), which are excited by electric field components parallel and perpendicular to the antenna symmetry axis, respectively. In the symmetric mode, the current and charge distributions in the two arms are mirror images of each other with respect to the antenna's symmetry plane, and the current vanishes at the joint that connects the two arms (Fig. 6.3c). This means that, in the symmetric mode, each arm behaves similarly to an isolated rod antenna of length h, and therefore the first-order antenna resonance occurs at $h \approx \lambda_{eff}/2$, where λ_{eff} is the effective wavelength [2]. In the antisymmetric mode, antenna current flows across the joint (Fig. 6.3d). The current and charge distributions in the two arms have the same amplitude but opposite sign, and they approximate those in the two halves of a straight rod antenna of length $2h$, and the condition for the first-order resonance of this mode is $2h \approx \lambda_{eff}/2$. The calculations of Figure 6.3e indeed show that the two modes differ by about a factor of 2 in resonance wavelength.

The calculated intensity $\left|E_{s,v}(\omega)\right|^2$ and phase $\phi(\omega)$ of the cross-polarized light $E_{s,v}(\omega) = \left|E_{s,v}(\omega)\right| e^{i\phi(\omega)}$ for a representative V-antenna are plotted in Figures 6.3e and 6.3f. The specific parameters $\Gamma_{a,i}, \Gamma_{s,i}, m_i, \omega_{0,i}$ $(i \in x, y)$ for its two modes correspond to antenna geometries specified in the caption of Figure 6.3. As can be clearly seen from the solid curve in Figure 6.3f, our two-oscillator element is able to span twice the range of phase of either single oscillator (dashed or dotted), even though the two oscillators are uncoupled. This phase extension, which can be seen as the $e^{i\pi}$ term in Equation 6.10, is due to the fact that the projections of the scattered fields from the spatially overlapped x- and y-oriented oscillators onto the v-axis are opposite in phase (Fig. 6.3b).

Equation 6.10 encodes the θ-dependence of the polarization conversion properties with a $\sin(2\theta)$ term. No cross-polarized light is generated for $\theta = 0°$ or $90°$ when the incident field is aligned along one of the two orthogonally oriented modes, and maximum polarization conversion is obtained for $\theta = 45°$. A feature of the two-oscillator element is that the rotation of both oscillators relative to the incident polarization allows for control of the scattering amplitude independent of the phase response or the line width of the resonances. Within the 0°–90° range, the θ-dependence in Equation 6.10 only affects the amplitude of $E_{s,v}$, so θ can be used as a degree of freedom to control the cross-polarized scattering amplitude of the two-oscillator element without altering its phase response. Due to the $\sin(2\theta)$ dependence, a rotation of the structure by 90° maintains the amplitude of cross-polarized scattering while adding an extra phase of π to the scattered light. This feature is used later to generate 8 distinct phase elements from 4 structures and allows us to construct antenna elements that are able to span the full 0-to-2π range in phase, while maintaining relatively large scattering amplitudes.

6.1.4 Analytical Models for V-Shaped Optical Antennas

In this section, we present the results of analytical models for solving the current distribution and scattered fields of V-shaped antennas and in doing so obtain a detailed picture of their near- and far-field properties. In particular, we are able to study the near-field coupling between the two antenna arms, and accurately map the amplitude, phase, and polarization of the antenna radiation in arbitrary directions. The convenient modeling tools enable us to select and assemble various V-shaped antennas into more complex optical systems.

Models describing the response of antennas have been extensively studied [28, 45–47]. One of the main challenges is that the integral equations governing the behavior of antennas have no exact analytical solutions. However, the integral equations can be solved in an approximate way by following an iteration procedure developed in 1950s by King [28]. Furthermore, with the development of numerical methods in the last few decades, we can obtain accurate numerical solutions of the integral equations in an efficient manner following the method of moments (MoM). We present in the Appendix of this chapter the integral equations for V-antennas, as well as the iteration method and the MoM used to calculate the antenna behavior. We study how the methods and approximations used for long wavelengths apply to the mid-infrared spectral range, where plasmonic properties play a significant role, by comparing the results of our analytical models with the results of FDTD simulations. Our main goal here was to propose fast and efficient methods as alternatives to full-wave simulations to probe a large design parameter space, bringing techniques commonly used in the microwave and radio frequency ranges to the attention of the optics community.

Our solutions follow the derivations presented in Reference 28 for solving Hallén's integral equation using the iteration method and in Reference 47 for solving Pocklington's integral equation using the MoM. We first derive the integral equations governing the behavior of V-shaped cylindrical antennas, reduce the 2D problem to one dimension, and implement a numerical solution based on the iteration method or the MoM. We obtain the current distribution driven at the surface of the antenna by an incident excitation field. The far field scattered by the antenna in any direction, with amplitude, phase, and polarization information, is then calculated as the coherent sum of the fields scattered by a series of infinitesimal current elements distributed along the antenna and having their amplitude and phase given by the current distribution, using an analytical expression for the radiation pattern of interfacial dipoles [48].

The main approximation used in our derivations is the thin-wire approximation ($a << \lambda_o$ and $a << h$, with a being the antenna radius and h its arm length) which enables us to consider the current distribution on the antenna to be purely axial and azimuthally invariant [46, 47]. While fully justified at long wavelengths, this approximation may seem crude for mid-infrared antennas for which typically $\lambda_o/a \sim 50$ and $h/a \sim 10$. Our first task is thus to validate our results by comparing them to the results of well-established simulation tools such as FDTD. We assume monochromatic light at $\lambda_o = 7.7$ μm coming at normal incidence with respect to the antenna plane. We calculated the amplitude and phase of the scattered light in the far field, in the direction normal to the plane of the antenna, for different antenna arm lengths h, ranging from 0.3 to 1.6 μm, and different opening angles Δ, ranging from 0° to 180°. The results are summarized in Figure 6.4 for the symmetric and antisymmetric plasmonic modes (see inset schematics). We observe a good agreement

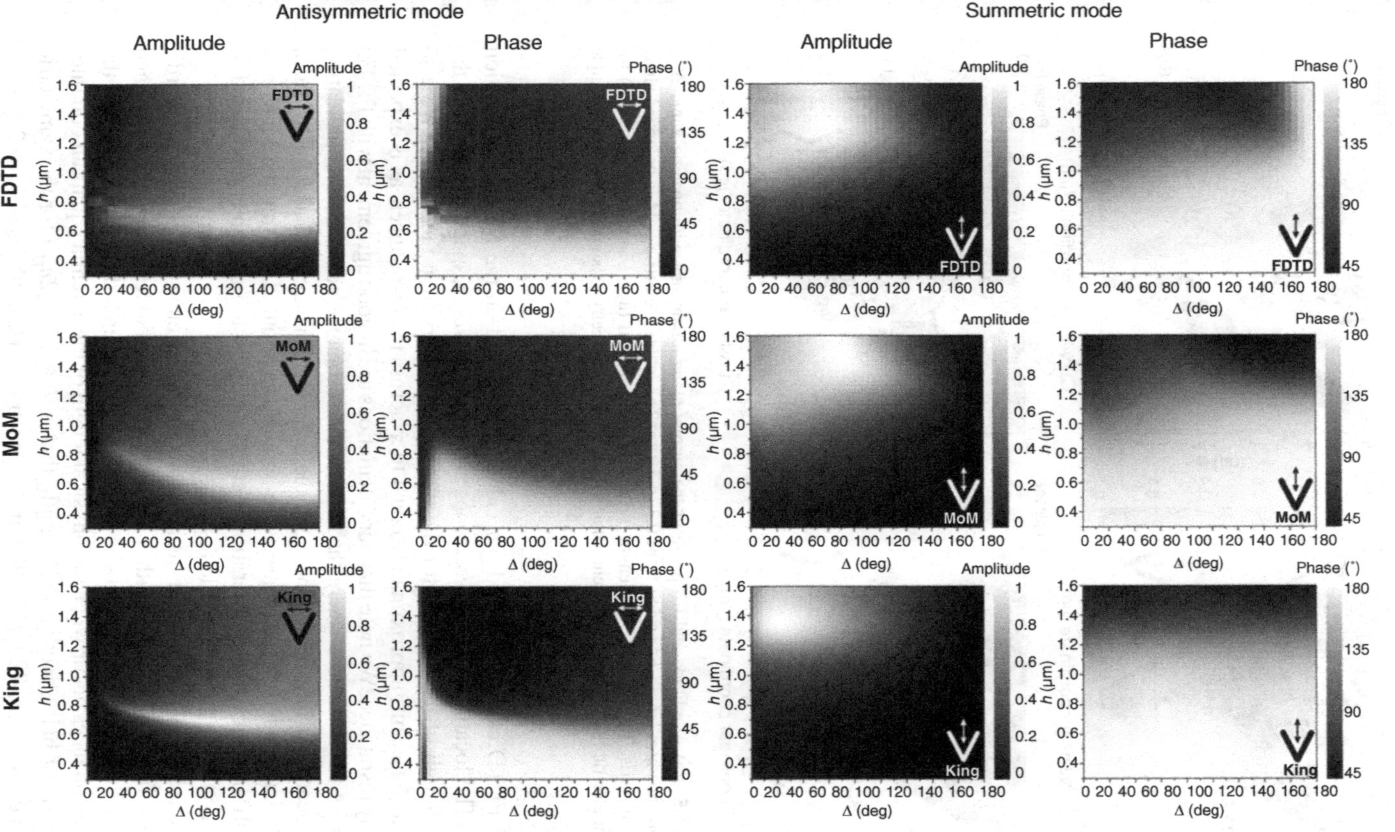

FIGURE 6.4 The first, second, and third rows are, respectively, FDTD simulations, method of moments (MoM) calculations, and calculations following King's iteration method of the amplitude and phase responses of V-antennas. The first and third columns are the amplitudes of the scattered light ($|E|$) as a function of the antenna arm length h and the angle between the antenna arms Δ, for the antisymmetric and the symmetric modes, respectively. The second and fourth columns are the phases of the scattered light for the antisymmetric and the symmetric modes, respectively. The arrows indicate the orientation of the two antenna modes.

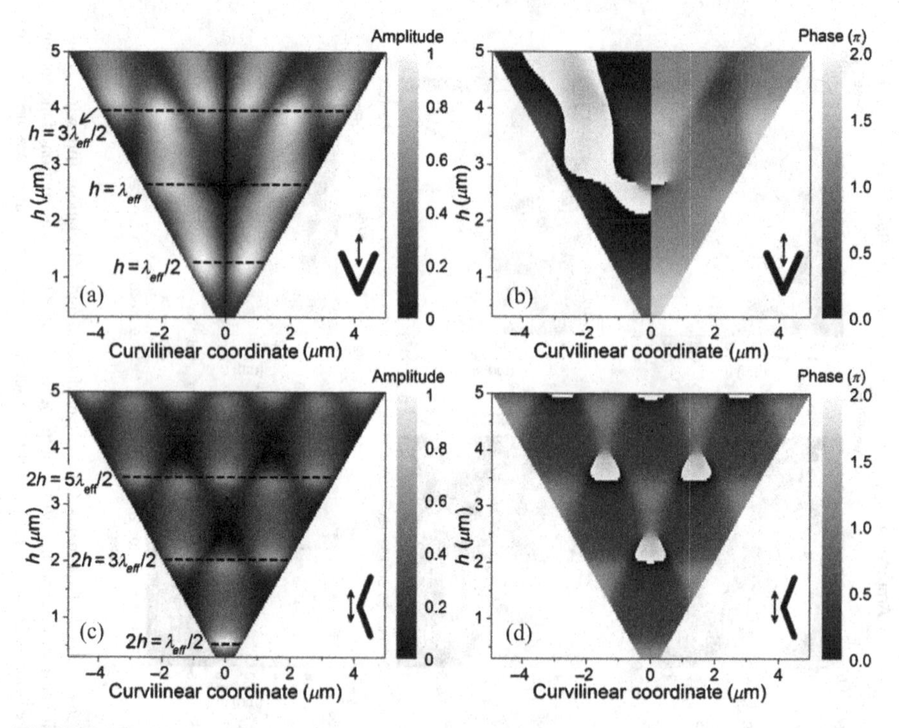

FIGURE 6.5 (a) Amplitude and (b) phase of the current along an antenna with opening angle $\Delta = 45°$ and antenna arm length h varying from 0.3 to 5 μm. The incident electric field is polarized along the symmetry axis. The incident wavelength is $\lambda_o = 7.7$ μm. (c) and (d) are similar to (a) and (b) for an incident electric field polarized perpendicular to the antenna symmetry axis and for an opening angle $\Delta = 135°$.

between FDTD calculations and our calculations based on the MoM and iteration method. The locations of the two modes are different with respect to h because the physical length of the antenna for the two modes differs by a factor of 2. We also observe a phase shift approximately equal to π across the resonances, as is expected across any resonance. We note that our calculations are in good agreement with FDTD simulations for $\Delta < 90°$, where the shift of the resonance peaks occurs, indicating that the near-field interactions between the antenna arms are well accounted for in our analytical models. These maps were obtained in about 20 s using the MoM and about half an hour using the iteration method on a desktop computer, compared to about 1 month for the FDTD calculations.

Figure 6.5 shows the current distributions on the antenna (in curvilinear coordinates) as a function of arm length h for the symmetric modes with $\Delta = 45°$ (a and b) and for the antisymmetric modes with $\Delta = 135°$ (c and d). We observe that the antenna resonances occur when the physical length of the antenna equals an odd integer multiple of half effective wavelength, that is, $h \approx N\lambda_o/(2n_{eff})$ for the symmetric modes and $2h \approx N\lambda_o/(2n_{eff})$ for the antisymmetric modes, where $N = 1, 3, 5, \ldots$

Interestingly, we also observe "forbidden" resonances when the antenna length equals an integer multiple of the effective wavelength, that is, $h \approx N\lambda_o/n_{eff}$, for the symmetric modes when Δ is small (Fig. 6.5a). In straight rod antennas, these resonances feature symmetric charge distributions (e.g., $+ - +$) along the antenna axis and cannot be excited by plane waves [49,50]. In the case of our V-antennas, the coupling between the two arms breaks this symmetry so these "forbidden" resonances can be optically addressed.

6.1.5 Optical Properties of V-Shaped Antennas: Experiments and Simulations

We characterize the spectral response of these antennas by Fourier transform infrared (FTIR) spectroscopy and numerical simulations. In Figure 6.6, we mapped the two oscillator modes of V-shaped antennas as a function of wavelength and opening angle Δ by showing the measured (b–d) and calculated (e–g) transmission spectra. The gold antennas fabricated on silicon wafers have arm length $h = 650$ nm, width $w = 130$ nm, thickness $t = 60$ nm, and opening angle ranging from 45° to 180°. The orientation of the incident polarization is shown in the upper right corner. Figures 6.6b and 6.6e correspond to excitation of only the x-oriented symmetric antenna mode, whereas Figures 6.6c and 6.6f correspond to the y-oriented antisymmetric mode and Figures 6.6d and 6.6g show both excited modes. The spectral position of these resonances are slightly different from the first-order approximation which would yield $\lambda_x \approx 2hn_{eff} \approx 3.4$ μm and $\lambda_y \approx 4hn_{eff} \approx 6.8$ μm, taking n_{eff} as 2.6 [15–17], with the differences attributed to the finite aspect ratio of the antennas and near-field coupling effects, which are especially strong for small Δ when the arms are in closer proximity to each other, leading to a significant resonance shift (Figs. 6.6c and 6.6f). All of the results of the experiment are reproduced very well in simulations, including the feature at 8–9 μm due to a phonon resonance in the 2 nm native silicon oxide layer on the silicon substrate, which is enhanced by the strong near fields formed around the metallic antennas. In Figures 6.6c, 6.6d, 6.6f), and 6.6g, a higher order antenna mode is clearly visible at $\lambda_o \approx 2.7$ μm for large Δ.

We measured the generated cross-polarization using our FTIR setup in transmission mode, inserting a polarizer after the sample at 90° to the incident polarization. The resulting spectra for incident polarization 45° from the x-axis (Fig. 6.6a), normalized to the light directly transmitted through the bare silicon substrate, are shown in Figure 6.7a. As expected, the polarization conversion peaks in the 3–8 μm range, in the vicinity of the two antenna resonances. The corresponding FDTD simulations are shown in Figure 6.7b and retain the same features as the experiment, though the simulated polarization conversion spectra are more clearly broken up into two resonances. The experimental data show less of this separation probably because of inhomogeneous broadening in the experiment due to fabrication imperfections, non-normal incident angle in our FTIR-microscope setup (numerical aperture of the microscope lens is ~0.4), and limited coherence area of our thermal source.

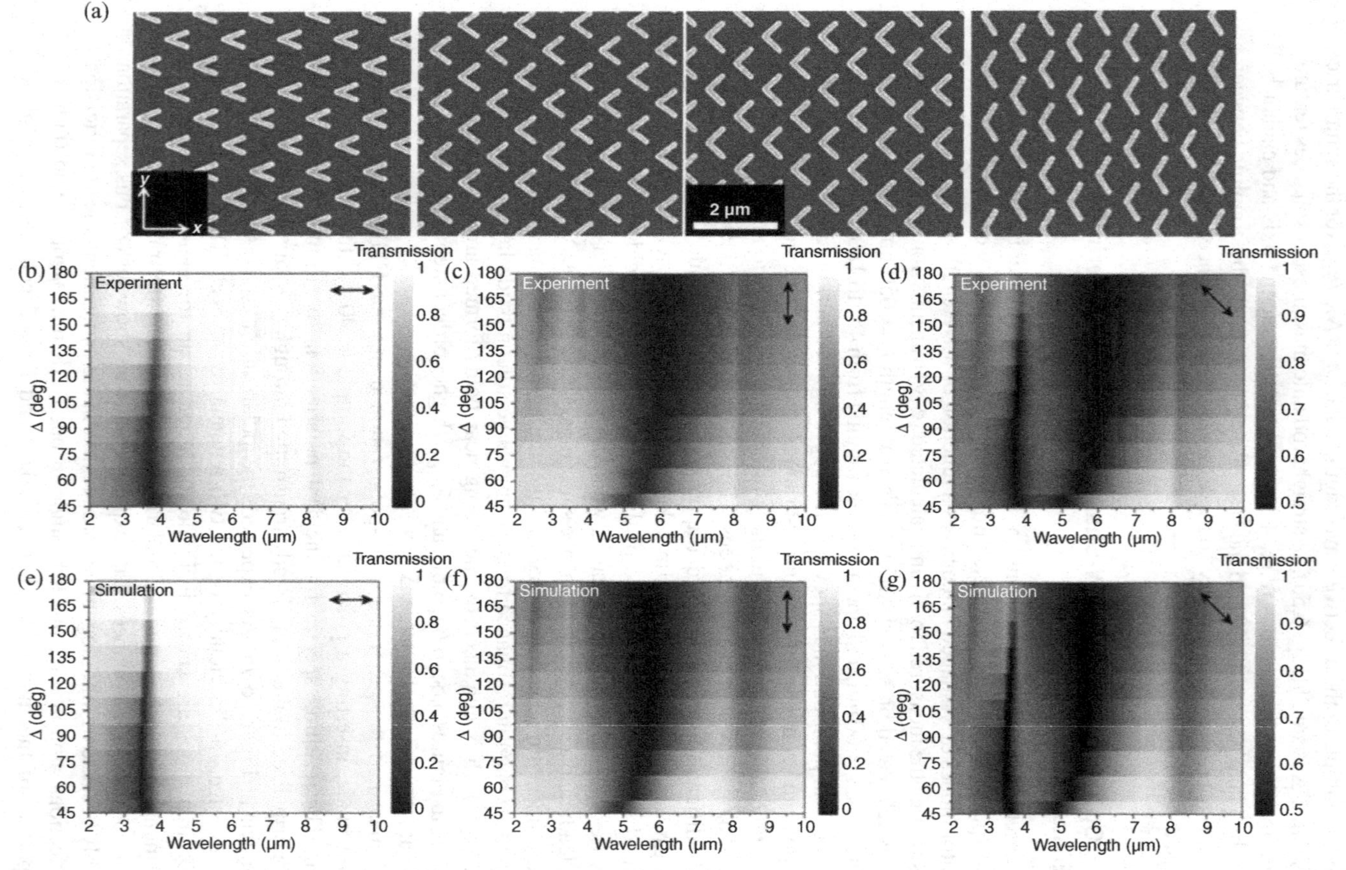

FIGURE 6.6 (a) SEM images of gold V-shaped antennas fabricated on a silicon substrate with opening angles $\Delta = 45°$, $75°$, $90°$, and $120°$ from left to right. (b–d) Measured transmission spectra through the V-antenna arrays at normal incidence as a function of wavelength and angle Δ for fixed arm length $h = 650$ nm. The incident light is polarized (b) along the symmetry axis of the antennas, (c) perpendicular to the symmetry axis, and (d) at a 45° angle. (e–g) FDTD simulations corresponding to the experimental spectra in (b–d), respectively. The feature at $\lambda = 8$–9 μm is due to the phonon resonance in the ~2 nm SiO_2 on the substrate.

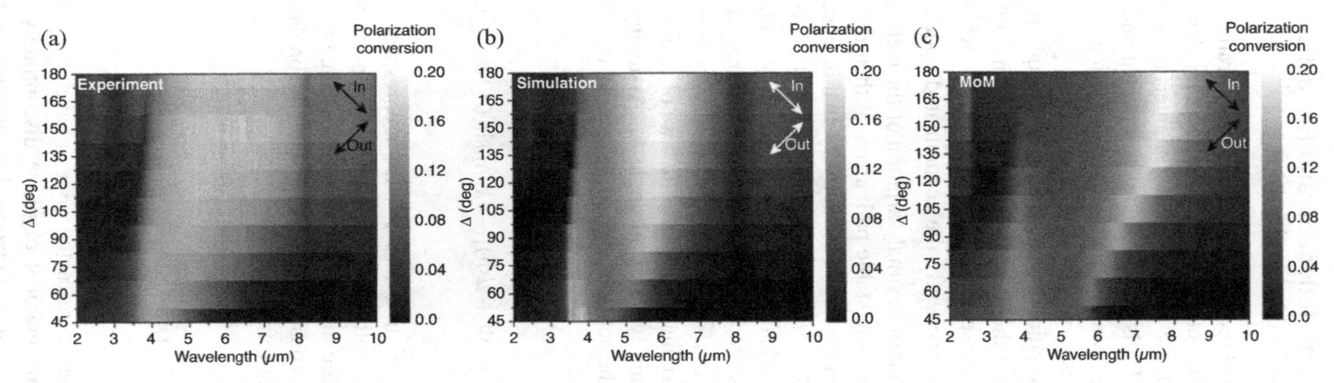

FIGURE 6.7 Experimental measurements (a), FDTD simulations (b), and MoM calculations (without considering the absorption of the SiO_2 layer on the silicon substrate) of the cross-polarized scattering for the V-antenna arrays in Figure 6.6. The arrows indicate the polarization of the incident and output light.

6.2 APPLICATIONS OF PHASED OPTICAL ANTENNA ARRAYS

6.2.1 Generalized Laws of Reflection and Refraction: Meta-Interfaces with Phase Discontinuities

In this section, we show one of the most dramatic demonstrations of controlling light using the tuneable phase shift between the emitted and incident radiation of optical resonators; that is, a linear phase variation along an interface introduced by an array of phased optical antennas leads to anomalously reflected and refracted beams in accordance with generalized laws of reflection and refraction.

The propagation of light is governed by Fermat's principle, which states that the trajectory taken between two points A and B by a ray of light is that of least optical path, $\int_A^B n(\vec{r})dr$, where $n(\vec{r})$ is the local index of refraction. Light chooses this least path from point A to B because it lies at an extremum, where the derivative of the optical path length, or equivalently the accumulated optical phase $\int_A^B d\varphi(\vec{r}) = \int_A^B k_o n(\vec{r})dr$, with respect to infinitesimal variation of the path, is zero. Light waves that stay close to the variationally stable path arrive at their destination with nearly the same phase and therefore interfere constructively. In this sense, Fermat's principle can be stated as the principle of stationary phase [51–53]. Now suppose an abrupt phase shift $\Phi(\vec{r}_s)$ is introduced in the optical path by suitably engineering the interface between two media; $\Phi(\vec{r}_s)$ depends on the coordinate $\vec{r}_s$ along the interface. Then the total phase shift $\Phi(\vec{r}_s) + \int_A^B k_o n(\vec{r})dr$ must be stationary for the actual path that light takes. This allows us to revisit and generalize the laws of reflection and refraction.

Consider an incident plane wave at an angle θ_i. Assuming that the two paths are infinitesimally close to the actual light path (Fig. 6.8), then the phase difference between them is zero:

$$[k_o n_i \sin(\theta_i)\, dx + (\Phi + d\Phi)] - [k_o n_t \sin(\theta_t)\, dx + \Phi] = 0, \qquad (6.11)$$

where θ_t is the angle of refraction, Φ and $\Phi + d\Phi$ are, respectively, the phase shifts at the locations where the two paths cross the interface, *dx* is the distance between the intersections, and n_i and n_t are the refractive indices of the two media. If the phase gradient along the interface is designed to be constant, the previous equation leads to the generalized Snell's law of refraction:

$$\sin(\theta_t)\, n_t - \sin(\theta_i)\, n_i = \frac{\lambda_o}{2\pi}\frac{d\Phi}{dx}. \qquad (6.12)$$

Equation 6.12 implies that the refracted ray can have an arbitrary direction, provided that a suitable constant gradient of phase discontinuity along the interface ($d\Phi/dx$) is introduced. Because of the nonzero phase gradient in this modified Snell's law, the two angles of incidence $\pm\theta_i$ lead to different values for the angle of refraction.

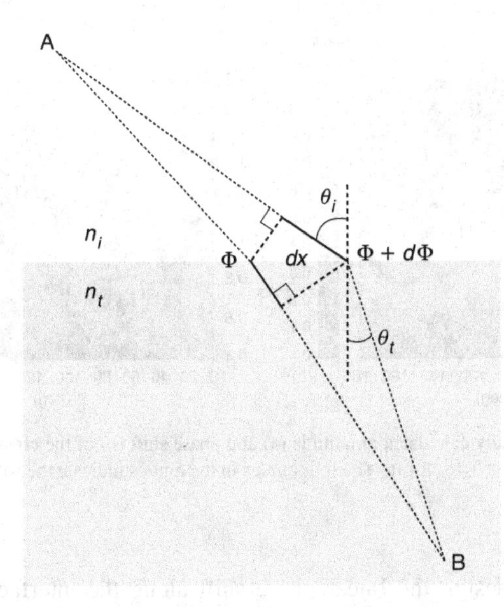

FIGURE 6.8 Schematics used to derive the generalized Snell's law of refraction. The interface between the two media is artificially structured in order to introduce an abrupt phase shift in the light path, which is a function of the position along the interface. Φ and $\Phi + d\Phi$ are the phase shifts where the two paths cross the boundary.

As a consequence, there are two possible critical angles for total internal reflection, provided that $n_t < n_i$:

$$\theta_c = \arcsin\left(\pm\frac{n_t}{n_i} - \frac{\lambda_o}{2\pi n_i}\frac{d\Phi}{dx}\right). \tag{6.13}$$

Similarly, for the reflected light we have

$$\sin(\theta_r) - \sin(\theta_i) = \frac{\lambda_o}{2\pi n_i}\frac{d\Phi}{dx}, \tag{6.14}$$

where θ_r is the angle of reflection. There is a nonlinear relation between θ_r and θ_i, which is dramatically different from conventional specular reflection. Equation 6.14 predicts that there is always a critical incidence angle

$$\theta_c' = \arcsin\left(1 - \frac{\lambda_o}{2\pi n_i}\left|\frac{d\Phi}{dx}\right|\right) \tag{6.15}$$

above which the reflected beam becomes evanescent.

As shown in previous sections, the phase shift between the scattered and the incident radiation of an optical antenna changes appreciably across a resonance. By spatially tailoring the geometry of the antennas in an array and hence their phase

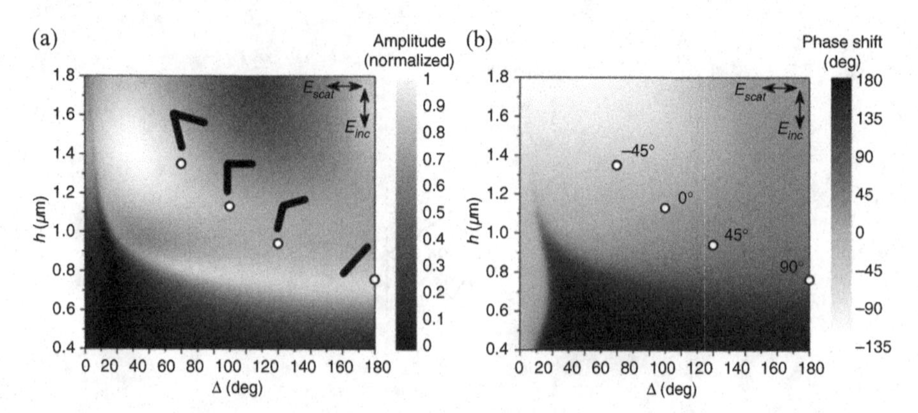

FIGURE 6.9 Analytically calculated amplitude (a) and phase shift (b) of the cross-polarized scattered light for gold V-antennas at $\lambda_o = 8$ μm. The four circles in the figures indicate the values of h and Δ used in experiments.

response, one can design the linear phase shift along the interface and mold the wavefront of the reflected and refracted beams in nearly arbitrary ways. The spacing between the antennas in the array should be subwavelength to provide efficient scattering and to prevent the occurrence of grating diffraction. However, it should not be too small to introduce strong near-field coupling between neighboring antennas that will perturb their phase responses. Note that due to the discreteness in our approach to approximate the linear phase distribution, in general there will always be regularly reflected and refracted beams, which follow conventional laws of reflection and refraction (i.e., $d\Phi/dx = 0$ in Eqs. 6.12 and 6.14). The antenna packing density controls the relative amount of energy in the anomalously reflected and refracted beams.

Figure 6.9 shows the amplitude and phase responses of gold V-antennas calculated using the iteration method previously discussed. The antenna symmetry axis is along 45° direction so that both the symmetric and antisymmetric plasmonic modes are excited, given a vertical incident polarization (Fig. 6.9a inset). As a result of the modal properties of the V-antennas and the degrees of freedom in choosing antenna geometry (h and Δ), the cross-polarized scattered light can have a large range of amplitudes and phases for a given wavelength $\lambda_o = 8$ μm. We chose four antennas indicated by circles in Figure 6.9, which provide an incremental phase of $\pi/4$ from left to right and almost equal scattering amplitudes for the cross-polarized scattered light.

Phase shifts covering the entire 0-to-2π range are needed to provide full control of the wavefront. We take the mirror structure (Fig. 6.10a, lower panel) of an existing V-antenna (Fig. 6.10a, upper panel) (or rotate the original antennas clockwise by 90°) so that the cross-polarized emission has an additional π phase shift. This is evident by observing that the currents leading to cross-polarized radiation are π out of phase in the two panels of Figure 6.10a. A set of eight antennas are thus created from the initial four antennas shown in Figure 6.9, and by periodically arranging these

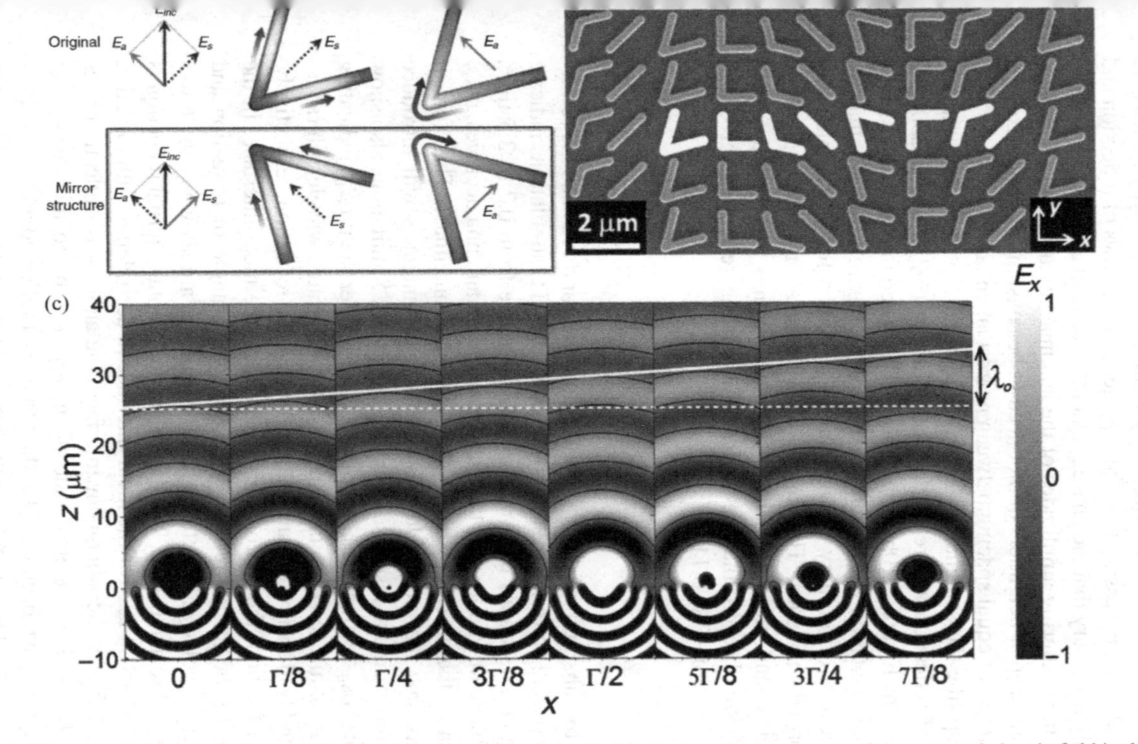

FIGURE 6.10 (a) V-antennas (upper panel) and their mirror structures (lower panel). The horizontal components of the scattered electric field in the two cases have a π phase difference. (b) SEM image of a meta-interface consisting of a phased optical antenna array fabricated on a silicon wafer. The meta-interface can introduce a linear phase distribution along the interface and is used to demonstrate the generalized laws of reflection and refraction. The unit cell of the structure (highlighted) comprises eight gold V-antennas of width ~220 nm and thickness ~50 nm and it repeats with a periodicity of $\Gamma = 11$ μm in the x-direction and 1.5 μm in the y-direction. The antennas are designed to have equal scattering amplitudes and constant phase difference $\Delta\Phi = \pi/4$ between neighbors. (c) FDTD simulations of the scattered electric field for the individual antennas composing the unit cell. Plots show the scattered electric field polarized in the x-direction for y-polarized plane wave excitation at normal incidence from the silicon substrate. The silicon substrate is located at $z \leq 0$. The antennas are equally spaced at a subwavelength separation $\Gamma/8$, where Γ is the unit cell length. The tilted white solid line is the envelope of the projection on the x–z plane of the spherical waves scattered by the antennas. On account of Huygens' principle, the anomalously refracted beam resulting from the superposition of these spherical waves is then a plane wave that satisfies the generalized Snell's law (Eq. 6.12) with a phase gradient $|d\Phi/dx| = 2\pi/\Gamma$ along the interface.

eight antennas we created meta-interfaces that can imprint a linear phase shift to the optical wavefronts. A representative fabricated sample with the densest packing of antennas is shown in Figure 6.10b. FDTD simulations confirm that the amplitudes of the cross-polarized radiation scattered by the eight antennas are nearly equal with phases in $\pi/4$ increments (Fig. 6.10c). The periodic antenna arrangement is used here for convenience, but is not necessary to satisfy the generalized laws of reflection and refraction. It is only necessary that the phase gradient is constant along the plasmonic interface and that the scattering amplitudes of the antennas are all equal. The phase increments between nearest neighbors do not need to be constant, if one relaxes the unnecessary constraint of equal spacing between nearest antennas.

We used a setup like the schematics shown in Figure 6.11a to demonstrate the generalized laws of reflection and refraction. Large arrays (~230 μm × 230 μm) like the one shown in Figure 6.10b were fabricated to accommodate the size of the plane-wave-like excitation (beam radius ~100 μm). Figure 6.11b summarizes the experimental results of ordinary and anomalous refraction for six samples with different Γ at normal incidence. The sample with the smaller Γ corresponds to larger phase gradient and more efficient light scattering into the cross-polarized anomalous beams. We observed that the angles of anomalous refraction agree well with theoretical prediction (Fig. 6.11b):

$$\theta_{t,\perp} = \arcsin\left[n_{Si}\sin(\theta_i) - \lambda_o/\Gamma\right], \tag{6.16}$$

which is obtained by substituting $-2\pi/\Gamma$ into Equation 6.12 for $d\Phi/dx$ and the refractive indices of silicon and air (n_{Si} and 1) for n_i and n_t. Our FDTD simulations indicate that the scattering cross-sections σ_{scat} of the antennas range from 0.7 to 2.5 μm^2, which is comparable to or smaller than the average area each antenna occupies, σ_{aver} (i.e., the total area of the array divided by the number of antennas). Therefore, it is reasonable to assume that near-field coupling between antennas will introduce only small deviations from the response of isolated antennas. Simulations also show that the absorption cross-sections σ_{abs} are ~5–7 times smaller than σ_{scat}, indicating comparatively small ohmic losses in the antennas at mid-infrared wavelength range.

Figures 6.11c and 6.11d show the angles of refraction and reflection, respectively, as a function of θ_i for both the silicon–air interface and the meta-interface. In the range of $\theta_i = 0°$–$9°$, the meta-interface exhibits negative angle of refraction and reflection for the cross-polarized scattered light (schematics are shown in the lower right insets of Figs. 6.11c and 6.11d). The critical angle for total internal reflection is modified to about –8° and +27° for the meta-interface in accordance with Equation 6.13 compared to ±17° for the silicon–air interface; the anomalous reflection does not exist beyond $\theta_i = -57°$. We note that antenna arrays in the microwave and millimeter-wave regime have been used for the shaping of reflected and transmitted beams in the so-called reflectarrays and transmitarrays. There is a connection between that body of work and our results in that both use abrupt phase changes associated with antenna resonances. However, the generalization of the laws of reflection and refraction we present is made possible by the deep-subwavelength thickness of our optical antennas and their subwavelength spacing. It is this meta-surface nature of the

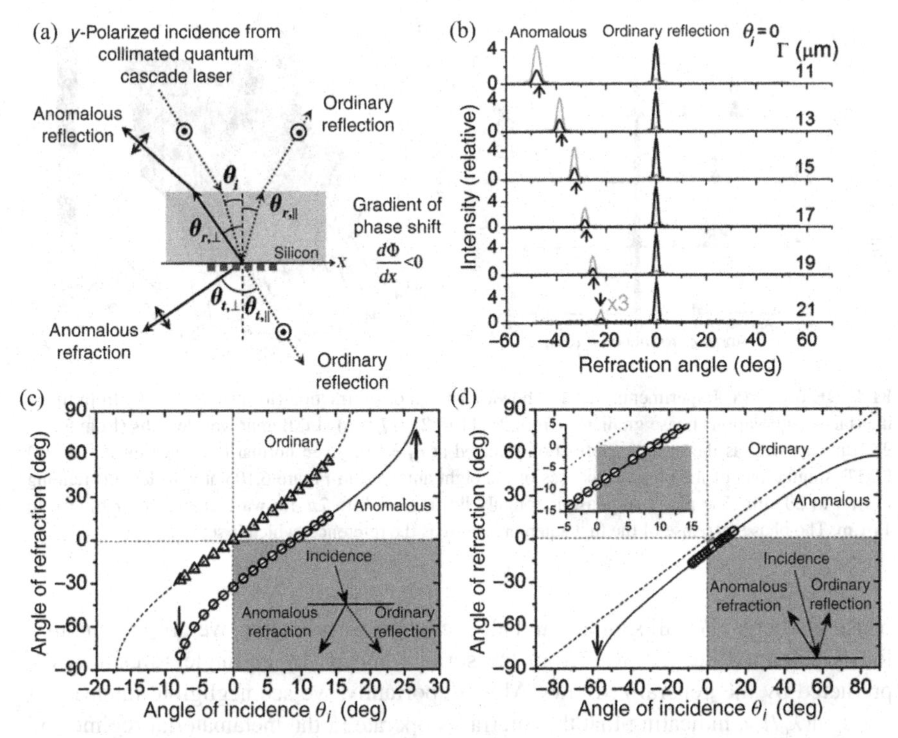

FIGURE 6.11 (a) Schematic experimental setup for y-polarized excitation (electric field normal to the plane of incidence). Angle of refraction and angle of reflection for the ordinary and anomalous beams, as well as angle of incidence, are labeled. (b) Measured far-field intensity profiles of the refracted beams for incidence normal to the interface. The gray and black curves are measured with and without a polarizer, respectively, for six samples with different Γ. The amplitude of the gray curves is magnified by a factor of three for clarity. The polarizer is used to filter out the anomalous refraction. The arrows indicate the calculated angles of anomalous refraction according to Equation 6.16. (c) Angle of refraction versus angle of incidence for the sample with $\Gamma = 15$ μm. The curves are theoretical calculations using the generalized Snell's law (Eq. 6.12) and the symbols are experimental data. The two arrows indicate the modified critical angles for total internal reflection. The shaded region represents "negative" refraction for the cross-polarized light as illustrated in the inset. (d) Angle of reflection versus angle of incidence for the sample with $\Gamma = 15$ μm. The upper left inset is the zoom-in view. The curves are theoretical calculations using Equation 6.14 and the symbols are experimental data. The arrow indicates the critical incidence angle above which the anomalously reflected beam becomes evanescent. The shaded region represents "negative" reflection for the cross-polarized light as illustrated in the lower right inset.

plasmonic interface that distinguishes it from reflectarrays and transmitarrays, which typically consist of a double-layer structure separated by a dielectric spacer of finite thickness, and the spacing between the array elements is not subwavelength.

The antenna designs in Figure 6.10b are broadband; that is, they provide approximately 0-to-2π phase coverage with approximately $\pi/4$ intervals over a wide range of wavelengths. Figure 6.12a shows experimental results of ordinary and anomalous refraction for normally incident light at five different wavelengths. There are four samples with unit cell length Γ ranging from 11 to 17 μm, corresponding to different

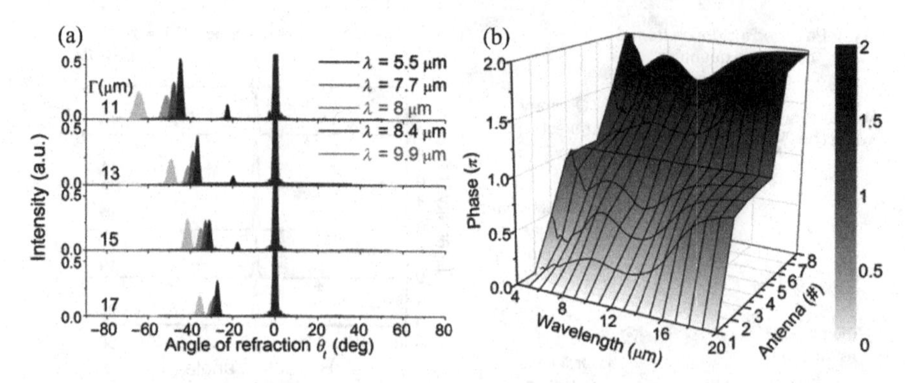

FIGURE 6.12 (a) Experimental results showing anomalous refraction (located at $\theta_t < 0$) from meta-interfaces with various phase gradients (from $2\pi/11$ to $2\pi/17$ μm) at different wavelengths (from 5.5 to 9.9 μm), as well as the ordinary refraction (located at $\theta_t = 0$), given normal incidence excitation. (b) FDTD simulations of the phase responses of the eight antennas in Figure 6.10b at wavelengths ranging from 4 to 20 μm. The phase response is roughly linear from 0 to 2π for wavelengths from $\lambda_o = 6$ to 14 μm. The phase response of the first antenna is used as the reference, which is set to be zero.

phase gradients. For all samples and all excitation wavelengths, we observe anomalously refracted beams away from the surface normal by an angle $-\arcsin(\lambda_o/\Gamma)$, predicted by the generalized laws. Most importantly, we see negligible intensity at $+\arcsin(\lambda_o/\Gamma)$, indicating that the interfaces operate in the metamaterial regime and do not function like a grating with periodicity Γ. The reason for this broadband behavior can be understood by looking at the phase responses of the eight constituent antennas (Fig. 6.12b). Although the antennas were designed to generate anomalous beams at $\lambda_o = 8$ μm, their phase responses, in terms of both the total phase coverage and the incremental phase between neighbors, do not vary significantly over a large range of wavelengths from $\lambda_o = 6$ to 14 μm.

6.2.2 Out-of-Plane Reflection and Refraction of Light by Meta-Interfaces

A key feature of the conventional laws of reflection and refraction is that the incident, reflected, and transmitted beams lie in the same plane, that is, the plane of incidence. Recent research on metamaterials and in particular on left-handed optical materials has shown that even if light can be refracted in unusual ways, the refraction angle is still described by the traditional form of Snell's law, albeit with a negative index of refraction [54–58]. In this section, we derive laws of reflection and refraction in three dimensions, applicable to the case in which the interfacial phase gradient does not lie in the plane of incidence, leading to the remarkable features of out-of-plane reflection and refraction [19].

We use the principle of stationary phase to derive the 3D reflection and refraction laws. Suppose the two paths in Figure 6.13a are infinitesimally close to the actual optical path. We shall have $\int_A \varphi(\vec{r})d\vec{r} = \int_B \varphi(\vec{r})d\vec{r}$, where the integrals are along

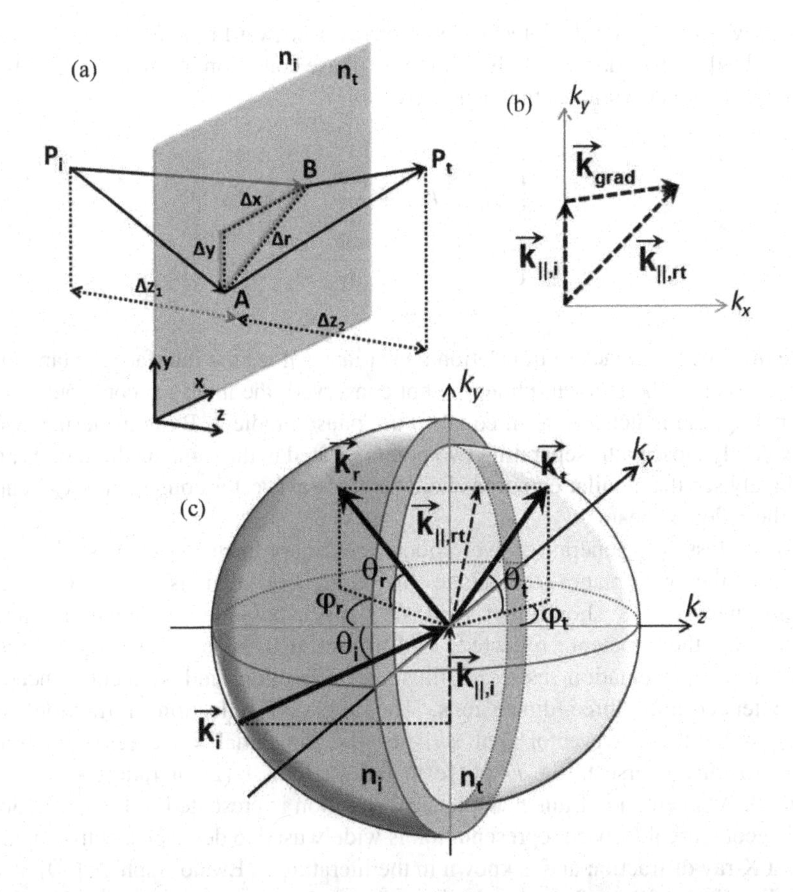

FIGURE 6.13 Schematics used to describe the generalized refraction at an interface with arbitrary orientation of the phase gradient. (a) The interface between two media of refractive index n_i and n_t is structured to introduce a constant interfacial phase gradient $d\vec{\Phi}/dr \equiv \vec{k}_{grad}$. Two points, P_i and P_t, are located, respectively, in mediums 1 and 2. The difference of the phase accumulated along two paths from P_i to P_t that are infinitesimally close to the actual optical path is zero according to Fermat's principle of stationary phase. (b and c) The projection of the incident wavevector on the interface $\vec{k}_{||,i}$ forms a nonzero angle with the direction of the interfacial phase gradient. As a result during the interaction with the meta-interface, the light beams acquire a k-vector component equal to the phase gradient ($\vec{k}_{||,rt} = \vec{k}_{||,i} + \vec{k}_{grad}$).

the paths through point A and B on the interface, respectively. The equation can be rewritten as

$$\left(\int_{P_i}^{A} \vec{k}_i \cdot d\vec{r} - \int_{P_i}^{B} \vec{k}_i \cdot d\vec{r}\right) + \left(\int_{A}^{P_t} \vec{k}_t \cdot d\vec{r} - \int_{B}^{P_t} \vec{k}_t \cdot d\vec{r}\right) + \frac{d\vec{\Phi}}{dr} \cdot (\vec{r}_A - \vec{r}_B) = 0, \tag{6.17}$$

where $\vec{k}_i$ ($\vec{k}_t$) is the wavevector of light in the medium of index n_i (n_t) and $\vec{r}_A$ ($\vec{r}_B$) is the position of A (B) on the x–y plane. For constant phase gradients, the accumulated phase of rays intersecting the interface is a convex downward function for all values of x and y [59], so we can rewrite the stationary phase condition in Equation 6.17 for the x and y spatial coordinates independently:

$$\begin{cases} k_{x,t} = k_{x,i} + \dfrac{d\Phi}{dx} \\ k_{y,t} = k_{y,i} + \dfrac{d\Phi}{dy} \end{cases} . \tag{6.18}$$

Note that due to the lack of translational invariance along the interface the tangential wavevector of the incident photon is not conserved; the interface contributes an additional "phase matching" term equal to the phase gradient. By considering two infinitesimally close paths separating two points located in the same medium one can immediately see that similar equations hold for the wavevector components $k_{x,r}$ and $k_{y,r}$ of the reflected beam.

Without loss of generality, we choose the coordinate system such that $\vec{k}_i$ lies in the y–z plane (the plane of incidence), that is, $k_{x,i} = 0$ (Fig. 6.13b). Equation 6.18 shows that it is no longer possible to define a plane that contains the incident, reflected, and transmitted beams. The usual planar k-space representation used to illustrate refraction and reflection needs to be extended into three dimensions. The angle of reflection (refraction) is now given by the wavevector that satisfies the tangential wavevector relation (Fig. 6.13b) and intersects the k-sphere in the medium 1 (2) of radius $k = n_i k_o$ ($k = n_t k_o$). A schematic of this new physical situation is presented in Figure 6.13c. This 3D geometrical k-space representation is widely used to describe electron, neutron, and X-ray diffraction and is known in the literature as Ewald's sphere [60].

The directions of the reflected and refracted wavevectors are characterized by the angles $\theta_{r(t)}$ (the angle between $\vec{k}_{r(t)}$ and its projection on the x–z plane) and $\varphi_{r(t)}$ (the angle formed by the projection of $\vec{k}_{r(t)}$ on the x–z plane and the z-axis), as defined in Figure 6.13c. With this choice of notation, we obtain the generalized law of reflection in 3D,

$$\begin{cases} \cos\theta_r \sin\varphi_r = \dfrac{1}{n_i k_o}\dfrac{d\Phi}{dx} \\ \sin\theta_r - \sin\theta_i = \dfrac{1}{n_i k_o}\dfrac{d\Phi}{dy} \end{cases} , \tag{6.19}$$

and the generalized law of refraction in 3D,

$$\begin{cases} \cos\theta_t \sin\varphi_t = \dfrac{1}{n_t k_o}\dfrac{d\Phi}{dx} \\ n_t \sin\theta_t - n_i \sin\theta_i = \dfrac{1}{k_o}\dfrac{d\Phi}{dy} \end{cases} . \tag{6.20}$$

Notice that when the phase gradient is oriented along the plane of incidence ($d\Phi/dx = 0$) the anomalous reflection and refraction are in plane and one recovers Equations 6.12 and 6.14. The nonlinear nature of these equations is such that two different critical angles now exist for both reflection and refraction. When a ray of light traverses an interface, it will propagate in the new medium as long as its longitudinal wavevector k_z remains real. This implies that the tangential components of the k-vector have to be smaller than the modulus of the k-vector in the medium. When a phase gradient along the interface provides an additional tangential component of the k-vector, the condition for the existence of a transmitted or reflected propagating beam is changed. From Equation 6.20, we can find the condition for $k_{z,t}$ to be zero ($\varphi_t = 90°$), leading to the expression for two critical angles for refraction:

$$\theta_i^{c,t} = \sin^{-1}\left[\pm\frac{1}{n_i}\sqrt{n_t^2 - \left(\frac{1}{k_o}\frac{d\Phi}{dx}\right)^2} - \frac{1}{n_i k_o}\frac{d\Phi}{dy}\right]. \tag{6.21}$$

Note that a critical angle for refraction may exist even when $n_i < n_t$ for some interfacial phase gradients. For reflection, the nonlinear relation between θ_r and θ_i yields two critical angles for reflection:

$$\theta_i^{c,r} = \sin^{-1}\left[\pm\sqrt{1 - \left(\frac{1}{n_i k_o}\frac{d\Phi}{dx}\right)^2} - \frac{1}{n_i k_o}\frac{d\Phi}{dy}\right]. \tag{6.22}$$

We experimentally observed out-of-plane refraction in accordance with the new 3D law (Eq. 6.20) using an interface patterned with phased optical antenna arrays oriented in such a way that the interfacial phase gradient forms an angle α with respect to the plane of incidence (Fig. 6.14).

We studied both the ordinary and anomalous refraction for various incidence angles and various phase gradient orientations. The magnitude of the phase gradient is fixed to $d\Phi/dr = 2\pi/15$ (radian/μm) for all the experiments. The experimental results are summarized in Figure 6.15b and unambiguously show out-of-plane refraction that agrees well with the prediction of Equation 6.20. Figure 6.15a shows measured far-field intensity distribution for a phase gradient perpendicular to the plane of incidence.

In summary, 3D laws of reflection and refraction are derived for optically thin meta-interfaces that impart to the incident wavefront a phase gradient arbitrarily oriented with respect to the plane of incidence. Due to the tangential wavevector provided by the anisotropic interface, the incident beam and the anomalously reflected and refracted beams are in general non-coplanar and two different critical angles exist for both reflection and refraction. The beams' direction can be controlled over a wide range by varying the angle between the plane of incidence and the phase gradient. Experiments on arrays of subwavelength optical antennas demonstrate out-of-plane refraction in excellent agreement with the 3D Snell's law, illustrating the unique beaming capabilities of meta-interfaces at optical frequencies.

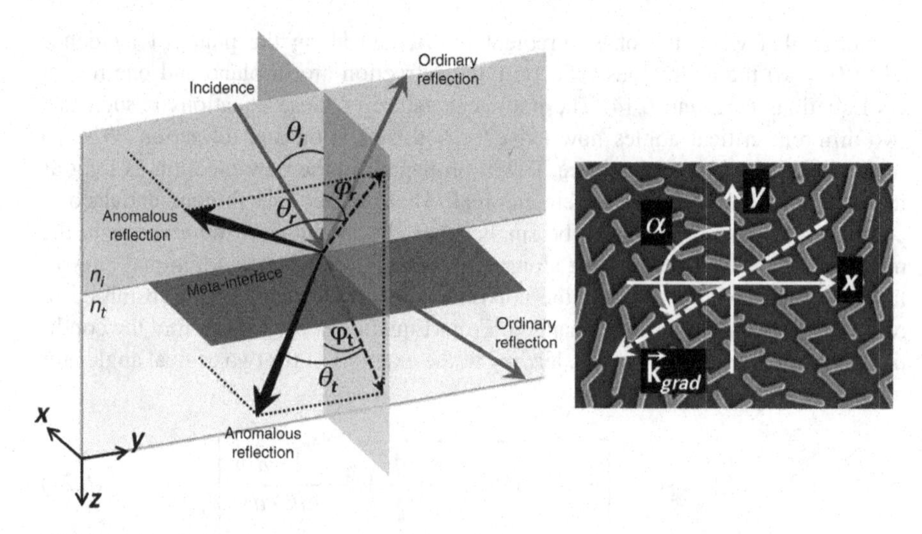

FIGURE 6.14 Schematic representation of the reflection and refraction of light in three dimension. The incident light from a collimated quantum cascade laser emitting at $\lambda_o = 8$ μm impinges at an angle θ_i with respect to the z-axis on an interface between silicon and air. The incidence is from silicon and the polarization of the incident light is maintained so that it forms an angle of 45° with respect to the two plasmonic modes of the antennas. The sample is the same as that used in Figure 6.10b. When the interfacial phase gradient $\vec{k}_{grad}$ imposed by the V-antenna arrays is not parallel to the y-axis, a component of the phase gradient out of the plane of incidence is created, resulting in out-of-plane anomalous beams satisfying the 3D laws of reflection and refraction (Eqs. 6.19 and 6.20).

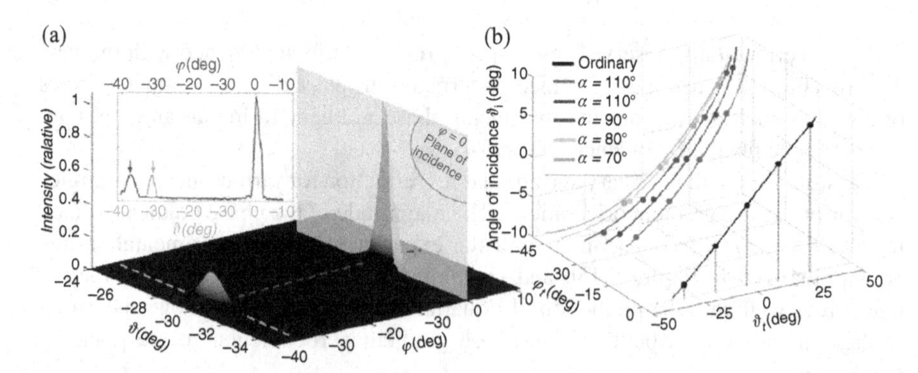

FIGURE 6.15 Experimental observation of out-of-plane refraction. (a) Measured far-field intensity as a function of the angular position of the detector θ and φ (defined in a similar way as θ_t and φ_t in Fig. 6.14) for a laser beam incident on the interface from the silicon side at an angle $\theta_i = -8.45°$ and for a phase gradient perpendicular to the plane of incidence. As expected, the ordinary refracted beam is in plane ($\varphi = 0°$) at an angle following the conventional Snell's law, that is, $\theta = -30°$. The anomalous beam is refracted out of plane at angles $\varphi = -38°$ and $\theta = -30°$. The inset shows the angular distribution of the intensity at a fixed angle $\varphi = -38°$ (light curve) and at a fixed angle $\theta = -30°$ (dark curve). (b) Angles of refraction versus angles of incidence and orientations of phase gradient α. Curves are theoretical calculations based on the 3D Snell's law (Eq. 6.20). Circles are experimental data.

6.2.3 Giant and Tuneable Optical Birefringence

We have shown that the optical properties of V-antennas, and in general any 2D plasmonic structures that support two charge-oscillation eigenmodes, can be captured by a simple model involving two independent, orthogonally oriented harmonic oscillators. We show in this section that meta-interfaces consisting of V- and Y-shaped plasmonic antennas exhibit widely tailorable birefringence, where the optical anisotropy can be controlled by interference between the light scattered by the two plasmonic eigenmodes of the antennas.

We studied the birefringence properties of the meta-interfaces shown in Figure 6.10b by changing the incident polarization at normal incidence so that the two orthogonal plasmonic modes are excited with different amplitudes, leading to a rotation of the polarization of the scattered light. Assume that the angle between the incident polarization and the x-axis is α and the symmetry axes of the eight antenna elements are 45° away from the vertical direction (Fig. 6.16a). We decompose the

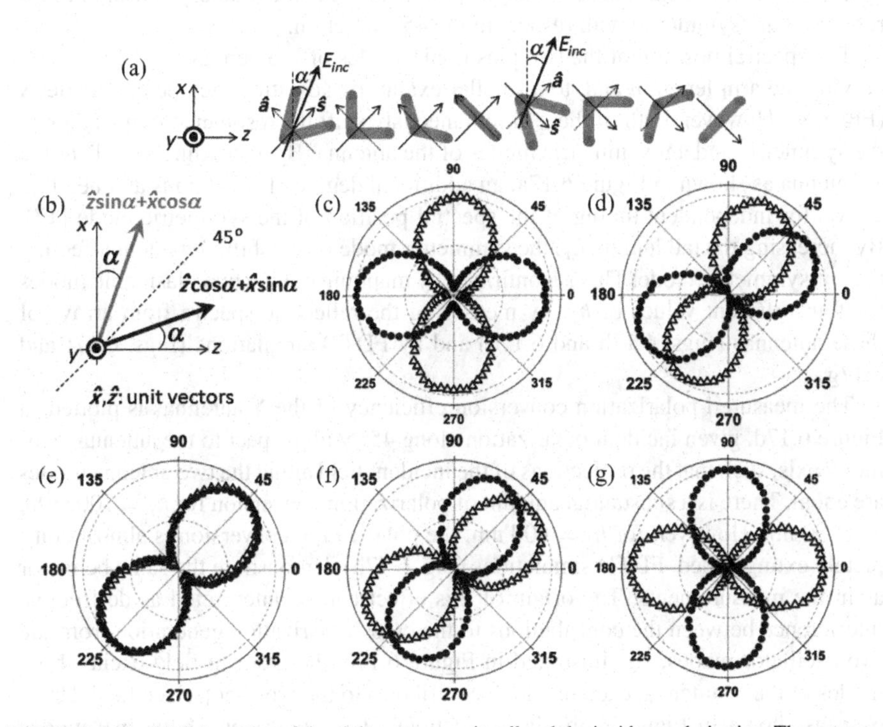

FIGURE 6.16 (a) Schematics of the eight-antenna unit cell and the incident polarization. The antenna symmetry axes, $\hat{s}$, are 45° from the vertical direction. (b) For incident polarization along the α-direction, the scattered fields from the antennas contain two components, directed along the gray and black arrows, respectively. Only the latter has the properly engineered phase response, that is, $\pi/4$ phase difference between neighbors, which leads to anomalously refracted beams polarized along the (90° – α)-direction. (c–g) Measured intensity of the ordinarily and anomalously refracted beams (triangles and dots, respectively) as a function of the rotation angle of a linear polarizer located in front of a detector (α = 0°, 30°, 45°, 60°, and 90° from (c) to (g)). The free-space wavelength is 8 μm.

incident electric field into components that drive the symmetric and antisymmetric modes, respectively (i.e., parallel and perpendicular to the antenna symmetry axis), and calculate the scattered light based on the amplitude and phase responses of the two modes. Theoretical analysis shows that the scattered light contains two contributions that are polarized along the α-direction and ($90° - \alpha$)-direction, respectively (Fig. 6.16b). The ($90° - \alpha$)-polarized components of all eight antennas have the same amplitude and incremental phase of $\pi/4$, which give rise to an anomalously refracted beam. The α-polarized components, however, do not have the same amplitude and the proper phase relation between neighboring antennas, which lead to a beam that propagates in the same direction as the ordinary refraction, as well as a small optical background over a large angular range. These birefringence properties of the meta-interface are experimentally confirmed at $\lambda_o = 8$ μm, and Figures 6.16c–6.16g show measured intensity of the ordinarily and the anomalously refracted beams as a function of the rotation angle of a linear polarizer located in front of the detector for different incident polarizations. The "eight" patterns in the figures indicate that both beams are linearly polarized, and that the polarizations of the ordinary and anomalous refraction are symmetric with respect to the 45° direction.

The spectral position of the two plasmonic modes of V-antennas can be tuned by varying the arm length h and, to a smaller extent, by adjusting the opening angle Δ (Fig. 6.9). However, both of these simultaneously shift the resonance frequencies of the symmetric and antisymmetric modes of the antenna. By appending a "tail" to the V-antenna as shown in Figure 6.17a, an additional degree of freedom is attained that allows for independent tuning of the spectral position of the symmetric mode [17]. By increasing the tail length, h_T, the symmetric mode is red-shifted without affecting the antisymmetric mode. This is confirmed by mapping out the two plasmonic modes for four different values of h_T, by measuring the reflection spectra from arrays of these antennas (Figs. 6.17b and 6.17c) and by FDTD simulations (Figs. 6.17f and 6.17g).

The measured polarization conversion efficiency of the Y-antennas is plotted in Figure 6.17d, given incident polarization along 45° with respect to the antenna symmetry axis, such that the projections of the incident field along the two antenna modes are equal. There is a substantial amount of polarization conversion for $h_T = 100, 300$, and 700 nm. However, for $h_T = 500$ nm, the polarization conversion is almost completely extinguished. FDTD simulations (Fig. 6.17h) demonstrate the same behavior as in the measurements. The origin of this effect can be interpreted as destructive interference between the contributions to the cross-polarization generation from the two oscillator modes. As illustrated in Figure 6.17e, the incident field excites both modes of the Y-antenna, each of which contributes to the cross-polarized field. However, as shown in Figure 6.3b, the projections of the emission of the two modes onto the y-axis are opposite in phase, so when the two modes are nearly identical in amplitude and phase responses (as is the case for $h_T = 500$ nm), their contributions to the polarization conversion are π out of phase, resulting in destructive interference. In general, arrays of resonant structures which support two orthogonal plasmonic eigenmodes can be viewed as meta-interfaces with large and tuneable birefringence: They can rotate the polarization of light over a thickness of just tens of nanometers at

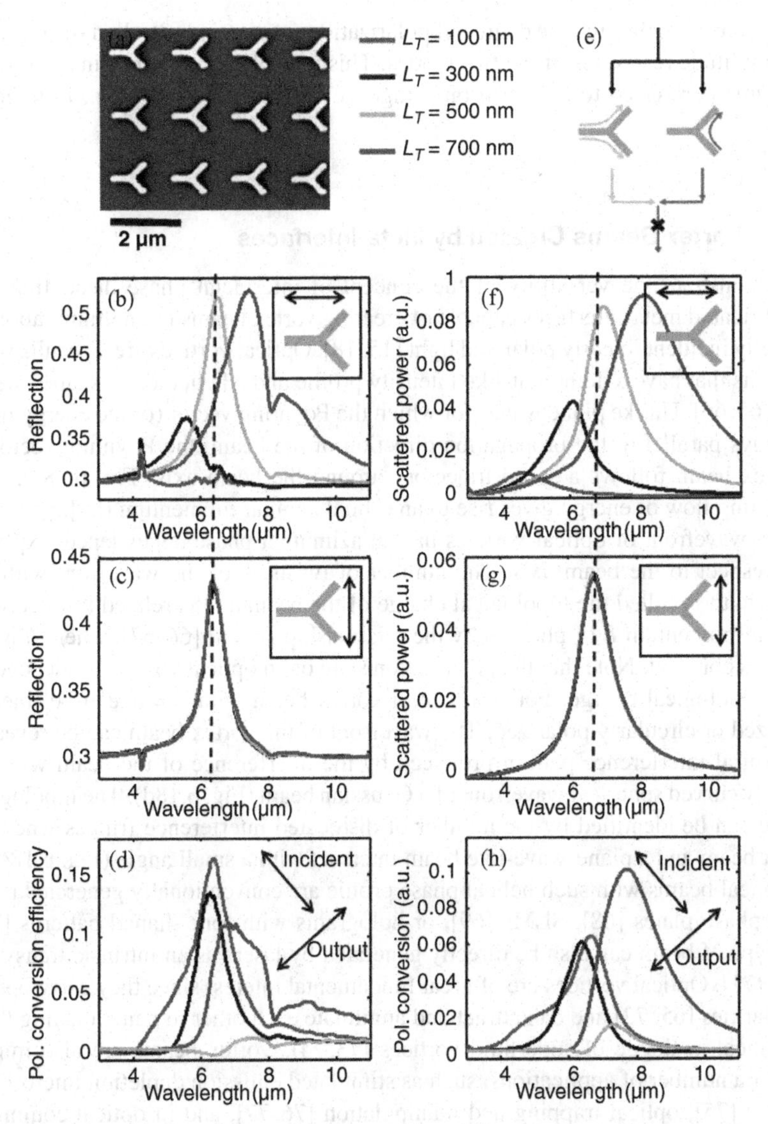

FIGURE 6.17 Y-shaped plasmonic antennas. (a) SEM image of the antenna array. (b and c) Measured normal incidence reflectivity spectra of the symmetric and antisymmetric antenna modes, respectively, as a function of tail length h_T, which varies from 100 nm to 700 nm by increments of 200 nm. The reflectivity of the bare silicon substrate is ~0.3. The vertical dashed line shows that for $h_T = 500$ nm the two modes are overlapping. The arrows in the insets indicate the polarization of the incident field. (d) Polarization conversion spectrum with incident polarization along 45°, with the incident and measured polarizations indicated with arrows. The polarization conversion is nearly extinguished for one intermediate value of h_T. (e) Diagram explaining the extinguishment of polarization conversion due to destructive interference between contributions of the two antenna modes when h_T is adjusted such that the two modes have the same resonant response. (f–h) FDTD simulations corresponding to the measurements in (b–d).

optical wavelengths, and the degree of polarization rotation is controlled by the phase and amplitude responses of the two modes. This type of birefringence in anisotropic structures is referred to as "form birefringence" in literature (see, e.g., References 61–64).

6.2.4 Vortex Beams Created by Meta-Interfaces

To demonstrate the versatility of the concept of interfacial phase discontinuities, we fabricated meta-interfaces capable of creating vortex beams upon illumination by normally incident linearly polarized light [15, 18]. Optical vortices are a peculiar type of beams that have a doughnut-like intensity profile and a helicoidally shaped wavefront [65, 66]. Unlike plane waves for which the Poynting vector (or the energy flow) is always parallel to the propagation direction of the beam, the Poynting vector of a vortex beam follows a spiral trajectory around the beam axis (Fig. 6.18a). This circulating flow of energy gives rise to an orbital angular momentum [66].

The wavefront of optical vortices has an azimuthal phase dependence, $\exp(il\theta)$, with respect to the beam axis. The number of twists, l, of the wavefront within a wavelength is called the topological charge of the beam and is related to the orbital angular momentum L of photons by the relationship $L = \hbar l$ [66, 67], where $\hbar$ is the Planck's constant. Note that the polarization state of an optical vortex is independent of its topological charge. For example, a vortex beam with $l = 1$ can be linearly polarized or circularly polarized. The wavefront of the vortex beam can be revealed by a spiral interference pattern produced by the interference of the beam with the strongly curved spherical wavefront of a Gaussian beam (Fig. 6.18d). The topological charge can be identified by the number of dislocated interference fringes when the vortex beam and a plane-wave-like beam intersect with a small angle (Fig. 6.18e).

Optical beams with such helical phase profile are conventionally generated using spiral phase plates [68], SLMs [69], or holograms with fork-shaped patterns [70]. This type of beam can also be directly generated by lasers as an intrinsic transverse mode [71]. Optical vortices are of great fundamental interest since they carry optical singularities [65, 72] and can attract and annihilate each other in pairs, making them the optical analogue of superfluid vortices [73, 74]. Vortex beams are also important for a number of applications, such as stimulated emission depletion microscopy (STED) [75], optical trapping and manipulation [76, 77], and in optical communication systems, where the quantized orbital angular momentum can carry additional information [78, 79].

Figure 6.19 shows the experimental setup used to generate and characterize the optical vortices. It consists of a Mach–Zehnder interferometer where the optical vortices are generated in one arm and their optical wavefronts are revealed by interference with a reference Gaussian beam propagating through the other arm. A laser beam from a distributed feedback QCL emitting monochromatic light at $\lambda_o = 7.75$ μm in continuous wave mode with power of ~10 mW is collimated and separated in two parts by a beam splitter. One part of the beam is rotated in polarization using a

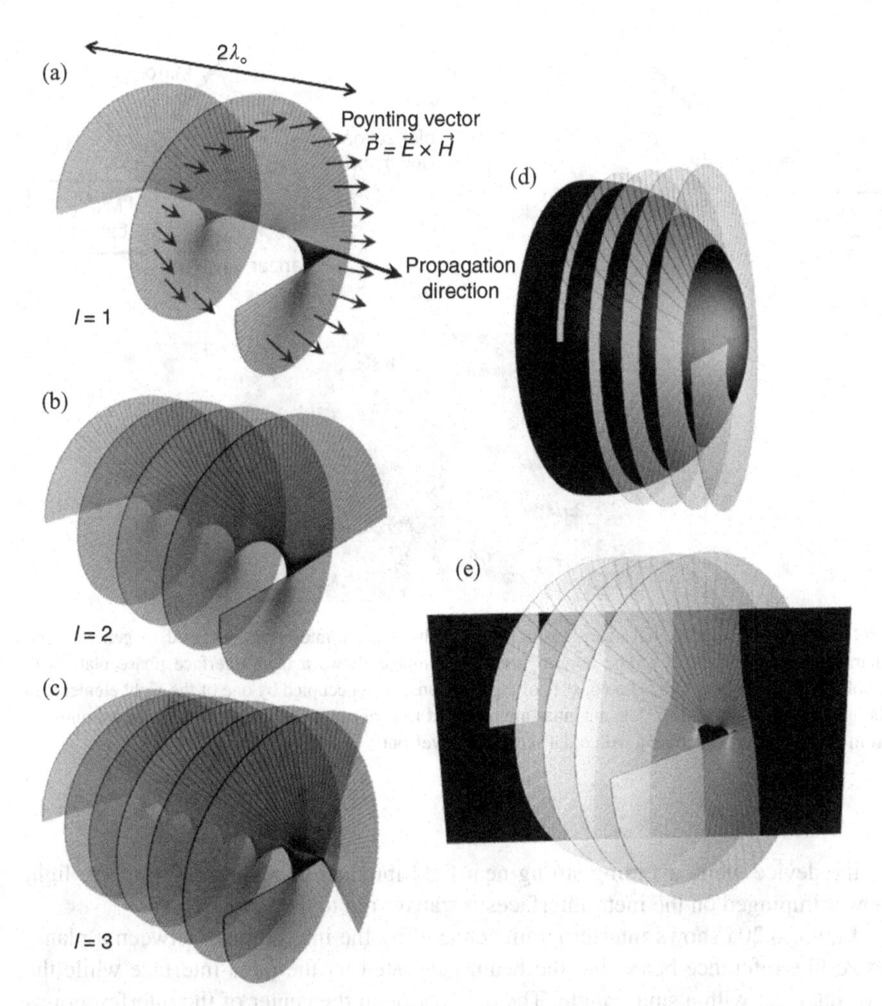

FIGURE 6.18 (a–c) Wavefronts of optical vortex beams with topological charge $l = 1$, 2, and 3; l represents the number of twists of wavefront within a wavelength. (d) and (e) show common methods for characterizing an optical vortex. One can interfere the vortex beam with a diverging or converging Gaussian beam, producing a spiral interference pattern, as demonstrated in (d) by the spiral intersection between the wavefronts of the two beams. Alternatively, one can interfere the vortex beam with a plane wave, producing an interference pattern with dislocated fringes, schematically shown in (e).

set of mirrors to serve as the reference beam. The second part is focused on a meta-interface phase mask using a ZnSe lens (20 inch focal length, 1 inch diameter). The phase mask comprises a silicon–air interface decorated with a 2D array of V-shaped gold plasmonic antennas designed and arranged so that it introduces a spiral shaped phase distribution to the scattered light cross-polarized with respect to the incident polarization. The inset SEM image of Figure 6.19 corresponds to $l = 1$. We chose a packing density of about 1 antenna per 1.5 μm^2 ($\sim\lambda_o^2/40$), to maximize the efficiency

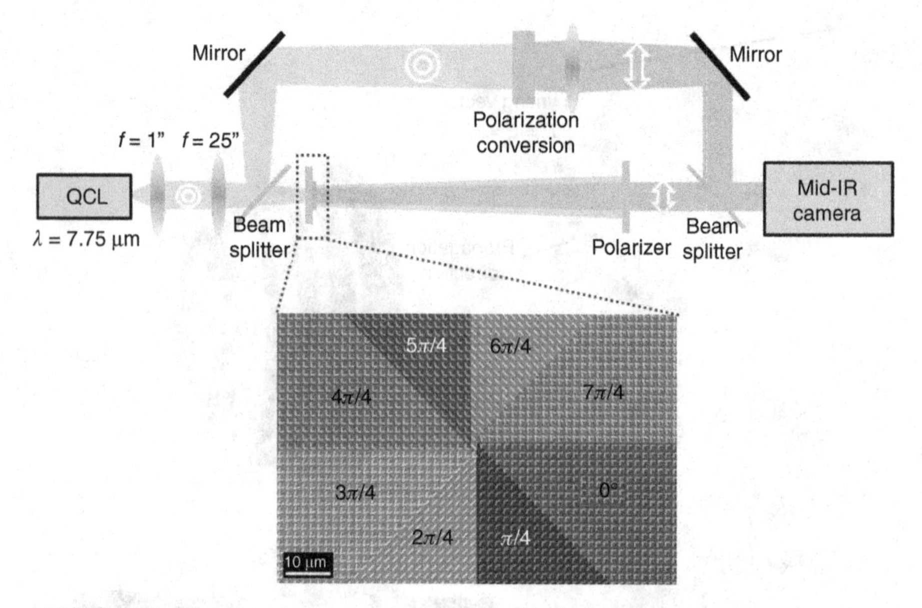

FIGURE 6.19 Experimental setup based on a Mach–Zehnder interferometer used to generate and characterize optical vortices. The bottom inset SEM image shows a meta-interface phase plate with topological charge one. The plate consists of eight regions, each occupied by one of the eight elements in the unit cell in Figure 6.10b. The antennas are arranged to generate a phase shift that varies azimuthally from 0 to 2π, thus producing a helicoidal scattered wavefront.

of the device while avoiding strong near-field interactions. About 30% of the light power impinged on the meta-interfaces is transferred to the vortex beams.

Figure 6.20a shows interferograms created by the interference between a plane-wave-like reference beam and the beam generated by the meta-interface while the two intersect with a small angle. The dislocation at the center of the interferograms confirms the presence of a phase defect at the core of the beam generated by the meta-interface. The orientation and the number of the dislocated fringes of the interferograms can be used to characterize the sign and the topological charge L of the vortex beam.

Figure 6.21 presents FDTD simulations of the evolution of the cross-polarized scattered field after a Gaussian beam at $\lambda_o = 7.7$ μm impinges normally on a 50×50 μm^2 meta-interface phase plate like the one shown in Figure 6.19. We observe that a phase singularity, where the phase is undefined, is established at the center of the phase distribution as close as 1 μm ($\sim\lambda_o/8$) away from the interface. The presence of such phase singularity produces the characteristic zero of intensity at the beam axis less than a wavelength away from the interface. The fact that a meta-interface molds the incident wavefront into an arbitrary shape almost instantaneously presents an advantage over conventional optical components, such as liquid-crystal SLM, which is optically thick, and diffractive optics, which require observers to be in the

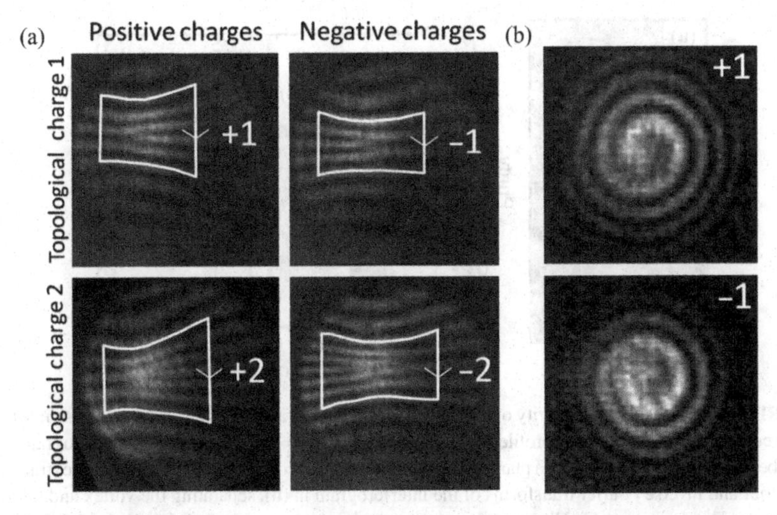

FIGURE 6.20 (a) Interferograms obtained from the interference between a plane-wave-like reference beam and optical vortices with different topological charges. The dislocation of the fringe pattern indicates the presence of a phase defect along the axis of the transmitted beam. Vortex beams with single (double) charge(s) are generated by introducing an angular distribution ranging from 0 to 2π (4π) using meta-interfaces. The azimuthal direction of the angular phase distribution defines the sign of the topological charge (also called chirality). (b) Spiral interferograms created by the interference of vortex beams and a co-propagating Gaussian beam.

far-field zone characterized by Fraunhofer distance $2D^2/\lambda_o$, where D is the size of the element [59].

Meta-interfaces with discretized spiral phase distribution (e.g., the one shown in Fig. 6.19) are used and thus we are concerned about the quality of the optical vortex beams. We conducted a quantitative analysis of the generated optical vortices in terms

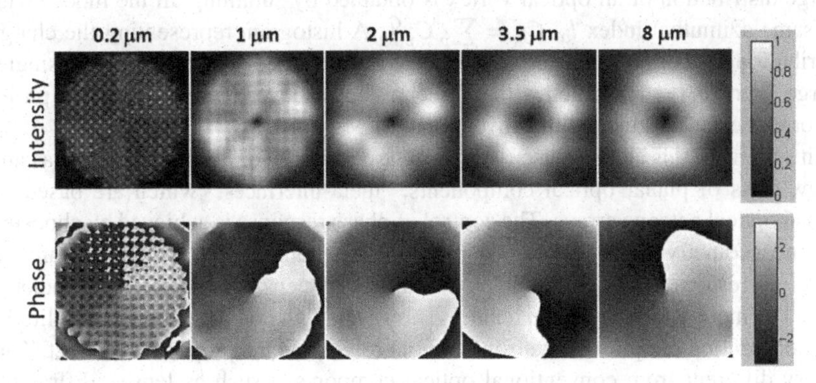

FIGURE 6.21 FDTD simulations of the cross-polarized scattered field as a function of distance from a meta-interface phase plate designed to create a single-charged optical vortex. The characteristic zero intensity at the center of the beam and the phase singularity develop as soon as the evanescent near-field components vanish, that is, about 1 μm or one-eighth of wavelength behind the interface.

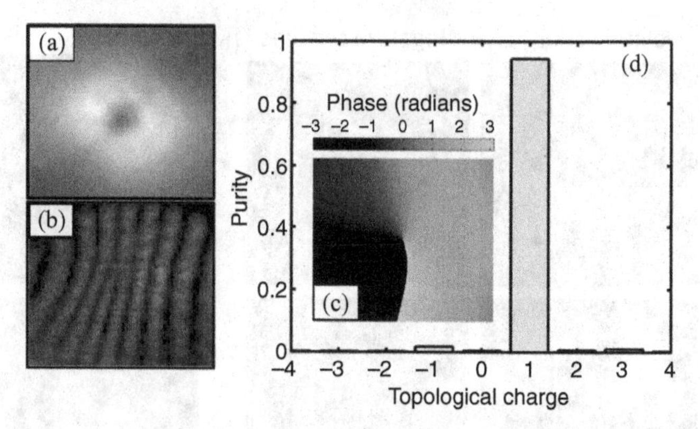

FIGURE 6.22 Analysis of the purity of the optical vortices generated by meta-interface phase masks. (a) Measured donut-shaped intensity profile of a vortex beam with $l = 1$. (b) Measured interferogram of the vortex beam when it interferes with a plane wave. The phase profile of the vortex beam in (c) is extracted via 2D Fourier and inverse Fourier transforms of the interferogram in (b), separating the vortex and reference beam in the Fourier space. (d) Histogram representing the decomposition of the single-charged vortex beam on a basis set of Laguerre–Gaussian modes with topological charge index l_o.

of the purity of their topological charge. The amplitude distribution of the optical vortex is obtained from the measured intensity distribution (Fig. 6.22a), and its phase profile (Fig. 6.22c) is obtained by the recorded interferogram (Fig. 6.22b). The purity of the optical vortex is calculated by decomposing its complex field on a complete basis set of optical modes with angular momentum, that is, the Laguerre–Gaussian (LG) modes ($E_{l,p}^{LG}$) [80]. The weight of a particular LG mode in the vortex beam is given by $C_{l,p}^{LG} = \int\int E_{vortex} E_{l,p}^{LG*} dxdy$, where E_{vortex} is the complex electric field of the vortex beam, the star denotes complex conjugate, and the integers l and p are the azimuthal and radial Laguerre–Gaussian mode indices, respectively. The relative charge distribution of an optical vortex is obtained by summing all the modes with the same azimuthal index l_o, $C_{l_o} = \sum_p C_{p,l_o}^{LG}$. A histogram representing the charge distribution for our vortex beam is plotted in Figure 6.22d. The purity of the single-charged vortex ($l = 1$) created with our technique is above 90%, similar to the purity of vortex beams obtained with conventional SLMs.

In summary, this chapter discusses the phase response of plasmonic antennas and a new class of planar optical components, "meta-interfaces," which are based on phased optical antenna arrays. The control of phase response is achieved by choosing antenna geometry so it operates at different positions on the antenna resonance curve, corresponding to different phase shifts between the scattered and incident light. The meta-interface design features subwavelength control of phase, amplitude, and polarization of optical wavefront without relying on the propagation effect. This is very different from conventional optical components such as lenses, diffractive optics, and wave plates. The subwavelength resolution of meta-interface will enable not only far-field engineering but also near-field and meso-field engineering of optical wavefronts.

Assume the V-antenna is composed of two rods with length h and radius a, located in the x–z plane (Fig. 6.A.1). The angle between the two rods is Δ. One antenna rod, denoted No. 1, is aligned with z-axis, while the other, denoted No. 2, is aligned with s-axis. The wavevector of the linearly polarized incident light, $\vec{\beta}$, lies in the symmetry plane of the V-antenna (light blue in Fig. 6.A.1A) and the angle between $\vec{\beta}$ and the x–z plane is θ. The antenna response can be decomposed into a linear combination of two modes characterized by symmetric and antisymmetric current distributions in the two rods as discussed in the main text. The currents in the two rods satisfy

$$I_{1z}^{a}(z) = -I_{2s}^{a}(s)\,|_{z=s} \text{ for the antisymmetric mode,} \tag{6.A.1}$$

$$I_{1z}^{s}(z) = I_{2s}^{s}(s)\,|_{z=s} \text{ for the symmetric mode,} \tag{6.A.2}$$

where the superscripts "a" and "s" are used to distinguish the two modes. The boundary conditions at the ends and the joint of the V-antenna for the antisymmetric mode are

$$I_{1z}^{a}(z = h) = I_{2s}^{a}(s = h) = 0, \tag{6.A.3}$$

$$\phi_1^{a}(z = 0) = \phi_2^{a}(s = 0) = 0. \tag{6.A.4}$$

The latter states that the scalar potential is zero where the two rods join. This is due to antisymmetry, that is, $\phi_1^a(z) = -\phi_2^a(s)$, and electrical contact, that is, $\phi_1^a(z = 0) = \phi_2^a(s = 0)$. For the symmetric mode, we have

$$I_{1z}^{s}(z = h) = I_{2s}^{s}(s = h) = 0, \tag{6.A.5}$$

$$A_{1z}^{s}(z = 0) = A_{2s}^{s}(s = 0) \approx 0, \tag{6.A.6}$$

The vector potential at the antenna joint is nearly zero because it is primarily determined by the current in the immediate vicinity of the joint (see Eqs. 6.A.8, 6.A.9, and 6.A.10), which is close to zero for the symmetric mode. For a better approximation, one can calculate the vector potential at the end of an isolated straight rod antenna of length h and use it to replace the zero at the right-hand side of Equation (6.A.6).

The boundary condition on the surface of the first rod reads

$$E_{1z} = -\frac{\partial \phi_1}{\partial z} - j\omega A_{1z} = -\frac{j\omega}{\beta^2}\left(\frac{\partial}{\partial z}\nabla \cdot \vec{A}_1 + \beta^2 A_{1z}\right) = 0, \tag{6.A.7}$$

where $\vec{A}_1$ and ϕ_1 are the vector and scalar potentials on the surface of the rod, respectively; A_{1z} is the z-component of $\vec{A}_1$; β is free-space wavevector; and ω is angular frequency. Note that we have used the Lorentz gauge $\nabla \cdot \vec{A}_1 = -j\beta^2\phi_1/\omega$. Equation 6.A.7 states that the tangential component of the electric field on the surface of the rod E_{1z} is zero, which is valid for antennas made of perfect electric conductors. It can be generalized to account for antennas made of realistic metals, which will be discussed later. There are three contributions to the vector potential on the surface of an antenna rod: (1) that due to the current distribution on the rod itself ("self-interaction"), (2) that due to the current distribution on the other rod ("mutual interaction"), and (3) that due to the incident light. For example, $A_{1z}^a = A_{11z}^a + A_{12z}^a + A_{13z}^a$ is the z-component of the vector potential on the surface of the first antenna rod for the antisymmetric mode, where the three terms on the right-hand side represent the three contributions. As a result, Equation 6.A.7 can be rewritten as

$$\left[\frac{\partial^2\left(A_{11z}^a + A_{12z}^a\right)}{\partial z^2} + \beta^2\left(A_{11z}^a + A_{12z}^a\right)\right] + \frac{\partial^2 A_{12x}^a}{\partial z \partial x} + \left(\frac{\partial}{\partial z}\nabla\cdot\vec{A}_{13}^a + \beta^2 A_{13z}^a\right) = 0, \tag{6.A.8}$$

where

$$A_{11z}^a = \frac{\mu_o}{4\pi}\int_0^h I_{1z}^a(z')\exp\left[-j\beta R_1\left(z,z'\right)\right]/R_1\left(z,z'\right)dz', \tag{6.A.9}$$

$$\begin{aligned} A_{12z}^a &= \cos(\Delta)\frac{\mu_o}{4\pi}\int_0^h I_{2s}^a(s')\exp\left[-j\beta R_{12}\left(z,s'\right)\right]/R_{12}\left(z,s'\right)ds' \\ &= -\cos(\Delta)\frac{\mu_o}{4\pi}\int_0^h I_{1z}^a(z')\exp\left[-j\beta R_{12}\left(z,z'\right)\right]/R_{12}\left(z,z'\right)dz', \end{aligned} \tag{6.A.10}$$

$$A_{12x}^a = -\sin(\Delta)\frac{\mu_o}{4\pi}\int_0^h I_{1z}^a(z')\exp\left[-j\beta R_{12}\left(z,z'\right)\right]/R_{12}\left(z,z'\right)dz' \tag{6.A.11}$$

are, respectively, contributions to the vector potential on the first rod from its own current and from the z- and x-components of the current in the second rod. $R_1(z,z') = \sqrt{(z-z')^2 + a^2}$ and $R_{12}(z,s') = \sqrt{z^2 - 2zs'\cos(\Delta) + s'^2 + a^2}$ are, respectively, the distances between the element of integration on the two rods (primed coordinates) and the monitor point on the first rod (unprimed coordinates) (see Fig. 6.A.1B). μ_o is the magnetic permeability of the surrounding dielectric. Note that in Equations 6.A.10 and 6.A.11, we have used the antisymmetric current condition (6.A.1). The contribution to the vector potential on the first rod from the incident light satisfies the following equation [33]:

$$\frac{\partial}{\partial z}\nabla\cdot\vec{A}_{13}^a + \beta^2 A_{13z}^a = \frac{j\beta^3}{\omega}\left(1 - \frac{q^2}{\beta^2}\right)U_{inc}^a\exp(jqz), \tag{6.A.12}$$

where

$$q = \beta \cos(\theta) \cos(\Delta/2), \tag{6.13}$$

$$U_{inc}^{a} = \frac{E_{inc}^{a} \sin(\Delta/2)}{\beta\left[\sin^{2}(\theta) + \cos^{2}(\theta)\sin^{2}(\Delta/2)\right]}, \tag{6.14}$$

and E_{inc}^{a} is the component of the incident electric field that drives the antisymmetric mode (Fig. 6.A.1A).

According to (6.A.11) it can be shown that

$$\xi_{x}^{a}(z) \equiv \frac{j\omega}{\beta^{2}}\frac{\partial A_{12x}^{a}}{\partial x} = -\sin^{2}(\Delta)\frac{j\omega\mu_{o}}{4\pi\beta^{2}}\int_{0}^{h} I_{1z}^{a}(z')\left[1 + j\beta R_{12}(z,z')\right] \times \exp\left[-j\beta R_{12}(z,z')\right]/R_{12}^{3}(z,z')z'dz'. \tag{6.A.15}$$

Substitute (6.A.12) and (6.A.15) into (6.A.8), we have

$$\frac{\partial^{2}(A_{11z}^{a} + A_{12z}^{a})}{\partial z^{2}} + \beta^{2}(A_{11z}^{a} + A_{12z}^{a}) = \frac{j\beta^{2}}{\omega}\frac{\partial\xi_{x}^{a}}{\partial z} - \frac{j\beta^{3}}{\omega}\left(1 - \frac{q^{2}}{\beta^{2}}\right)U_{inc}^{a}\exp(jqz). \tag{6.A.16}$$

Solution to Equation 6.A.16 is

$$A_{11z}^{a} + A_{12z}^{a} = \frac{-j}{c}\left[C_{1}\cos(\beta z) + C_{2}\sin(\beta z) - V^{a}(z) + U_{inc}^{a}\exp(jqz)\right], \tag{6.A.17a}$$

where

$$V^{a}(z) = \int_{0}^{z}\frac{\partial\xi_{x}^{a}(w)}{\partial w}\sin\left[\beta(z-w)\right]dw = -\sin(\beta z)\,\xi_{x}^{a}(z=0) + \beta\int_{0}^{z}\xi_{x}^{a}(w)\cos\left[\beta(z-w)\right]dw. \tag{6.A.17b}$$

The scalar potential on antenna arm No. 1 is

$$\phi_{1}^{a}(z) = \frac{j\omega}{\beta^{2}}\nabla\cdot\vec{A}_{1}^{a} = \frac{j\omega}{\beta^{2}}\frac{\partial(A_{11z}^{a} + A_{12z}^{a})}{\partial z} + \frac{j\omega}{\beta^{2}}\frac{\partial A_{12x}^{a}}{\partial x} + \frac{j\omega}{\beta^{2}}\nabla\cdot\vec{A}_{13}^{a} = \frac{j\omega}{\beta^{2}}\frac{\partial(A_{11z}^{a} + A_{12z}^{a})}{\partial z} + \xi_{x}^{a}(z), \tag{6.A.18}$$

where we have used Equation 6.A.15 and $\nabla \cdot \vec{A}_{13}^a = 0$. With Equation 6.A17a, Equation 6.A.18 can be expressed explicitly as

$$\phi_1^a(z) = \frac{1}{\beta}\left[-C_1\beta\sin(\beta z) + C_2\beta\cos(\beta z) - \frac{\partial V^a(z)}{\partial z} + jqU_{inc}^a \exp(jqz)\right] + \xi_x^a(z). \tag{6.A.19}$$

With boundary condition (6.A.4), one can solve for one of the unknown constants in Equation 6.A.19:

$$C_2 = -\xi_x^a(z=0) - \frac{jq}{\beta}U_{inc}^a. \tag{6.A.20}$$

Using (6.A.17b) and (6.A.20), (6.A.17a) reduces to

$$\begin{aligned} A_{11z}^a + A_{12z}^a &= \frac{-j}{c}\left\{C_1\cos(\beta z) - \left[\xi_x^a(z=0) + \frac{jq}{\beta}U_{inc}^a\right]\sin(\beta z) - V^a(z) + U_{inc}^a \exp(jqz)\right\} \\ &= \frac{-j}{c}\left\{C_1\cos(\beta z) - \frac{jq}{\beta}U_{inc}^a \sin(\beta z) - \beta\int_0^z \xi_x^a(w)\cos[\beta(z-w)]\,dw + U_{inc}^a \exp(jqz)\right\}. \end{aligned} \tag{6.A.21}$$

Substitute Equations 6.A.9 and 6.A.15 into Equation 6.A.21, we can derive the integral equation for the current in the first antenna rod for the antisymmetric mode:

$$\begin{aligned} \int_0^h I_{1z}^a(z')K^a(z,z')dz' &= \frac{-4\pi j}{\eta_o}\left[C_1\cos(\beta z) - \frac{jq}{\beta}U_{inc}^a\sin(\beta z)\right] \\ &+ J^a(z) - \frac{4\pi j}{\eta_o}U_{inc}^a\exp(jqz), \end{aligned} \tag{6.A.22}$$

where

$$\begin{aligned} K^a(z,z') &= \exp\left[-j\beta R_1(z,z')\right]/R_1(z,z') \\ &- \cos(\Delta)\exp\left[-j\beta R_{12}(z,z')\right]/R_{12}(z,z'), \end{aligned} \tag{6.A.23}$$

$$J^a(z) = \frac{4\pi j}{\eta_o}\beta\int_0^z \xi_x^a(w)\cos[\beta(z-w)]\,dw. \tag{6.A.24}$$

η_o is the impedance of the surrounding medium, and C_1 in Equation 6.A.22 is a constant to be determined.

We then follow an iteration procedure as discussed in Reference 33 and use boundary condition (6.A.3) to eliminate C_1 and solve the integral equation (6.A.22). The current distribution in the rod No.1 to the nth-order correction for the antisymmetric mode is

$$I_{1z}^a(z) = \frac{-4\pi j}{\eta_o \psi^a}$$

$$\frac{U_{inc}^a\left[\sum_{m=0}^{n} H_{mz}^a/(\psi^a)^m \sum_{m=0}^{n} F_m^a(h)/(\psi^a)^m - \sum_{m=0}^{n} H_m^a(h)/(\psi^a)^m \sum_{m=0}^{n} F_{mz}^a/(\psi^a)^m\right] + \frac{jq}{\beta}U_{inc}^a\left[\sum_{m=0}^{n} G_m^a(h)/(\psi^a)^m \sum_{m=0}^{n} F_{mz}^a/(\psi^a)^m - \sum_{m=0}^{n} G_{mz}^a/(\psi^a)^m \sum_{m=0}^{n} F_m^a(h)/(\psi^a)^m\right]}{\sum_{m=0}^{n} F_m^a(h)/(\psi^a)^m},$$

(6.A.25)

in which

$$\psi^a = \left|\int_0^h \frac{\cos(\beta z') - \cos(\beta h)}{\cos(\beta z) - \cos(\beta h)} K^a\left(z, z'\right) dz'\right|, \quad (6.A.26)$$

$$F_{mz}^a = F_m^a(z) - F_m^a(h),\ G_{mz}^a = G_m^a(z) - G_m^a(h),\ H_{mz}^a = H_m^a(z) - H_m^a(h), \quad (6.A.27)$$

$$F_0^a(z) = \cos(\beta z),\ G_0^a(z) = \sin(\beta z),\ H_0^a(z) = \exp(jqz), \quad (6.A.28)$$

and

$$X_m^a(z) = X_{m-1,z}^a \psi^a - \int_0^h X_{m-1,z'}^a K^a(z, z')dz' + J^a\left(I_{1z}^a(z')\right.$$
$$\left. = X_{m-1,z'}^a, z\right) \text{ for } m \geq 1, \quad (6.A.29)$$

where X can be F, G, or H and $J^a(I_{1z}^a(z') = X_{m-1,z'}^a, z)$ is the same as $J^a(z)$ in Equation 6.A.24 but with $X_{m-1,z'}^a$ substituted for $I_{1z}^a(z')$. According to Equation 6.A.25, the solution with the first-order correction is

$$I_{1z,1st}^a(z) =$$

$$\frac{-4\pi j}{\eta_o \psi^a} \frac{-U_{inc}^a\left[\exp(jqh)\cos(\beta z) - \exp(jqz)\cos(\beta h)\right] + \frac{jq}{\beta}U_{inc}^a \sin[\beta(h - z)] + M_1^a(z)/\psi^a}{\cos(\beta h) + A_1^a/\psi^a},$$

(6.A.30)

where

$$M_1^a(z) = U_{inc}^a \left[H_{0z}^a F_1^a(h) + H_{1z}^a F_0(h) - H_0(h) F_{1z}^a - H_1^a(h) F_{0z}^a\right] + \frac{jq}{\beta} U_{inc}^a \left[G_0(h) F_{1z}^a + G_1^a(h) F_{0z}^a - G_{0z}^a F_1^a(h) - G_{1z}^a F_0(h)\right] \tag{6.A.31}$$

and

$$A_1^a = F_1^a(h). \tag{6.A.32}$$

The zeroth-order current for normal incidence ($q = 0$) is

$$I_{1z,0,normal}^a(z) = \frac{4\pi j U_{inc}^a}{\eta_o \psi^a} \frac{\cos(\beta z) - \cos(\beta h)}{\cos(\beta h)}, \tag{6.A.33}$$

which is exactly the same as the zeroth-order solution of the current in an infinitely thin straight receiving antenna of length $2h$ [33]. This makes sense because in the zeroth-order approximation the electromagnetic interactions between different parts of a V-antenna are not considered. Therefore its optical response is only determined by the antenna length as shown in Equation 6.A.33.

We can similarly derive the integral equation for current in the first rod for the symmetric mode:

$$\int_0^h I_{1z}^s(z') K^s(z, z') dz' = \frac{-4\pi j}{\eta_o} \left[-U_{inc}^s \cos(\beta z) + C_2 \sin(\beta z)\right] + J^s(z) - \frac{4\pi j}{\eta_o} U_{inc}^s \exp(jqz), \tag{6.A.34}$$

where

$$K^s(z, z') = \exp\left[-j\beta R_1(z, z')\right] / R_1(z, z') + \cos(\Delta) \exp\left[-j\beta R_{12}(z, z')\right] / R_{12}(z, z'), \tag{6.A.35}$$

$$J^s(z) = \frac{4\pi j}{\eta_o} \left\{\beta \int_0^z \xi_x^s(w) \cos[\beta(z - w)] dw - \sin(\beta z) \xi_x^s(z = 0)\right\}, \tag{6.A.36}$$

$$\xi_x^s(z) = \sin^2(\Delta) \frac{j\omega\mu_o}{4\pi\beta^2} \int_0^h I_{1z}^s(z') \left[1 + j\beta R_{12}(z, z')\right] \times \exp\left[-j\beta R_{12}(z, z')\right] / R_{12}^3(z, z') z' dz', \tag{6.A.37}$$

$$U_{inc}^s = \frac{E_{inc}^s \sin(\theta) \cos(\Delta/2)}{\beta \left[\sin^2(\theta) + \cos^2(\theta) \sin^2(\Delta/2)\right]}, \tag{6.A.38}$$

in which E^s_{inc} is the component of the incident electric field driving the symmetric mode (Fig. 6.A.1A).

The current distribution in the rod No.1 to the nth-order correction for the symmetric mode is

$$I^s_{1z}(z) = \frac{-4\pi j}{\eta_o \psi^s} \frac{U^s_{inc}\left[\sum_{m=0}^{n} F^s_m(h)/(\psi^s)^m \sum_{m=0}^{n} G^s_{mz}/(\psi^s)^m - \sum_{m=0}^{n} F^s_{mz}/(\psi^s)^m \sum_{m=0}^{n} G^s_m(h)/(\psi^s)^m\right] + U^s_{inc}\left[\sum_{m=0}^{n} H^s_{mz}/(\psi^s)^m \sum_{m=0}^{n} G^s_m(h)/(\psi^s)^m - \sum_{m=0}^{n} H^s_m(h)/(\psi^s)^m \sum_{m=0}^{n} G^s_{mz}/(\psi^s)^m\right]}{\sum_{m=0}^{n} G^s_m(h)/(\psi^s)^m}, \tag{6.A.39}$$

where

$$\psi^s = \left| \int_0^h \frac{\sin\left[\beta(h-z')\right] - \left[\sin(\beta z') - \sin(\beta h)\right]}{\sin\left[\beta(h-z)\right] - \left[\sin(\beta z) - \sin(\beta h)\right]} K^s\left(z, z'\right) dz' \right| \tag{6.A.40}$$

and F, G, and H are similarly defined as in Equations 6.A.27, 6.A.28, and 6.A.29.

The current distribution with the first-order correction is

$$I^s_{1z,1st}(z) = \frac{-4\pi j}{\eta_o \psi^s} \frac{-U^s_{inc}\sin\left[\beta(h-z)\right] - U^s_{inc}\left[\exp(jqh)\sin(\beta z) - \exp(jqz)\sin(\beta h)\right] + M^s_1(z)/\psi^s}{\sin(\beta h) + A^s_1/\psi^s}, \tag{6.A.41}$$

where

$$\begin{aligned} M^s_1(z) = {} & U^s_{inc}\left[F_0(h)G^s_{1z} + F^s_1(h)G^s_{0z} - F^s_{0z}G^s_1(h) - F^s_{1z}G_0(h)\right] \\ & + U^s_{inc}\left[H^s_{0z}G^s_1(h) + H^s_{1z}G_0(h) - H_0(h)G^s_{1z} - H^s_1(h)G^s_{0z}\right] \end{aligned} \tag{6.A.42}$$

and

$$A^s_1 = G^s_1(h). \tag{6.A.43}$$

The zeroth-order current for normal incidence ($q = 0$) is

$$I^s_{1z,0,normal}(z) = \frac{4\pi j U^s_{inc}}{\eta_o \psi^s} \frac{\sin\left[\beta(h-z)\right] + \left[\sin(\beta z) - \sin(\beta h)\right]}{\sin(\beta h)}. \tag{6.A.44}$$

Let $h = 2h'$ and $z = z' + h'$ in Equation 6.A.44. We have

$$I^{s}_{1z,0,normal}(z)|_{\substack{h=2h' \\ z=z'+h'}} = \frac{4\pi j U^{s}_{inc}}{\eta_o \psi^{s}} \frac{\cos(\beta z') - \cos(\beta h')}{\cos(\beta h')}. \qquad (6.A.45)$$

Thus we recover the zeroth-order current distribution in an infinitely thin *straight* receiving antenna of length $2h' = h$. One can further check that according to Equation 6.A.44,

$$I^{s}_{1z,0}(z = 0) = 0. \qquad (6.A.46)$$

That is, there is no current at the joint of the V-antenna for the symmetric mode.

The above derivation assumes that the antenna is made of a perfect electric conductor. To model real metals, one has to add a term to the right-hand side of Equation 6.A.7 and the new boundary condition on the surface of the rod No. 1 reads

$$E_{1z} = -\frac{j\omega}{\beta^2}\left(\frac{\partial}{\partial z}\nabla \cdot \vec{A}_1 + \beta^2 A_{1z}\right) = z^i I_z, \qquad (6.A.47)$$

where z^i is the surface impedance per unit length of the antenna [33]:

$$z^i = \frac{\eta_m}{2\pi a} = \frac{1}{2\pi a}\sqrt{\frac{\mu}{\varepsilon_r - j\sigma/\omega}}. \qquad (6.A.48)$$

The concept of surface impedance is valid for mid-infrared and longer wavelengths [37]. If the displacement current is much smaller than the conduction current as in the case of highly conductive metals ($\omega\varepsilon_r \ll \sigma$) the impedance per unit length of antenna will be

$$z^i \approx \frac{1+j}{2\pi a}\sqrt{\frac{\omega\mu}{2\sigma}}. \qquad (6.A.49)$$

As a result of the correction in Equation 6.A.47, one additional term should be added to Equation 6.A.29 for the antisymmetric mode and the similar expression for the symmetric mode (other equations previously derived will not change):

$$\begin{aligned} X_m(z) = {} & \frac{j4\pi z^i}{\eta_o}\int_0^z X_{m-1,s}\sin[\beta(z-s)]\,ds + X_{m-1,z}\psi \\ & - \int_0^h X_{m-1,z'}K(z,z')dz' + J(I_{1z}(z') = X_{m-1,z'}, z). \end{aligned} \qquad (6.A.50)$$

Combining the results of current distributions for the antisymmetric and symmetric modes, Equations 6.A.25 and 6.A.39, the currents in the two antenna arms can be written as

$$I_1(z) = I_{1z}^a(z) + I_{1z}^s(z), \tag{6.A.51}$$

$$I_2(s) = -I_{1z}^a(s) + I_{1z}^s(s), \tag{6.A.52}$$

based on which one can calculate the amplitude and phase shift of antenna radiation in the far field. Corrections up to the second order were used in our calculations (Figs. 6.2d and 6.2e).

REFERENCES

1. Grober RD, Schoelkopf RJ, Prober DE (1997) Optical antenna: towards a unity efficiency near-field optical probe. *Appl. Phys. Lett.* 70: 1354–1356.
2. Novotny L, van Hulst N (2011) Antennas for light. *Nat. Photon.* 5: 83–90.
3. Xu Q, Bao J, Rioux RM, Perez-Castillejos R, Capasso F, Whitesides GM (2007) Fabrication of large-area patterned nanostructures for optical applications by nanoskiving. *Nano Lett.* 7: 2800–2805.
4. Sukharev M, Sung J, Spears KG, Seideman T (2007) Optical properties of metal nanoparticles with no center of inversion symmetry: observation of volume plasmons. *Phys. Rev. B* 76: 184302.
5. Biagioni P, Huang JS, Duò L, Finazzi M, Hecht B (2009) Cross resonant optical antenna. *Phys. Rev. Lett.* 102: 256801.
6. Ginn J, Shelton D, Krenz P, Lail B, Boreman G (2010) Polarized infrared emission using frequency selective surfaces. *Opt. Express* 18: 4557–4563.
7. Watanabe T, Fujii M, Watanabe Y, Toyama N, Iketaki Y (2004) Generation of a doughnut-shaped beam using a spiral phase plate. *Rev. Sci. Instrum.* 75: 5131.
8. McLeod JH (1954) The axicon—a new type of optical element. *J. Opt. Soc. Am.* 44: 8.
9. Casasent D (1977) Spatial light modulators. *Proc. IEEE* 65: 1.
10. Goodman JW (2004) *Introduction to Fourier Optics*. Roberts & Co.
11. Pendry JB, Schurig D, Smith DR (2006) Controlling electromagnetic fields. *Science* 312: 1780–1782.
12. Leonhardt U (2006) Optical conformal mapping. *Science* 312: 1777–1780.
13. Cai W, Shalaev V (2009) *Optical Metamaterials: Fundamentals and Applications*. Springer.
14. Engheta N, Ziolkowski RW (2006) *Metamaterials: Physics and Engineering Explorations*. Wiley-IEEE Press.

15. Yu N, Genevet P, Kats MA, Aieta F, Tetienne JP, Capasso F, Gaburro Z (2011) Light propagation with phase discontinuities: generalized laws of reflection and refraction. *Science* 334: 333–337.

16. Kats MA, Yu N, Genevet P, Gaburro Z, Capasso F (2011) Effect of radiation damping on the spectral response of plasmonic components. *Opt. Express* 19: 21749.

17. Kats MA, Genevet P, Aoust G, Yu N, Blanchard R, Aieta F, Gaburro Z, Capasso F. (2012) Giant birefringence in optical antenna arrays with widely tailorable optical anisotropy. Proceedings of the National Academy of Sciences of the United States of America 109: 12364–12368

18. Genevet P, Yu N, Aieta F, Lin J, Kats MA, Blanchard R, Scully MO, Gaburro Z, Federico Capasso (2012) Ultra-thin plasmonic optical vortex plate based on phase discontinuities. *Appl. Phys. Lett.* 100: 013101.

19. Aietal F, Genevet P, Yu N, Kats MA, Gaburro Z, Capasso F. Out-of-plane reflection and refraction of light by plasmonic interfaces with phase discontinuities. Submitted to *Nat. Phys.*

20. Blanchard R, Aoust G, Genevet P, Yu N, Kats MA, Gaburro Z, Capasso F (2012) Modelling nanoscale V-shaped antennas for the design of optical phased arrays. *Phys. Rev. B* 85: 155457.

21. Miyazaki HT, Kurokawa Y (2006) Controlled plasmon resonance in closed metal/insulator/metal nanocavities. *Appl. Phys. Lett.* 89: 211126.

22. Fattal D, Li J, Peng Z, Fiorentino M, Beausoleil RG (2010) Flat dielectric grating reflectors with focusing abilities. *Nat. Photonics* 4: 466–470.

23. Fanl JA, Wu C, Bao K, Bao J, Bardhan R, Halas NJ, Manoharan VN, Nordlander P, Shvets G, Capasso F (2010) Self-assembled plasmonic nanoparticle clusters. *Science* 328: 1135–1138.

24. Luk'yanchuk B, Zheludev NI, Maier SA, Halas NJ, Nordlander P, Giessen H, Chong CT (2010) The Fano resonance in plasmonic nanostructures and metamaterials. *Nat. Mater.* 9: 707–715.

25. Hu X, Chan CT, Ho KM, Zi J (2011) Negative effective gravity in water waves by periodic resonator arrays. *Phys. Rev. Lett.* 106: 174501.

26. Liu Z, Zhang X, Mao Y, Zhu YY, Yang Z, Chan CT, Sheng P (2000) Locally resonant sonic materials. *Science* 289: 1734–1736.

27. Jackson JD (1999) *Classical Electrodynamics*. 3rd ed. John Wiley & Sons, Inc. p 665.

28. King RWP (1956) *The Theory of Linear Antennas*. Harvard University Press.

29. Maier SA (2007) *Plasmonics: Fundamentals and Applications*. Springer.

30. Novotny L, Bert H (2011) *Principles of Nano-Optics*. Cambridge University Press.

31. Zuloaga J, Nordlander P (2011) On the energy shift between near-field and far-field peak intensities in localized plasmon systems. *Nano Lett.* 11: 1280.

32. Zhang S, Genov DA, Wang Y, Liu M, Zhang X (2008) Plasmon-induced transparency in metamaterials. *Phys. Rev. Lett.* 101: 047401.

33. Liu N, Langguth L, Weiss T, Kastel J, Fleischhauer M, Pfau T, Giessen H (2009) Plasmonic analogue of electromagnetically induced transparency at the Drude damping limit. *Nat. Mater.* 8: 758.

34. Griffiths DJ (1999) *Introduction to Electrodynamics*. 3rd ed. Benjamin Cummings.

35. Heitler W (1954) *The Quantum Theory of Radiation*. 3rd ed. London: Oxford University Press.

36. Kelly KL, Coronado E, Zhao LL, Schatz GC (2003) The optical properties of metal nanoparticles: the influence of size, shape, and dielectric environment. *J. Phys. Chem. B* 106: 668.

37. Grady NK, Halas NJ, Nordlander P (2004) Influence of dielectric function properties on the optical response of plasmon resonant metallic nanoparticles. *Chem. Phys. Lett.* 399: 167.

38. Bruzzone S, Malvaldi M, Arrighini GP, Guidotti C (2004) Light scattering by gold nanoparticles: role of simple dielectric models. *J. Phys. Chem. B* 108: 10853.

39. Grimault AS, Vial A, de la Chapelle ML (2006) Modeling of regular gold nanostructures arrays for SERS applications using a 3D FDTD method. *Appl. Opt. B* 84: 111.

40. Bryant GW, de Abajo FJG, Aizpurua J (2008) Mapping the plasmon resonances of metallic nanoantennas. *Nano Lett.* 8: 2.

41. Ross BM, Lee LP (2009) Comparison of near- and far-field measures for plasmon resonance of metallic nanoparticles. *Opt. Lett.* 34: 7.

42. Palik ED (1997) *Handbook of Optical Constants of Solids*. Vol. 3. Academic Press.

43. Ashcroft NW, Mermin ND (1976) *Solid State Physics*. Thomson.

44. van Driel HM (1984) Optical effective mass of high density carriers in silicon. *Appl. Phys. Lett.* 44: 6.

45. Stutzman WL, Thiele GA (1981) *Antenna Theory and Design*. New York: John Wiley & Sons, Inc.

46. Balanis CA (1982) *Antenna Theory, Analysis and Design*. New York: John Wiley & Sons, Inc.

47. Orfanidis SJ. (2010) online publication *Electromagnetic Waves and Antennas*. http://www.ece.rutgers.edu/~orfanidi/ewa.

48. Engheta N, Papas CH, Elachi C (1982) Radiation patterns of interfacial dipole antennas. *Radio Sci.* 17: 1557.

49. Cubukcu E, Capasso F (2009) Optical nanorod antennas as dispersive one-dimensional Fabry-Perot resonators for surface plasmons. *Appl. Phys. Lett.* 95: 201101.

50. Huang J-S, Kern J, Geisler P, Weinmann P, Kamp M, Forchel A, Biagioni P, Hecht B (2010) Mode imaging and selection in strongly coupled nanoantennas. *Nano Lett.* 10: 2105–2110.

51. Brorson SD, Haus HA (1988) Diffraction gratings and geometrical optics. *J. Opt. Soc. Am. B* 5: 247–248.

52. Feynman RP, Hibbs AR (1965) *Quantum Mechanics and Path Integrals*. New York: McGraw-Hill.

53. Hecht E (1997) *Optics*. 3rd ed. Addison Wesley Publishing Company.

54. Veselago VG (1968) The electrodynamics of substances with simultaneously negative values of ε and μ. *Sov. Phys. Usp.* 10: 509–514.

55. Pendry JB (2000) Negative refraction makes a perfect lens. *Phys. Rev. Lett.* 85: 3966–3969.
56. Shelby RA, Smith DR, Schultz S (2001) Experimental verification of a negative index of refraction. *Science* 292: 77–79.
57. Parazzoli CG, Greegor RB, Li K, Koltenbah BEC, Tanielian M (2003) Experimental verification and simulation of negative index of refraction using Snell's law. *Phys. Rev. Lett.* 90: 107401.
58. Houck AA, Brock JB, Chuang IL (2003) Experimental observations of a left-handed material that obeys Snell's law. *Phys. Rev. Lett.* 90: 137401.
59. Born M, Wolf E (1999) *Principles of Optics*. 7th ed. Cambridge University Press.
60. Ewald PP (1969) Introduction to the dynamical theory of X-ray diffraction. *Acta Crystallogr. A* 25: 103–108.
61. Kikuta H, Ohira Y, Iwata K (1997) Achromatic quarter-wave plates using the dispersion of form birefringence. *Appl. Opt.* 36: 7.
62. Sung J, Sukharev M, Hicks EM, Van Duyne RP, Seideman T, Spears KG (2008) Nanoparticle spectroscopy: birefringence in two-dimensional arrays of L-shaped silver nanoparticles. *J. Phys. Chem. C* 112: 3252.
63. Feng L, Mizrahi A, Zamek S, Liu Z, Lomakin V, Fainman Y (2011) Metamaterials for enhanced polarization conversion in plasmonic excitation. *ACS Nano* 5: 6.
64. Shcherbakov MR, Dobynde MI, Dolgova TV, Tsai D-P, Fedyanin AA (2010) Full Poincare sphere coverage with plasmonic nanoslit metamaterials at Fano resonance. *Phys. Rev. B* 82: 193402.
65. Nye JF, Berry MV (1974) Dislocations in wave trains. *Proc. R. Soc. Lond. A* 336: 165–190.
66. Padgett M, Courtial J, Allen L (2004) Light's orbital angular momentum. *Phys. Today* 57: 35–40.
67. Allen L, Beijersbergen MW, Spreeuw RJC, Woerdman JP (1992) Orbital angular momentum of light and the transformation of Laguerre-Gaussian laser modes. *Phys. Rev. A* 45: 8185–8189.
68. Beijersbergen MW, Coerwinkel RPC, Kristensen M, Woerdman JP (1994) Helical-wavefront laser beams produced with a spiral phaseplate. *Opt. Commun.* 112: 321–327.
69. Heckenberg NR, McDuff R, Smith CP, White AG (1992) Generation of optical phase singularities by computer-generated holograms. *Opt. Lett.* 17: 221–223.
70. Yu V, Vasnetsov MV, Soskin MS (1990) Laser beams with screw dislocations in their wavefronts. *JETP Lett.* 52: 429.
71. White AG, Smith CP, Heckenberg NR, Rubinsztein-Dunlop H, McDuff R, Weiss CO, Tamm C (1991) Interferometric measurements of phase singularities in the output of a visible laser. *J. Mod. Opt.* 38: 2531.
72. M., Berry, Nye J, Wright F (1979) The elliptic umbilic diffraction catastrophe. *Philos. Trans. R. Soc. Lond.* 291: 453.
73. Coullet P, Gil L, Rocca F (1989) Optical vortices. *Opt. Commun.* 73: 403.
74. Genevet P, Barland S, Giudici M, Tredicce JR (2010) Bistable and addressable localized vortices in semiconductor lasers. *Phys. Rev. Lett.* 104: 223902.

75. Hell SH (2007) Far-field optical nanoscopy. *Science* 316: 1153.

76. He H, Friese MEJ, Heckenberg NR, Rubinsztein-Dunlop H (1995) Direct observation of transfer of angular momentum to absorptive particles from a laser beam with a phase singularity. *Phys. Rev. Lett.* 75: 826–829.

77. Padgett M, Bowman R (2011) Tweezers with a twist. *Nat. Photonics* 5: 343.

78. Leach J, Padgett MJ, Barnett SM, Franke-Arnold S, Courtial J (2002) Measuring the orbital angular momentum of a single photon. *Phys. Rev. Lett.* 88: 257901.

79. Gibson G, Courtial J, Padgett M, Vasnetsov M, Pas'ko V, Barnett S, Franke-Arnold S (2004) Free-space information transfer using light beams carrying orbital angular momentum. *Opt. Express* 12: 5448–5456.

80. Siegman AE (1986) *Lasers*. Sausalito, CA: University Science Books. Chapter 16.

7

Integrated Plasmonic Detectors

PIETER NEUTENS AND PAUL VAN DORPE
Imec, Kapeldreef 75 3001 Leuven, Belgium, and Physics Dept., KULeuven, Leuven, Belgium

7.1 INTRODUCTION

Surface plasmons are known to have the potential to guide, concentrate, and scatter light at the nanoscale. These properties are particularly interesting in emerging fields that rely on either small-footprint and fast devices like in on-chip optical communication, optical absorption in small volumes such as cost-effective solar cells, or biosensing, where the extent of the field is similar to the size of the molecules that are probed. In all these cases, surface plasmon-based devices have been investigated extensively to improve the performance or to create new possibilities. In one important application, surface plasmons can directly enhance the absorption in (thin film) solar cells. In optical communication schemes and biosensing, surface plasmons have been studied extensively using optical techniques, while it is highly desirable to detect them locally using integrated plasmon detectors.

To understand how surface plasmons can be applied for enhancing the response of semiconductor detectors, the detection of biochemical reactions, or signal transmission, one has to gain insight into the nature of surface plasmons. Surface plasmons are collective electron oscillations at the interface between a metal and a dielectric. As the reader will have discovered while wandering through this book, one of the most amazing properties of surface plasmons is the way in which they can focus and channel light using subwavelength-sized metallic structures. For this reason, surface plasmons have the potential of merging the dimensions of fast photonics with nowadays electronic components.

For the development of plasmon-enhanced photodetectors, small-area photodetectors are combined with a plasmonic antenna structure, which can collect light over a large area and focus it on the much smaller semiconductor area. Also metal

Active Plasmonics and Tuneable Plasmonic Metamaterials, First Edition. Edited by Anatoly V. Zayats and Stefan A. Maier.

particle resonances can be applied to maximize the overlap of the plasmonic mode and the small detector active area. In this way it is possible to combine the advantages of small photodetectors, like a low noise level and high-speed operation, with the advantages of large-area photodetectors, such as the high detector photoresponse.

In the study of optical communication on chip, plasmonic waveguides have been studied extensively in the last decade [1–7]. The delay caused by electrical interconnects increases when their dimensions are scaled down, opposite to the transistors, for which the performance increases with scaling. Researchers today are investigating if the use of light as the information carrier can supply an integrated circuit with a better performance as their copper counterparts. At optical frequencies, it is possible to communicate with a high bandwidth, and as a consequence an optical waveguide can carry digital signals with a capacity exceeding that of its electrical counterpart with a factor of more than one thousand. Unlike diffraction-limited dielectric waveguides, metallic waveguides can be fabricated with subwavelength dimensions and benefit from the same properties of optical signal transmission and can be used for both optical and electrical signal transmission. For this reason, they have been suggested to function as optical interconnects on chip. In order to integrate these metallic waveguides on chip, electro-optical transducers need to be developed to realize a fast and efficient coupling between electrical and optical signals, leading to the design of plasmon sources and detectors.

In the field of solar cell research, the properties of surface plasmons are applied to increase the conversion efficiency. Material cost is the main factor determining the price of solar cells today. As a consequence, great interest has arisen in both solid state and organic thin film solar cells, where thin semiconductor layers are deposited on a cheap substrate. A lower material cost asides, thin semiconductor junctions offer an increased collection efficiency compared to wafer-based solar cells because the minority carrier diffusion length is larger than the film thickness. However, the absorption length of the incident radiation can exceed the semiconductor film thickness, leading to only a partial conversion of the optical power in the semiconductor junction. Both localized surface plasmons on metallic nanoparticles and propagating surface plasmon polaritons (SPPs) on metallic films can offer a way to increase the absorption of the incident light by channeling and concentrating the electromagnetic fields inside the thin semiconductor layer.

SPPs are intrinsically very sensitive to small changes in the refractive index near the metal/dielectric interface in a confined region whose spatial extent is dictated by the evanescent tail of the SPP. At visible wavelengths, this is in the order of a few hundred nanometers or below. Therefore SPPs provide excellent means to probe biochemical events occurring at the metal surface. Bio-recognition is achieved by using localized surface plasmon resonance (LSPR) biosensors through the detection of LSPR changes caused by adsorbate-induced changes in local dielectric constant. Usually the LSPR spectrum of a nanoparticle is studied with an optical detector in the far field. In order to develop lab-on-chip plasmonic biosensors, the integration of plasmonic sources and detectors is of paramount importance.

In general we can divide the plasmon detection mechanisms into three large groups. The underlying physical principle is presented in Figure 7.1. More than two decades ago, the study of surface plasmon detection by means of tunnel junction

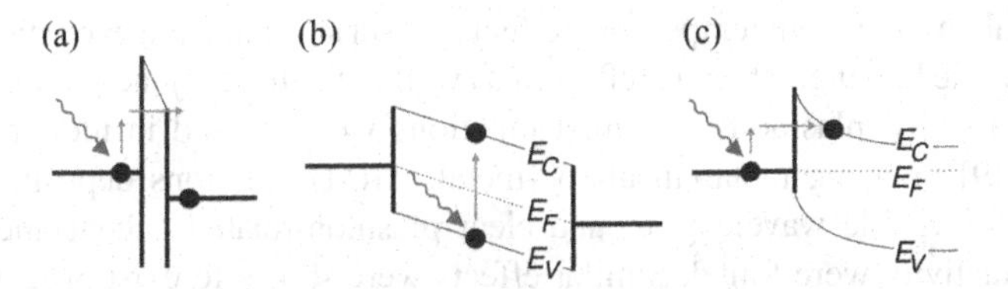

FIGURE 7.1 Plasmon detection mechanisms. (a) A tunnel junction detector. (b) Standard semiconductor photodetectors. (c) A Schottky diode photodetector.

detectors started. In Figure 7.1a the detection process is visualized. If plasmons are focused inside the tunnel junction, it will bias the junction and induce tunneling of electrons back and forth at the plasmon frequency, even when no external bias voltage is applied. Due to the rectification property caused by the nonlinear and asymmetric nature of the IV characteristics of the tunnel junction, a DC bias develops across the junction under illumination and the system will behave as a detector. Such bias voltage can be detected either as a light-induced DC current or as an additional bias if the tunnel junction is placed under an externally applied bias voltage. The second detection mechanism is the integration of standard semiconductor detectors with plasmonic nanostructures, shown in Figure 7.1b. Among these detectors we can find photoconductors, MSM photodetectors, and photodiodes. MSM photodetectors are mainly studied for plasmonic enhancement of ultrafast photodetectors while photodiodes mainly find their applications in plasmon-enhanced solar cells. A third detection method is the application of Schottky photodiodes, depicted in Figure 7.1c. In Schottky photodetectors, two detection modes are possible. For plasmon energies above the bandgap energy of the semiconductor, electron–hole pair generation across the bandgap is the main contributor to the photocurrent. For energies below the bandgap but larger than the work function of the junction, direct excitation of electrons to the conduction band is impossible, but emission of carriers can occur from the metal to the semiconductor over the Schottky barrier, leading to a measurable photocurrent. In this chapter we will discuss the combination of these three types of detectors with plasmonic waveguides and nanostructures.

7.2 ELECTRICAL DETECTION OF SURFACE PLASMONS

7.2.1 Plasmon Detection with Tunnel Junctions

The first demonstrations of electrical detection of surface plasmons were performed using electrical tunnel junctions. Indeed, as shown in Figure 7.1a, the rectifying nature of an asymmetric tunnel barrier results in a DC component upon AC excitation, similar to the rectification in a pn-diode. The particular advantage of a tunnel junction compared to a pn-diode is the intrinsic speed of the tunnel junction. As the speed is determined by the tunnel time, which can go down to 1 fs or below, bandwidths up to the NIR and visible are feasible [8]. However, these types of detectors have very low efficiencies, which can be strongly enhanced using plasmon excitations. Substantial experimental evidence for this has been shown already in the 1980s, although care has to be taken to rule out temperature effects, as surface plasmon excitations usually involve sample heating, which is reflected as well in the tunnel junction conductance. More specifically, plasmons in tunnel junctions were excited in the Kretschmann geometry [9] or using metal–insulator–metal (MIM) junctions deposited on gratings [10] for visible wavelengths, and clear plasmon-related enhancements of the photoconductivity were found. Similar effects were shown to exist in scanning tunneling microscopy (STM) on a metal film, where a clear light-induced tunneling current was observed upon the excitation of surface plasmons using the Kretschmann geometry [11].

Tunnel junctions are particularly interesting for the ultrafast detection of terahertz or IR radiation as little fast alternatives are available. In order to decrease the response time, and hence increase the bandwidth, the junctions should exhibit a relatively low resistance-area product. A well-established tunnel junction is the Ni/NiO/Ni junction, which has a high-quality NiO barrier, but exhibits a very small barrier height. Its properties have been investigated thoroughly by Hobbs et al., who optimized the fabrication technique and completely characterized the electrical properties [12]. The same group at IBM reported later on a waveguide-integrated version of their detector, which is depicted in Figure 7.2 [13].

In this work a Au antenna is fabricated on top of a Si photonic waveguide. The Au antenna captures and concentrates the light into the Ni/NiO tunnel junction that is deposited on top of and in-between the Au antenna (that serves as an electrical contact simultaneously). The device operates at 1.6 μm and exhibits a quantum efficiency of 6%, which is a remarkably high number compared to older results.

Next to tunnel junctions, gold quantum point contacts also exhibit photoconductivity through a process that is called photo-assisted transport (PAT). This was shown earlier by Ittah et al. by direct laser irradiation of a $1G_0$ Au point contact, where photo-assisted processes were observed and attributed to interactions between the photons and the ballistic electrons in the point contact [14]. The point contacts were fabricated in a double process. First, two Au electrodes with a nanometer-sized gap were deposited using a shadow masking technique. Subsequently, by applying a relatively large voltage across the gap, the mobile Au atoms restructure in the zone of high field until a contact is established. This proved to be a rather reliable method for fabricating quantum point contacts. In a next paper, the same group integrated this sensor in a plasmonic waveguide, where free-space radiation was converted to running SPPs using a grating on a Au SPP waveguide strip. A polarization and distance-dependent analysis demonstrated the electrical conversion of the optically excited SPPs [15].

7.2.2 Plasmon-Enhanced Solar Cells

Conventional photovoltaic solar cells presently employ a range of optical techniques to couple incident sunlight into the solar cell. For example, random surface textures are used in commercial mono- and multi-crystalline silicon solar cells and highly ordered inverted pyramid designs used in the highest efficiency silicon solar cells. These textured surfaces play two roles: One is to reduce the overall surface reflection

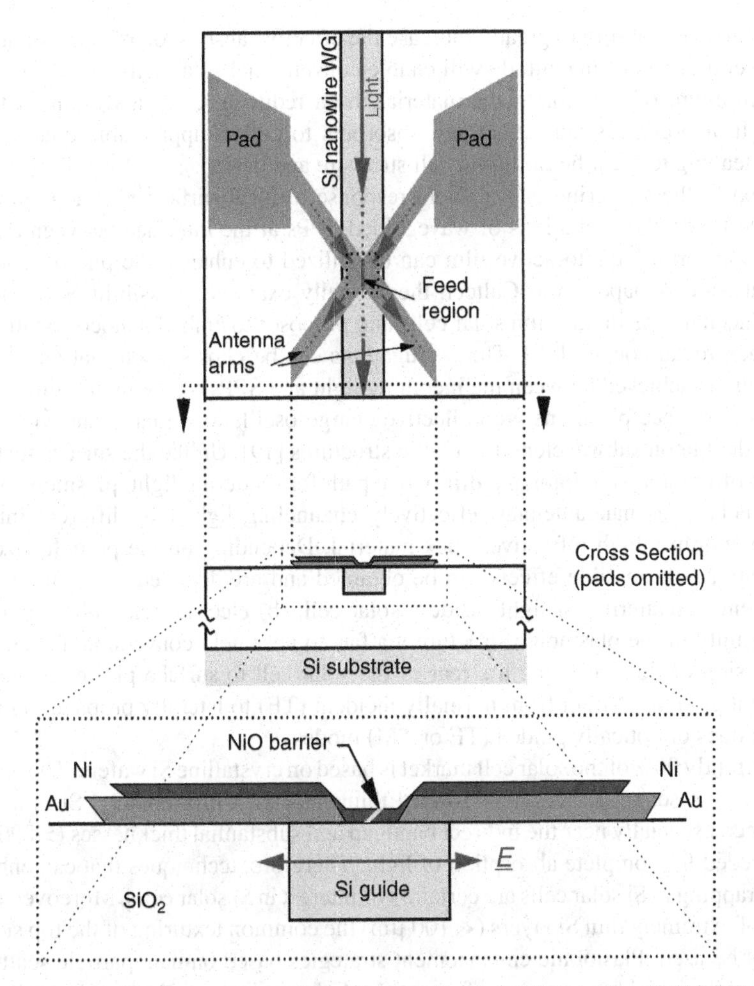

FIGURE 7.2 Schematic representation of the waveguided integrated detector from Hobbs et al. On top of the silicon waveguide there is a Au antenna, connected to metal connection pads. In the gap between the two leads, the Ni/NiO tunnel junction is deposited.

and the second is to scatter the light internally within the solar cell where it becomes trapped to a significant extent. These textured surface light-trapping techniques can be highly effective in practice [16] and approach the ergodic limit for optical path length extension [17]. The technology is mature and is commonly used in commercial solar cell devices. However, the texturing process, to be efficient, consumes typically 10 μm of active material and becomes therefore practically unusable for solar cells with thickness much smaller than 100 μm. Plasmonic structures have the ability to enhance the electric field locally, making them suitable for ultrathin or nanostructured active regions.

The consequences of plasmonics in photovoltaics are profound, not only in terms of fundamental physics but also for enabling the manufacture of new, highly efficient solar cells. The ability to greatly increase the effective absorption of semiconductors by several orders of magnitude will enable conventional solar cells to be made with the minimum of semiconductor material, both reducing costs and manufacturing time. It also enables nanostructured absorbers to collect appreciable quantities of light, leading to great flexibility in cell structure and design.

Next to the scattering or near-field response of local surface plasmons in metal nanoparticles, the excitations of waveguided SPPs at the interface between a metal and an absorbing photo-active film can be utilized to enhance the photoabsorption in that film. A paper from Caltech theoretically examines possibilities to include such mechanisms in thin film solar cells and proposes to embed nanocorrugations in the back metal contact [18]. The corrugations can be chosen such that an efficient coupling is achieved between the incoming light and SPPs on the metal film.

Local surface plasmons are collective charge oscillations that occur when light is incident upon subwavelength metallic structures [19]. Unlike the surface textures, which offer only a randomizing diffractive path for scattering light, plasmonic structures behave as nanoantennas, effectively channeling light into different internal modes within a high refractive index material. Depending on the particle size and geometry, four possible effects can be obtained and are depicted in Figure 7.3: (a) preferential scattering of light inside a solar cell, (b) electric field enhancement in the vicinity of the plasmonic structure via far- to near-field conversion, (c) efficient conversion of light reaching the rear of the solar cell to surface plasmons, and (d) efficient guiding of light from normally incident (TE) to laterally propagating (TM) SPP modes or optically guided (TE or TM) modes.

Currently 90% of the solar cell market is based on crystalline Si wafers. The general trend in the sector, hence, goes toward thinner cells. Unfortunately, Si is a weak absorber, especially near the indirect bandgap and substantial thicknesses (>200 μm) are needed for complete absorption of light. Therefore, techniques that can enhance light trapping in Si solar cells are certainly of interest in Si solar cells. Moreover, in the limit of extremely thin Si layers ($\ll 100$ μm) the common texturing of the top surface cannot be used. Plasmonic enhancement strategies based on nanoparticle scattering have been applied by a number of groups both theoretically and experimentally, and in particular, photocurrent enhancements up to 19% for planar wafers have been achieved [20,21]. These experiments, partially inspired by a seminal paper by Stuart

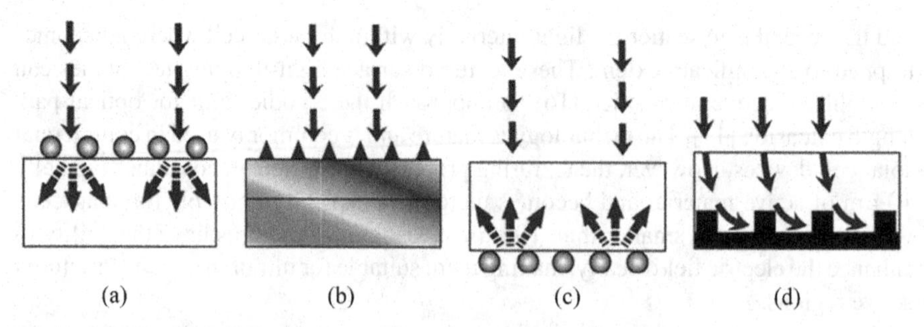

FIGURE 7.3 Four possible configurations for a plasmonic structure: (a) scattering of light, (b) local field enhancement, (c) (back) scattering of light that reaches the rear into surface plasmons, and (d) conversion of light into surface plasmon polariton resonance.

and Hall [22], indicated that the most promising enhancement mechanism based on plasmonic nanoparticles is given by their large scatter cross sections. Recent papers suggest possible geometries for optimized scatter cross sections [20, 23]. There are now also strong indications that particle scattering is probably most effective at the rear of the solar cell, as instead of broadband scattering, the absorption enhancement can be engineered in a narrow spectra range around the semiconductor bandgap, where the absorption coefficient drops significantly [24]. Although some experiments have made use of Au nanoparticles, cost and contamination issues typically favor Ag for the plasmonic metal. Ag also suffers less from absorption losses in the metal itself.

Next to silicon cells, a lot of effort goes into thin film solar cells, such as dye-sensitized, CIGS, or organic cells. In the case of organic cells, most of the light is already absorbed in a very thin (tens to a few hundred nanometers) film compared to conventional silicon cells. However, the required exciton splitting and the limited exciton diffusion lengths in these materials favor thinner cells. Also issues related to the series resistance require relatively thin cells. Due to this inherent trade-off that exists between optical and electrical properties, the ability to enhance the absorption of organic thin films could provide a significant boost to the efficiencies of solar cells. Electrical properties can be maintained, such that devices have a high fill factor, while absorption is increased, resulting in a direct increase of the short circuit current. The use of surface plasmons seems particularly well suited to fill this role. In an early work, it was shown that near-field plasmon-enhanced absorption was responsible for a 15% increase in photocurrent for a series-connected tandem cell compared with that of a single cell [25]. There, it was also shown that the near-field enhancement extended for distances up to ~10 nm and covered a broad range of wavelengths, even those significantly red-shifted from the plasmon energy.

Since then, there have been numerous reports in the literature where metal NPs have been incorporated into bulk heterojunction solar cells [26, 27]. In both cases, the Ag NPs were not embedded within the organic semiconductor layer, but either underneath [26] or within [27] the hole-injecting layer that is positioned between the anode indium tin oxide layer and the bulk heterojunction. In each of these reports, a photocurrent enhancement was seen, but the origin of the effect could not be explicitly identified, particularly due to the fact that the metal NPs were separated from the organic chromophores by distances too large such that near-field effects could play a significant role. Other studies focused on investigating the possibility of exploiting SPPs, via either nanohole arrays or grating structures [28–31].

In these studied SPPs enhanced effects were clearly shown. However, as usually a relatively simple model system was chosen, the reached efficiencies were moderate. An experiment where the efficiency of an already optimized cell can still be improved using plasmonic effects has yet to be demonstrated.

7.2.3 Plasmon-Enhanced Photodetectors

Next to solar cells, a lot of work has been published on the enhancement of photodetectors. While the main objective of plasmon-enhanced solar cells is to reduce the cost per cell and increase the conversion efficiency, the goal of plasmon-enhanced detectors is determined by its application. We will categorize the different

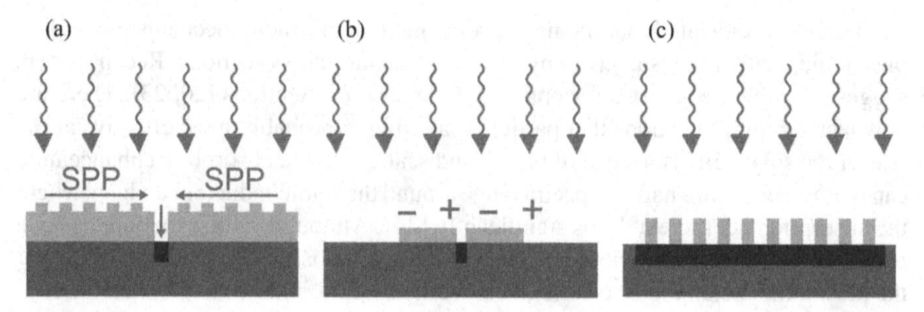

FIGURE 7.4 Geometries for plasmon-enhanced detectors. (a) Grating couplers to focus the generated SPPs inside a small-scale detector. (b) Particle antenna on a small-scale detector. (c) Metallic photonic crystal structures.

plasmon-enhanced detectors by the method used to focus the incident light in the active region of the detector. Different geometries are presented in Figure 7.4: (a) grating couplers to convert incident light to SPPs which are focused inside a small-scale detector, (b) particle antenna on a small-scale detector, and (c) metallic photonic crystal structures to enhance the photoresponse. The inclusion of an antenna or resonator can be used to enhance the photoresponse or to make the detector wavelength- and polarization-specific.

In the first geometry, a nanoscale semiconductor photodetector is discussed. Small-area photodetectors benefit from low noise levels, a low junction capacitance, and a possible high-speed operation. However, under the same optical power density, a lower output is obtained due to the decrease of the active area of the semiconductor detector. In order to increase the responsivity, and maintaining the advantages of nanoscale photodetectors, metallic antenna structures are fabricated on top of the small active semiconductor region. When the correct geometrical parameters are chosen, the plasmonic antennas convert the incident optical radiation into SPPs propagating in the sample plane, which are guided toward the active region of the photodetector, generating a photocurrent. This way, one can realize a small detector with a large collection area.

One of the first designs realizing a large photocurrent enhancement by means of an SPP antenna structure was realized by Ishi et al. [32]. They studied the photocurrent enhancement of a 300 nm diameter silicon Schottky photodiode equipped with a silver surface plasmon antenna. A schematic presentation of this so-called bull's-eye antenna detector is shown in Figure 7.5a. When the grating geometrical parameters are designed correctly so the corrugations can provide the correct momentum difference, coupling between incident photons and propagating surface plasmons on the silver surface is achieved. Under illumination of 840 nm laser light, a photocurrent enhancement of several 10-folds was measured compared with an unpatterned silver film. As the authors mention in Reference 32, at this wavelength the absorption length in silicon is approximately 10 μm and therefore propagating light cannot generate carriers efficiently within the thin absorption layer, leading to very low quantum efficiency. In future work, a great improvement could be made by resonantly coupling

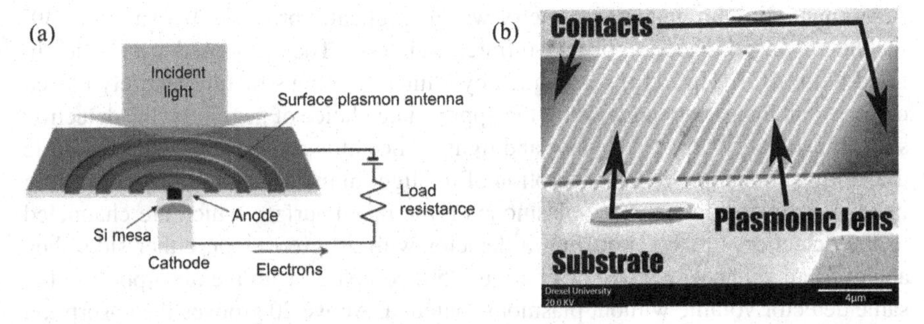

FIGURE 7.5 Plasmonic SPP antennas. (a) Circular bull's-eye antenna enhancing the photocurrent in a nanoscale Si Schottky diode detector. (b) Linear Bragg grating on a planar MSM detector.

the incident SPPs to a resonant cavity in which the resonant mode strongly overlaps with the nanoscale-active region of the photodetector. The limiting cutoff frequency due to the transit time of the carriers across the 200 nm thick depletion layer was estimated to be 80 GHz. This strongly suggests that plasmonic antennas can effectively increase the responsivity of a nanoscale photodetector while conserving the high-speed operation. The largest enhancements can be achieved with resonant systems, indicating that this approach only works at one specific excitation wavelength for which the metallic antenna is optimized. Later Laux et al., however, experimentally demonstrated that it was possible to let multiple antennas of this type overlap in space in order to separate light according to its wavelength and polarization [33].

In several designs, linear Bragg gratings were used to excite, guide, and focus plasmons on metal surfaces [34–36]. Shackleford et al. reported the experimental investigation of a metal–semiconductor–metal (MSM) photodetector where the metallic cathode and anode are each modified by the deposition of a series of 10 parallel linear corrugations [37]. These corrugations are designed to selectively couple 830 nm light to propagating SPPs, which in turn are guided to and concentrated in a slit MSM detector with an aperture of approximately 1 μm. A colored SEM picture of the device can be seen in Figure 7.5b. The planar structure of this device makes it very simple to fabricate and integrate in photonic circuitry. By performing current transient measurements for wavelengths between 800 and 870 nm, a maximum enhancement of approximately 1.9 was found for a wavelength of 830 nm. The FWHM of the transient response for both the MSM detector with and without corrugations was measured to be 15 ps. Also the decay time constant was found to be approximately equal, indicating that the integration of the plasmonic lens provides an increase of about 90% in responsivity without negatively affecting the speed of the detector. Also here, a larger increase in photocurrent enhancement could be obtained by designing the detector active area inside a plasmonic resonant cavity to which the incident SPPs can efficiently couple. White et al. calculated that a 10-fold enhancement of the optical absorption can be obtained by fabricating a detector inside a single slit by tuning the geometry in such a way that the resonant slit mode almost completely overlaps with the active semiconductor material, both spatially and spectrally [38].

A resonant slit with antenna geometry was theoretically proposed by Yu et al. [39] for the enhanced detection of mid-infrared radiation. They proposed a metallic slit filled with MCT (HgCdTe) surrounded by a linear grating structure, entirely placed on top of an insulating oxide. With the appropriate choice of geometry, the detection slit forms a Fabry–Perot resonator and light in the slit is resonantly enhanced, while the grating slits improve the absorption of the light in the detector slit by converting incident radiation into surface plasmons on the metal surface which are channeled to the detection slit. By modeling a detector with 20 grooves on either side, they achieved an enhancement factor of about 250, compared with the absorption in the same detector volume without plasmonic antenna. Above 20 grooves the absorption cross section saturates because of SPP propagation losses which are dominated by reradiation to free space.

As depicted in Figure 7.4b, the second way we presented in the introduction to enhance the photocurrent response is integrating a nanoscale structure on the detector that has a local plasmon resonance. Resonant antennas can confine strong optical fields inside a subwavelength volume. This was previously shown for dipole antennas [40] and bow-tie antennas [41]. By designing the structure in such a way that the region with highly confined optical fields overlaps with the active region of the photodetector, a strong enhancement of the photocurrent can be achieved. Both LSPR and local SPP resonances are used for this purpose. De Vlaminck et al. developed a technique for the local electrical transduction of a plasmon resonance in a single metal nanostructure [42]. A gold strip with nanoscale-transverse dimensions ranging between 200 and 500 nm was deposited on a free-standing GaAs photoconductor with a thickness of 100 nm. A SEM picture of the device can be found in Figure 7.6a. The combination of the Au strip on top of a thin GaAs layer results in a Fabry–Perot resonator for SPPs at the Au/GaAs interface where the edges of the particle act as reflective mirrors. Through comparison of the photoresponse of this detector with a reference detector without Au strip, they were able to study the effect of the resonance on the response of the photoconductor. The GaAs layer was explicitly chosen such that it probed the near field of the Au strip. It was found that the resonance wavelength of the metal structures scales linearly with the metal–semiconductor cavity size, which offers wide range tuneability. The signal enhancement for devices with widths between 230 and 400 nm was experimentally measured and varied between 1 and 3.2. If the devices and measurement setup discussed by De Vlaminck et al. could be optimized, devices like these photoconductors can be very promising for applications like integrated biosensors for refractive index sensing. Later, Barnard et al. also studied similar resonances in a tapered metal nanostrip antenna [43]. On a silicon-on-insulator (SOI) wafer, they fabricated an MSM detector, where the plasmonic nanoantenna was inserted in-between the two electrodes. A schematic overview of the sample is given in Figure 7.6b. Also in this case the (top) Si layer is very thin (40 nm), enabling to electrically probe the near field of the tapered gold nanostrip in function of its width. By scanning a monochromatic focused laser beam across the antenna, a photocurrent enhancement was observed each time the resonance condition for the Fabry–Perot resonator was satisfied. By performing photocurrent scans along the lengths of the tapered gold nanostrip, it was possible to match the experimental

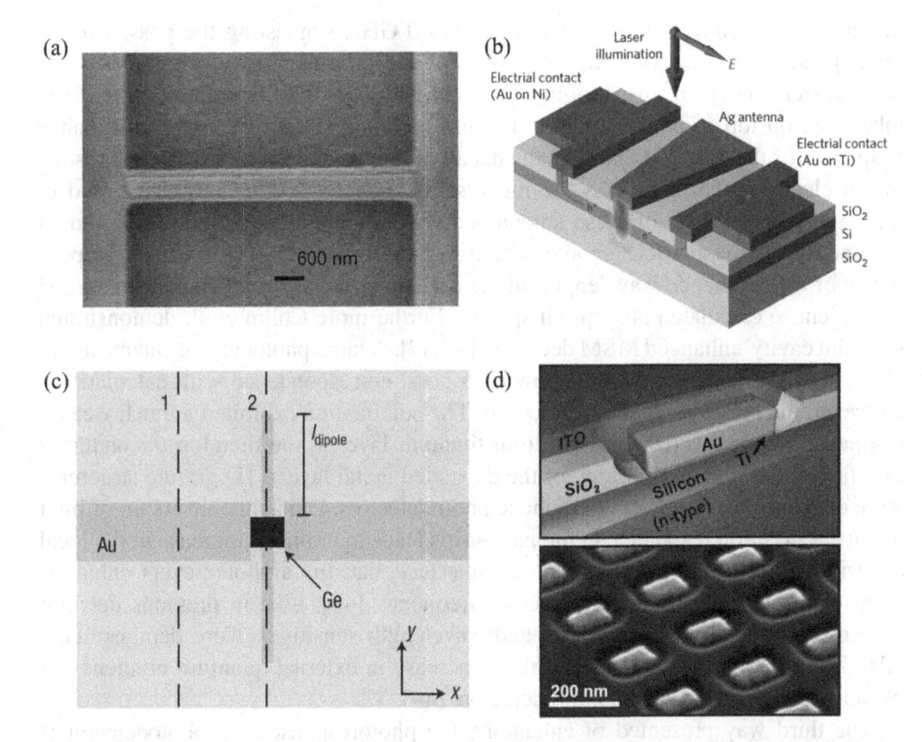

FIGURE 7.6 Photodetector enhancement by local surface plasmon resonances. (a) GaAs photoconductor with gold nanostrip. (b) Nanostrip resonance measurement by silicon MSM diode. (c) Half-wave antenna for photocurrent enhancement of a nanoscale Ge detector. (d) Nanoparticle plasmon-enhanced Schottky detector.

photocurrent enhancement map with the theoretical predictions based on a simple Fabry–Perot model. The work performed by Barnard et al. clearly showed that large micron-scale photodetectors are excellent tools to investigate the optical near-field properties of nanoscale plasmonic particles.

While the two previous detectors were mainly fabricated to perform an electrical near-field study of the properties of metal nanoparticles, nanoparticle resonators were—just like SPP antennas—also used to increase the photoresponse of small photodetectors. In 2008, Tang et al. demonstrated a very small germanium MSM detector with an active region of 150 × 60 × 80 nm which was equipped with a half-wave Hertz dipole antenna [44]. A schematic presentation is depicted in Figure 7.6c. The half-wave antenna was designed to concentrate the electric field inside the high-index germanium of the detector in the antenna gap. Polarization-dependent photocurrent measurements confirmed the half-wave behavior of the antenna. For a given device, the authors measured the photocurrent for the polarization along the antenna axis to be 20 times that for the perpendicular polarization for a very low bias voltage. Also spectral measurements showed a clear dipole peak at 1390 nm which was characteristic of the half-wavelength antenna. A rough approximation of the cutoff frequency was found to be over 100 GHz, suggesting the possibility of high-speed operation. Also Knight et al. reported on a nanoparticle antenna-based photodetector [45]. In this design, illustrated in Figure 7.6d, a nanorod antenna is fabricated on top of a silicon layer, forming a Schottky contact. At the resonance frequency of the nanorod antenna, the decay of the excited surface plasmons results in hot electrons that can cross the metal/semiconductor barrier and hence lead to a photocurrent. Arrays of these antenna-equipped Schottky detectors were studied in function of the particle geometry and light polarization. The expected dependency of the resonance wavelength on the particle size was found and showed good agreement to calculated absorption spectra. Furthermore Collin et al. demonstrated resonant cavity-enhanced MSM detectors [46,47]. A large photocurrent enhancement was found for TE polarization, showing a good correspondence with calculations. However, the plasmon-enhanced mode in TM polarization exhibited a much weaker resonance than expected, due to a thin titanium layer at the metal–semiconductor interface and to the morphology of the deposited metal layers. 1D grating structures were also integrated on top of graphene photodetectors. Efficient conversion of light to a local plasmon resonance in the nanostrips leads to dramatic increase in the local electric field at the electrode–graphene interface, causing a photocurrent enhancement of more than 20 at the resonance frequency [48]. Like in previous detector designs, here also the polarization and wavelength sensitivity were demonstrated. Also Liu et al. demonstrated a dramatic increase in external quantum efficiency of plasmon-enhanced graphene photodetectors [49].

The third way presented of enhancing the photoresponse of a photodetector is the inclusion of a metallic photonic crystal on the detector area or arranging the detector structures in a periodic way, forming a photonic crystal structure. Like other approaches, the idea of integrating a resonant structure on the detector is increasing the interaction length between the incoming light and the active semiconductor region. This is interesting for thin film semiconductor detectors with a large absorption length like, for example, silicon. Instead of passing only once through the absorbing layer, at resonance the light passes multiple times before being absorbed in the semiconductor and the metal or being reradiated, increasing the probability of being absorbed in the active layer. One way of making use of metallic photonic crystal structures is introducing defect structures inside a 1D or 2D metallic photonic crystal, creating a resonant cavity with a high field enhancement. A more common approach is to use photonic crystal structures to efficiently couple incident light to an in-plane confined resonant mode which overlaps with the underlying active area of the photodetector. When designing a photonic crystal-enhanced photodetector, several aspects have to be considered: The resonance wavelength has to match the maximum absorption region of the active semiconductor material, the crystal needs both excellent incident light coupling and in-plane SPP confinement properties and the plasmonic fields at resonance need to spatially match with the active region of the photodetector. A thorough investigation of these concepts was theoretically described by Rosenberg et al. [50] and also realized experimentally [51]. The calculated band structure of the square crystal lattice of square holes in the metal layer revealed no band gaps but several flat-band regions. A SEM picture of the device under study can be seen

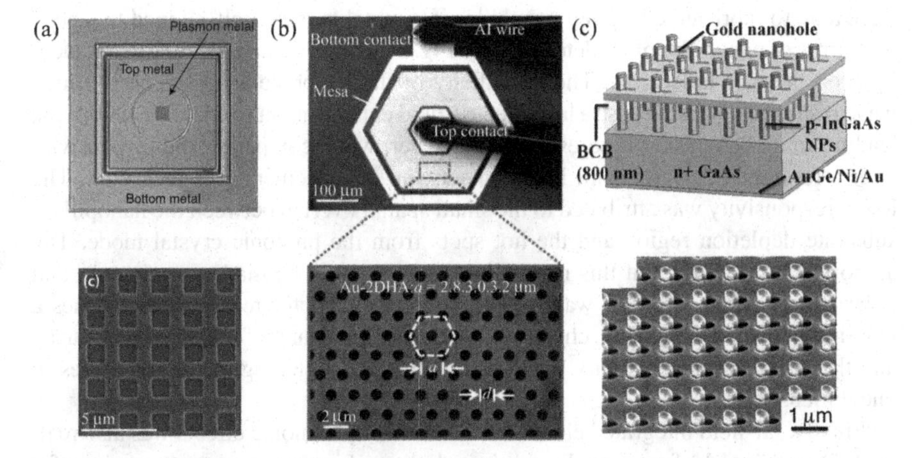

FIGURE 7.7 Photonic crystal-based enhanced photodetectors. Top row: the entire device, bottom row: a zoom of the photonic crystal lattice. (a) Square lattice photonic crystal for MIR quantum dot detector. (b) Hexagonal lattice for MIR quantum dot detector. (c) Periodically ordered array of nanopillar photodetectors with self-aligned hole array.

in Figure 7.7a. These band-edge modes have a group velocity close to zero and consequently the light propagates very slowly and is confined within the patterned region. The authors realized the developed concept on a dots-in-a-well photodetector material [51]. FDTD simulations revealed a fundamental plasmonic dipole-like mode at 7.8 μm and higher-order modes of the in-plane square lattice photonic crystal around 5.5 μm. By varying the lattice constant between 1.83 and 2.38 μm, the peak wavelength could be shifted from 5.5 to 7.2 μm. The sensor could be made polarization-dependent by stretching the square holes in one dimension, splitting the resonance in two well-defined peaks. By comparing the responsivity of the photonic crystal photodetector with an unpatterned detector, an enhancement of 5 was achieved at $T = 77$ K. A similar detector was developed by Chang et al. [52]. In this publication a hexagonal lattice of etched circles was processed in a 50 nm thick gold film on an InGaAs quantum well detector. The detector was optimized for dual band detection at 5 and 9 μm wavelengths. The resonance wavelength was found to increase linearly with the lattice constant, while the size of the holes determines the light transmission via evanescent tunneling through the thin metallic 2D holes. A maximum photoresponse enhancement of 2.3 was found at 8.8 μm for a lattice constant of 3.2 μm and a hole diameter of 1.6 μm ($T = 77$ K). Several other research groups also studied metallic photonic crystal-enhanced photodetectors based on InAs/InGaAs quantum wells or quantum dots [53–55].

In the previous papers, a large-scale metallic photonic crystal was fabricated on top of a large-volume detector for mid-infrared light detection. Another way to use a photonic crystal structure is to fabricate small-scale detectors and position them in a periodic pattern. Senanayake et al. demonstrated a nanopillar-based plasmon-enhanced photodetector array operating in the NIR spectral range [56]. The authors were able to fabricate elongated nanoholes in a metal surface, self-aligned to a periodic array of nanopillar photodetectors (see Fig. 7.7c). SPPs can be excited by incident light on the nanohole array. The periodicity of the nanohole array supports Bloch wave resonances which cause a high electric field enhancement inside the nanopillars, leading to an enhanced photoresponse. However, when comparing the responsivity with a reference detector with ITO contacts, no enhancement was measured. The lower responsivity was attributed to the small spatial overlap between the nanopillar–substrate depletion region and the hot spots from the photonic crystal mode. The authors demonstrated that this nanopillar detector could be made wavelength- and polarization-dependent. The wavelength where the detector responsivity reaches a maximum can be tuned by changing the lattice constant of the nanohole array and the polarization sensitivity is determined by the stretching of the nanoholes in one dimension.

Finally, far-field integrated electrical detection of plasmonic effects was also realized. Dunbar et al., for example, fabricated slit and groove structures on top of a standard CMOS camera and achieved an eight times enhancement in the optical transmission [57]. Also several other applications electrically detected plasmons in the far field; for example, in biosensor applications, Jonsson et al. studied supported lipid bilayers by detecting LSPR changes [58].

7.2.4 Waveguide-Integrated Surface Plasmon Polariton Detectors

Plasmonic waveguides have been studied for a variety of applications, ranging from detection of biological properties to optical communication. For most applications, it would be useful to be able to excite and detect SPPs locally in the metallic waveguide by means of an integrated plasmon source and detector, respectively. Depending on the application, the important aspects of these electro-optical transducers are the efficiency, speed, and scalability. Here we will present various ways of detecting SPPs on chip. Direct electrical detection of SPPs propagating on metallic waveguides has been demonstrated with several types of detectors and waveguides. First, we will discuss SPP detection with both organic and solid-state semiconductors. Later we will present SPP detection with a superconducting detector and last the detection of waveguide modes in dielectric waveguides with plasmon-assisted photodetectors.

The first demonstration of direct electrical detection of waveguided SPPs was by Ditlbacher et al. [59]. The authors fabricated a $150 \times 500\ \mu m^2$ large organic photodiode on top of a 100 nm thick silver film that serves as both the plasmonic waveguide and the bottom contact of the diode. A schematic overview can be seen in Figure 7.8a. In the silver film, a grating structure and a slit were etched as SPP excitation structures. SPPs are generated by coupling incident laser light to SPPs at the slit or grating. They propagate along the silver waveguide and excite excitons in the organic materials when entering the diode, giving rise to an electric current that can be assessed directly. By scanning the focused laser beam across the device while measuring the photocurrent, photocurrent maps were generated, showing a strongly enhanced photocurrent when the laser spot scans across the SPP excitation structures, due to the excitation, propagation, and subsequent electrical detection of guided SPPs.

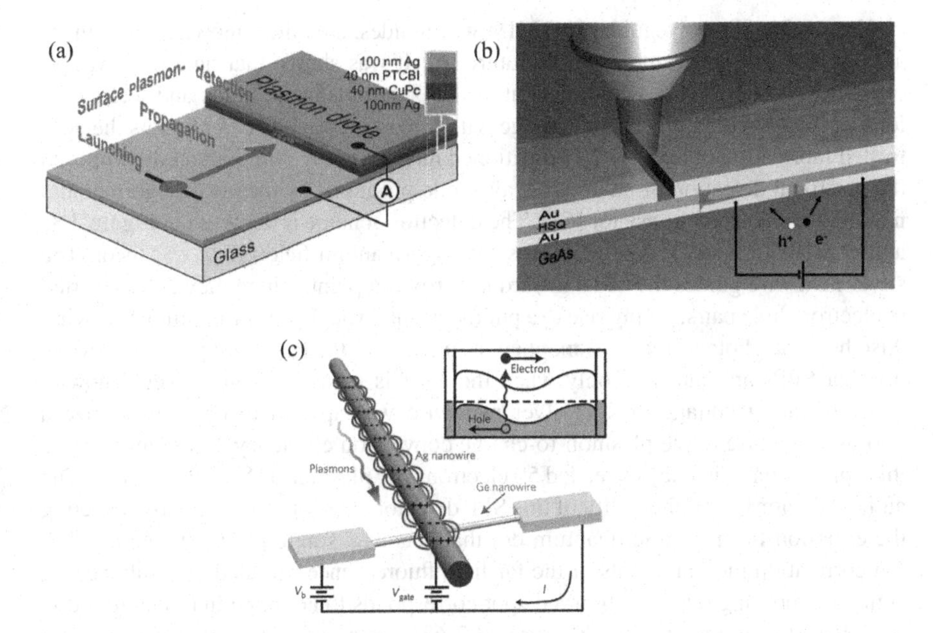

FIGURE 7.8 Integrated SPP detection methods. (a) Organic photodiode SPP detector. (b) GaAs MSM detector inside MIM waveguide. (c) Ge nanowire detector for plasmons propagating on silver nanowires.

Blocking regions were inserted between the excitation and detection region to exclude that the photocurrent generation was due to scattered light. Organic photodiodes offer a large processing flexibility and can be deposited on cheap substrates. However, in terms of speed, efficiency, and scalability, they cannot compete with solid-state photodetectors at this time.

In 2009, several research groups started working on integrating solid-state photodetectors inside plasmonic waveguides. We demonstrated the electrical detection of SPPs in MIM waveguides by means of a nanoslit GaAs MSM detector [60]. MIM waveguides offer the prospects of combining a high confinement of the plasmonic mode with a reasonably long propagation distance [7]. A schematic representation of the device under study can be seen in Figure 7.8b. In the bottom layer of the MIM waveguide, a subwavelength detection slit is processed, forming the two contacts of the nanoscale MSM detector. SPPs are generated by focusing a laser beam on the subwavelength slit in the top metal layer; they propagate through the MIM waveguide and couple to a local mode at the detection slit, which is strongly absorbed by the GaAs layer. 2D photocurrent scans were performed for both TE and TM polarization, revealing only a high photocurrent for TM polarization, consistent with theory. Photocurrent line scans were acquired for different distances between the excitation slit and the detection slit in order to study the SPP decay along the waveguide. For a gold waveguide with 100 nm thick cured HSQ insulating layer, 1/e decay lengths of 3.5–9.5 μm were found for, respectively, 660–870 nm excitation wavelengths. Because of their extremely low capacitance and short transit times, these detectors are very suitable for high-speed integrated electrical detection of SPPs.

Next to flat film waveguides and MIM waveguides, metallic nanowire waveguides also were extensively studied in literature [6]. It was shown that nanowire waveguides too can have a high confinement of the electromagnetic fields and maintain a long propagation length. The challenge with nanowire waveguides remains the controlled fabrication of horizontally positioned nanowires on chip. Falk et al. proposed the electrical detection of nanowire waveguide plasmons by means of a germanium nanowire field-effect transistor [61]. The detection scheme is depicted in Figure 7.8c and consists of a silver nanowire crossing a germanium field-effect transistor. The silver nanowire guides the SPPs toward the crossing point, where they are converted to electron–hole pairs, giving rise to a photocurrent through the germanium nanowire. Also here the photocurrent was mapped in function of the laser spot position, revealing that SPPs are launched only when the laser is focused on the silver nanowire ends. A gating voltage on the silver nanowire was applied in order to realize a 300-fold increase in the plasmon-to-charge conversion efficiency. For some devices this conversion efficiency exceeded 50 electrons per plasmon at 1 V bias voltage. The authors demonstrated the utility of this SPP detector concept by electrically detecting the emission from a CdSe quantum dot that acts as a single-plasmon source. Photon correlation measurements of the far-field fluorescence revealed an antibunching signature, proving that the observed spot corresponds to an individual quantum dot. Together with an integrated SPP source [62, 63], the previous two approaches could lead to a plasmonic optoelectronic circuit. Also for plasmon biosensing, for example, like described by Dostalek et al. [64], integrated SPP sources and detectors would be of great interest in order to achieve a fully integrated plasmon biosensor.

The last direct semiconductor SPP detector which will be discussed in this chapter was demonstrated by Akbari et al. [65, 66]. The authors propose detection of SPPs with a Schottky photodetector. A gold stripe is deposited on an n-type silicon layer, functioning simultaneously as the waveguide and a Schottky contact. An ohmic contact is fabricated at the bottom of the wafer. The sample was cleaved perpendicular to the stripe axis and the $as_b{}^0$ SPP mode at the gold–silicon interface is excited by end-fire coupling with a polarization maintaining tapered optical fiber at the side of the sample. The excitation energy is chosen to be smaller than the bandgap energy so no direct electron–hole pair generation across the bandgap occurs. Absorption of the SPPs in the metal waveguide causes hot electrons which can cross the Schottky barrier if the excitation energy is large enough. These electrons can be collected, giving rise to a photocurrent. To prove the $as_b{}^0$ SPP mode was excited, photocurrent measurements were performed for TE and TM polarization, revealing that only a high photocurrent was obtained for TM-polarized light. This type of detector, however, is not confined to a small area, since a distance of several times the propagation length is necessary if one wants to achieve complete absorption of the optical power in the injected plasmon pulse.

Also in 2009, a completely different approach to integrated detection of SPPs was published by Heeres et al. [67]. The authors coupled a plasmon waveguide to a superconducting single-photon detector, thereby demonstrating on-chip electrical detection of single plasmons. The photodetector consisted out of a meandering NbN wire. The wire enters the superconductive state below approximately 9 K. If a bias current close to the critical current is applied, a single photon can create a local region in the normal resistive state, which is detected as a voltage pulse at the terminals of the wire. A flat polycrystalline gold waveguide was deposited on top of the detectors, separated by a thin dielectric, and again grating structures were used for achieving SPP excitation. Measuring the pulse frequency in function of the laser spot position confirmed plasmon excitation, propagation, and subsequent electrical detection by the superconductor detector. Time correlation measurements using a single-photon source proved the single-plasmon sensitivity of this type of detector. This publication demonstrates a very sensitive plasmon detector with an excellent time resolution which can be widely applied for research purposes. However, for commercial applications, the low-temperature operation still forms a large barrier.

Not only are plasmon detectors useful for the detection of free-space radiation or waveguided SPPs, but they can also be employed for integrating dielectric waveguides with small, high-speed detectors, as plasmonic detectors can considerably shorten the length over which the full optical power of the light pulse is absorbed. Ren et al. proposed a Ge-on-SOI MSM photodetector with interdigitated electrodes to obtain a large TM enhancement at a wavelength of 1550 nm [68]. In Figure 7.9a both a schematic presentation of the device and a SEM picture are shown. A TE/TM-polarized optical mode propagates through the silicon core layer (220 nm thick). By calculating the correct grating pitch, the detector can be optimized for 1550 nm light detection. Under TE and TM illumination, the quantum efficiencies are found to be 14.1% and 86.7%, revealing a very high responsivity for TM polarization. FDTD simulations were performed to confirm that the interdigitated electrodes function as a 1D grating coupler above the detector's depletion layer. Also the time response of

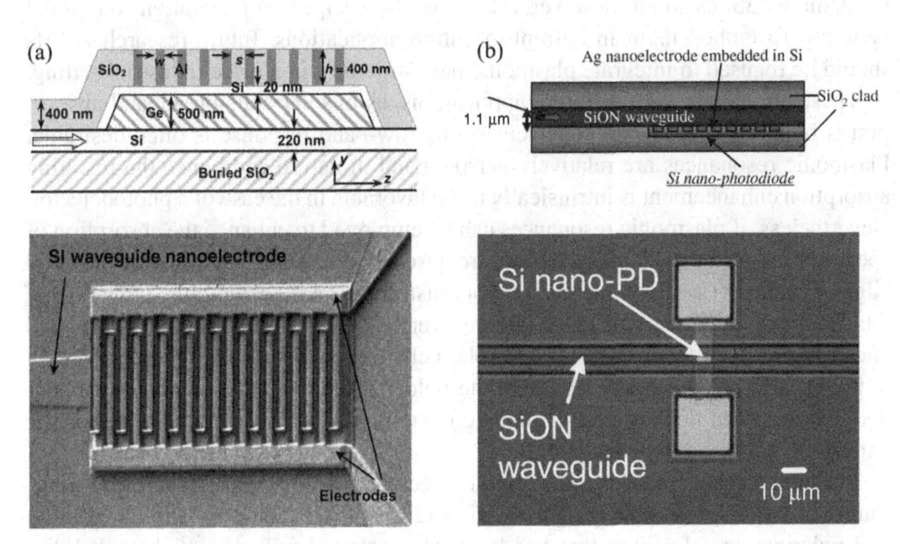

FIGURE 7.9 Plasmon-enhanced detectors for detection of dielectric waveguide optical modes. Top row: device cross-sections. Bottom row: microscope pictures. (a) Plasmon-enhanced germanium detector fabricated on a silicon waveguide. (b) Plasmon-enhanced silicon detector on a SiON waveguide.

the detector was studied by the authors. A 3 dB bandwidth of 11.4 and 15.6 GHz was found for the TM and TE modes, respectively, meaning there is a trade-off between the enhancement in responsivity and the response speed. Also Fujikata et al. presented a similar detection scheme [69]. A low-loss SiON waveguide in-between SiO_2 cladding layers was studied. On the silicon-active photodetector surface, silver grating MSM detector-interdigitated electrodes were processed. The resulting device can be seen in Figure 7.9b. An 850 nm laser light was introduced in the waveguide via butt-coupling with an optical fiber. A 10% quantum efficiency was achieved for TM polarization, which was two to three times larger than for TE polarization due to the surface plasmon resonance-enhanced near-field optical coupling between the waveguide and the Si layer. Also here the time response of the detector was measured, reaching a 17 ps FWHM of the generated electrical pulse following a 780 nm optical pulse smaller than 2 ps. The authors applied their design to a prototype of a large-scale integration on-chip optical clock system. They demonstrated a 5 GHz optical clock circuit operation with a four-branching H-tree structure, validating that plasmon-enhanced integrated photodetection for dielectric waveguide has very promising prospects for future on-chip optical circuits based on photonics and plasmonics.

7.3 OUTLOOK

The unique properties of plasmonic nanostructures to guide and focus light into deep-subwavelength volumes have led to a number of demonstrations where plasmonic nanoantennas strongly enhance the absorption in photovoltaic cells and photodetectors for free-space radiation or integrated in plasmonic or photonic waveguide. Most of the demonstrations so far, however, are proof-of-principle demonstrations on model systems. To embed them in current or future applications, future research efforts should be focused to integrate plasmonic nanostructures in a more industrial setting.

These are clearly very different applications, as in solar cells broadband absorption is required while in photodetectors a narrow-band response is often desirable. Plasmonic resonances are relatively narrow band in nature and hence the resulting absorption enhancement is intrinsically more favorable in the case of a photodetector. Nevertheless, if plasmonic resonances can be employed to enhance the absorption of specific parts of the solar spectrum, where absorption is weak, such as near the band edge of (indirect) semiconductors, without disturbing or shadowing the other part of the spectrum, they can still be considered very useful in the quest for thinner and cheaper cells. The field of plasmonic solar cells has been very active recently, with a number of companies also active in the field. In order to be successfully embedded in a solar cell flow, issues related to contamination and cost-effective plasmonic patterning will need to be tackled.

Plasmon (or antenna)-enhanced photodetectors have a bright future in ultrafast, multi- or hyperspectral imaging in a wide range of wavelengths. The wavelength and polarization selectivity that has been demonstrated will with high probability find its way in several applications. Moreover, recent results on plasmonic-enhanced absorption in graphene show a route toward highly confined and atomically thin photodetectors that can be supported by an arbitrary substrate. More problematic is the integration of plasmonic effects in conventional silicon photodetectors, where contamination issues and the limited temperature budget are a potential road block.

Integrating plasmonic detectors with plasmonic and photonic waveguides is a very promising application, as the waveguided light can be captured in extremely small volumes, leading to potentially very fast, sensitive, and dense photodetectors. This is interesting for a number of applications, such as highly multiplexed integrated surface plasmon resonance-based biosensing and fast detection in integrated optical communication schemes. In the latter application, however, the lack of electrically excitable coherent plasmon sources and the large losses present in plasmonic waveguides prevent a pure plasmonic information routing scheme. However, when exploiting the nearly loss-free propagation in silicon or silicon nitride photonic waveguides, the ability of plasmonic nanostructures to reduce the interaction length between a photodetector and the waveguide will lead to faster detectors. However, in this case also, issues related to integration in an industrial fabrication platform need to be considered.

REFERENCES

1. Boltasseva A, Nikolajsen T, Leosson K, Kjaer K, Larsen MS, Bozhevolnyi SI (2005) Integrated optical components utilizing long-range surface plasmon polaritons. *J. Lightw. Technol.* 23(1): 413–422.
2. Berini P, Charbonneau R, Lahoud N, Mattiussi G (2005) Characterization of long-range surface-plasmon-polariton waveguides. *J. Appl. Phys.* 98: 043109.
3. Bozhevolnyi SI, Volkov VS, Devaux E, Ebbesen T (2005) Channel plasmon-polariton guiding by subwavelength metal grooves. *Phys. Rev. Lett.* 95: 046802.
4. Moreno E, Rodrigo SG, Bozhevolnyi SI, Martin-Moreno L, Garcia-Vidal FJ (2008) Guiding and focusing of electromagnetic fields with wedge plasmon polaritons. *Phys. Rev. Lett.* 100: 023901.
5. Maier SA, Kik PG, Atwater HA, Meltzer S, Harel E, Koel BE, Requicha AG (2003) Local detection of electromagnetic energy transport below the diffraction limit in metal nanoparticle plasmon waveguides. *Nature* 2: 229–232.
6. Oulton RF, Sorger VJ, Genov DA, Pile DFP, Zhang X (2008) A hybrid plasmonic waveguide for subwavelength confinement and long-range propagation. *Nat. Photonics* 2: 496–500.
7. Dionne JA, Lezec HJ, Atwater HA (2006) Highly confined photon transport in subwavelength metallic slot waveguides. *Nano Lett.* 6: 1928–1932.
8. Nagae M (1972) Response time of metal-insulator-metal tunnel junctions. *Jpn. J. Appl. Phys.* 11: 1611–1621.
9. Soole JBD, Lamb RN, Hughes HP, Apsley N (1986) Surface plasmon enhanced photoconductivity in planar metal-oxide-metal tunnel junctions. *Solid State Commun.* 59: 607–611.

10. Berthold K, Höpfel RA, Gornik E (1985) Surface plasmon polariton enhanced photoconductivity of tunnel junctions in the visible. *Appl. Phys. Lett.* 46(7): 626–628.

11. Möller R, Albrecht U, Boneberg J, Koslowski B, Leiderer P, Dransfeld K (1991) Detection of surface plasmons by scanning tunneling microscopy. *J. Vac. Sci. Technol. B* 9: 506.

12. Hobbs PCD, Laibowitz RB, Libsch FR (2005) Ni-NiO-Ni tunnel junctions for terahertz and infrared detection. *Appl. Opt.* 44: 6813.

13. Hobbs PCD, Laibowitz RB, Libsch FR, LaBianca NC, Chiniwalla PP (2007) Efficient waveguide-integrated tunnel junction detectors at 1.6 μm. *Opt. Express* 15: 16376.

14. Ittah N, Gilad N, Yutsis I, Selzer Y (2009) Measurement of electronic transport through 1G0 gold contacts under laser irradiation. *Nano Lett.* 2009: 1615.

15. Ittah N, Selzer Y (2011) Electrical detection of surface plasmon polaritons by 1G0 gold quantum point contacts. *Nano Lett.* 11: 529.

16. Campbell P, Green MA (1987) Light trapping properties of pyramidally textured surfaces. *J. Appl. Phys.* 62(1): 243–249.

17. Yablonovitch E (1982) Statistical ray optics. *J. Opt. Soc. Am.* 72(7): 899–907.

18. Ferry VE, Sweatlock LA, Pacifici D, Atwater HA (2008) Plasmonic nanostructure design for efficient light coupling into solar cells. *Nano Lett.* 8(12): 4391–4397.

19. Maier SA (2007) *Plasmonics: Fundamentals and Applications*. New York: Springer.

20. Hägglund C, Zäch M, Petersson G, Kasemo B (2008) Electromagnetic coupling of light into a silicon solar cell by nanodisk plasmons. *Appl. Phys. Lett.* 92: 053110.

21. Rockstuhl C, Fahr S, Lederer F (2008) Absorption enhancement in solar cells by localized plasmon polaritons. *J. Appl. Phys.* 104: 123102.

22. Stuart HR, Hall DG (1998) Island size effects in nanoparticle-enhanced photodetectors. *Appl. Phys. Lett.* 73: 3815.

23. Catchpole K, Polman A (2008) Design principles for particle enhanced solar cells. *Appl. Phys. Lett.* 93(19): 191113.

24. Ouyang Z, Pillai S, Beck F, Kunz O, Varlamov S, Catchpole KR, Campbell P, Green MA (2010) Effective light trapping in polycrystalline silicon thin-film solar cells by means of rear localized surface plasmons. *Appl. Phys. Lett.* 96(26): 261109.

25. Rand BP, Peumans P, Forrest SR (2004) Long-range absorption enhancement in organic tandem thin-film solar cells containing silver nanoclusters. *J. Appl. Phys.* 96(12): 7519.

26. Morfa AJ, Rowlen KL, Reilly TH, Romero MJ, van de Lagemaat J (2008) Plasmon-enhanced solar energy conversion in organic bulk heterojunction photovoltaics. *Appl. Phys. Lett.* 92(1): 013504.

27. Kim S, Na S, Jo J, Kim D, Nah Y (2008) Plasmon enhanced performance of organic solar cells using electrodeposited Ag nanoparticles. *Appl. Phys. Lett.* 93(7): 073307.

28. Mapel JK, Singh M, Baldo MA, Celebi K (2007) Plasmonic excitation of organic double heterostructure solar cells. *Appl. Phys. Lett.* 90(12): 121102.

29. Tvingstedt K, Persson N, Inganas O, Rahachou A, Zozoulenko IV (2007) Surface plasmon increase absorption in polymer photovoltaic cells. *Appl. Phys. Lett.* 91(11): 113514.

30. Reilly TH, van de Lagemaat J, Tenent RC, Morfa AJ, Rowlen KL (2008) Surface-plasmon enhanced transparent electrodes in organic photovoltaics. *Appl. Phys. Lett.* 92(24): 243304.
31. Lindquist NC, Luhman WA, Oh S, Holmes RJ (2008) Plasmonic nanocavity arrays for enhanced efficiency in organic photovoltaic cells. *Appl. Phys. Lett.* 93(12): 123308.
32. Ishi T, Fujikata J, Makita K, Baba T, Ohashi K (2005) Si nano-photodiode with a surface plasmon antenna. *Jpn. J. Appl. Phys.* 44(12): 364–366.
33. Laux E, Genet C, Skauli T, Ebbesen TW (2008) Plasmonic photon sorters for spectral and polarimetric imaging. *Nat. Photonics* 2: 161–164.
34. Ropers C, Neacsu CC, Elsaesser T, Albrecht M, Raschke MB, Lienau C (2007) Grating-coupling of surface plasmon onto metallic tips: a nanoconfined light source. *Nano Lett.* 7(9): 2784–2788.
35. Weeber J-C, Gonzalez MU, Baudrion A-L, Dereux A (2005) Surface plasmon routing along right angle bent metal strips. *Appl. Phys. Lett.* 87: 221101.
36. Chen C, Verellen N, Lodewijks K, Lagae L, Maes G, Borghs G, Van Dorpe P (2010) Groove-gratings to optimize the electric field enhancement in a plasmonic nanoslit-cavity. *Appl. Phys. Lett.* 108: 034319.
37. Shackleford JA, Grote R, Currie M, Spanier JE, Nabet B (2009) Integrated plasmonic lens photodetector. *Appl. Phys. Lett.* 94: 083501.
38. White JS, Veronis G, Yu Z, Barnard ES, Chandran A, Fan S, Brongersma ML (2009) Extraordinary optical absorption through subwavelength slits. *Opt. Lett.* 34(5): 686–688.
39. Yu Z, Veronis G, Fan S, Brongersma ML (2006) Design of midinfrared photodetectors enhanced by surface plasmons on grating structures. *Appl. Phys. Lett.* 89: 151116.
40. Muhlschlegel P, Eisler H-J, Martin OJF, Hecht B, Pohl DW (2005) Resonant optical antennas. *Science* 308(5728): 1607–1609.
41. Kinkhabwala A, Yu Z, Fan S, Avlasevich Y, Mullen K, Moerner WE (2009) Large single-molecule fluorescence enhancements produced by a bowtie nanoantenna. *Nat. Photonics* 3: 654–657.
42. De Vlaminck I, Van Dorpe P, Lagae L, Borghs G (2007) Local electrical detection of single nanoparticle plasmon resonance. *Nano Lett.* 7(3): 703–706.
43. Barnard ES, Pala RA, Brongersma ML (2011) Photocurrent mapping of near-field optical antenna resonances. *Nat. Nanotechnol.* 6: 588–593.
44. Tang L, Kocabas SE, Latif S, Okyay AK, Ly-Gagnon D-S, Saraswat KC, Miller DAB (2008) Nanometre-scale germanium photodetector enhanced by a near-infrared dipole antenna. *Nat. Photonics* 2: 226–229.
45. Knight MW, Sobhani H, Nordlander P, Halas NJ (2011) Photodetection with active optical antennas. *Science* 332: 702–704.
46. Collin S, Pardo F, Teissier R, Pelouard J-L (2004) Efficient light absorption in metal-semiconductor-metal nanostructures. *Appl. Phys. Lett.* 85(2): 194–196.
47. Collin S, Pardo F, Pelouard J-L (2003) Resonant-cavity-enhanced subwavelength metal-semiconductor-metal photodetector. *Appl. Phys. Lett.* 83(8): 1521–1523.
48. Echtermeyer TJ, Britnell L, Jasnos PK, Lombardo A, Gorbachev RV, Grigorenko AN, Geim AK, Ferrari AC, Novoselov KS (2011) Strong plasmonic enhancement of photovoltage in graphene. *Nat. Commun.* 2(458): 1–5.

49. Liu Y, Cheng R, Liao L, Zhou H, Bai J, Liu G, Liu L, Huang Y, Duan X (2011) Plasmon resonance enhanced multicolour photodetection by graphene. *Nat. Commun.* 2: 579.

50. Rosenberg J, Shenoi RV, Krishna S, Painter O (2010) Design of plasmonic photonic crystal resonant cavities for polarization sensitive infrared photodetectors. *Opt. Express* 18(4): 3672–3686.

51. Rosenberg J, Shenoi RV, Vandervelde TE, Krishna S, Painter O (2009) A multispectral and polarization-selective surface-plasmon resonant midinfrared detector. *Appl. Phys. Lett.* 95: 161101.

52. Chang C-C, Sharma YD, Kim Y-S, Bur JA, Shenoi RB, Krishna S, Huang D, Lin S-Y (2010) A surface plasmon enhanced infrared photodetector based on InAs quantum dots. *Nano Lett.* 10(5): 1704–1709.

53. Lee SC, Krishna S, Brueck SRJ (2011) Plasmonic-enhanced photodetectors for focal plane arrays. *IEEE Photonics Technol. Lett.* 23(14): 935–937.

54. Lee SC, Krishna S, Brueck SRJ (2009) Quantum dot infrared photodetector enhanced by surface plasma wave excitation. *Opt. Express* 17(25): 23160–23168.

55. Wu W, Bonakdar A, Mohseni H (2010) Plasmonic enhance quantum well infrared photodetector with high detectivity. *Appl. Phys. Lett.* 96: 161107.

56. Senanayake P, Hung C-H, Shapiro J, Lin A, Liang B, Williams BS, Huffacker DL (2011) Surface plasmon-enhanced nanopillar photodetectors. *Nano Lett.* 11(12): 5279–5283.

57. Dunbar LA, Guillaumee M, De Leon-Perez F, Santschi C, Grenet E, Eckert R, Lopez-Tejeira F, Garcia-Vidal FJ, Martin-Moreno L, Stanley RP (2009) Enhanced transmission from a single subwavelength slit aperture surrounded by grooves on a standard detector. *Appl. Phys. Lett.* 95: 011113.

58. Jonsson MP, Jonsson P, Dahlin AB, Hook F (2007) Supported lipid bilayer formation and lipid-membrane-mediated biorecognition reactions studied with a new nanoplasmonic sensor template. *Nano Lett.* 7(11): 3462–3468.

59. Ditlbacher H, Aussenegg FR, Krenn JR, Lamprecht B, Jakopix B, Leising G (2006) Organic diodes as monolithically integrated surface plasmon polariton detectors. *Appl. Phys. Lett.* 89: 161101.

60. Neutens P, Van Dorpe P, De Vlaminck I, Lagae L, Borghs G (2009) Electrical detection of confined gap plasmons in metal–insulator–metal waveguides. *Nat. Photonics* 3: 283–286.

61. Falk AL, Koppens FHL, Yu CL, Kang K, de Leon Snapp N, Akimov AV, Jo M-H, Lukin MD, Park H (2009) Near-field electrical detection of optical plasmons and single-plasmon sources. *Nat. Phys.* 5: 475–479.

62. Neutens P, Lagae L, Borghs G, Van Dorpe P (2010) Electrical excitation of confined surface plasmon polaritons in metallic slot waveguides. *Nano Lett.* 10(4): 1429–1432.

63. Walters RJ, van Loon RVA, Brunets I, Schmitz J, Polman A (2009) A silicon-based electrical source of surface plasmon polaritons. *Nat. Mater.* 9: 21–25.

64. Dostalek J, Ctyroky J, Homola J, Brynda E, Skalsky M, Nekvindova P, Spirkova J, Skvor J, Schrofel J (2001) Surface plasmon resonance biosensor based on integrated optical waveguide. *Sens. Act. B* 76: 8–12.

65. Akbari A, Berini P (2009) Schottky contact surface-plasmon detector integrated with an asymmetric metal stripe waveguide. *Appl. Phys. Lett.* 95: 021104.

66. Akbari A, Tait RN, Berini P (2010) Surface plasmon waveguide Schottky detector. *Opt. Express* 18(8): 8505–8514.

67. Heeres RW, Dorenbos SN, Koene B, Solomon GS, Kouwenhoven LP, Zwiller V (2010) On-chip single plasmon detection. *Nano Lett.* 10: 661–664.

68. Ren F-F, Ang K-W, Song J, Yu M, Lo G-Q, Kwong D-L (2010) Surface plasmon enhanced responsivity in a waveguided germanium metal-semiconductor-metal photodetector. *Appl. Phys. Lett.* 97: 091102.

69. Fujikata J, Nose K, Ushida J, Nishi K, Kinoshita M, Shimizu T, Ueno T, Okamoto D, Gomyo A, Mizuno M, Tsuchizawa T, Watanabe T, Yamada K, Itabashi S, Ohashi K (2008) Waveguide-integrated Si nano-photodiode with surface-plasmon antenna and its applications to on-chip optical clock distribution. *Appl. Phys. Express* 1: 022001.

8

Terahertz Plasmonic Surfaces for Sensing

STEPHEN M. HANHAM
Centre for Terahertz Science and Engineering Imperial College London, London, UK

STEFAN A. MAIER
Experimental Solid State Group, Department of Physics, Imperial College London, London, UK

The terahertz (THz) band occupies the region in the electromagnetic spectrum between microwaves and infrared, lying at the intersection of the electronic and optical regimes. It is broadly defined as the electromagnetic energy with wavelength range between 1000 and 100 μm (300 GHz and 3 THz). The band has attracted significant attention in recent years due to its scientific richness and promise of many exciting new applications which exploit the unique properties of this submillimeter-wave radiation. Potential applications span the gamut from the detection of chemical, biological, and explosive agents, security imaging and material science, to biomedical applications [1, 2].

The adaption of plasmonic concepts from the optical region of the spectrum to the THz band offers the exciting potential to realize new applications as well as overcome some of the limitations of the band. The inherently long wavelength of THz radiation poses a challenge for sensing and imaging objects of interest with feature sizes below the diffraction limit of hundreds of microns. The deep subwavelength confinement offered by plasmonics allows objects well below this limit to be sensed and imaged. Furthermore, the weakness of THz sources and the deficiency of its detectors can be somewhat compensated for by the use of resonant plasmonic structures to locally confine and enhance the electromagnetic field, allowing for sensitive detection.

This chapter first briefly examines some of the sensing applications of THz radiation that make the band so interesting. We then show how THz plasmonics can be

Active Plasmonics and Tuneable Plasmonic Metamaterials, First Edition. Edited by Anatoly V. Zayats and Stefan A. Maier.

realized by way of semiconductor and engineered metal surfaces and how they can be applied toward sensing. The use of THz plasmonic antennas for receiving and concentrating THz radiation and the phenomena of extraordinary transmission (ET) in the THz band are described in Sections 8.5 and 8.6, respectively. The chapter concludes with a description of some of the possibilities offered by the exciting new material graphene.

8.1 THE TERAHERTZ REGION FOR SENSING

THz radiation has a number of unique properties which make it suitable for a range of applications, which are often not possible elsewhere in the spectrum. It shares the ability of radio waves to penetrate a large range of nonmetallic, nonpolar substances such as cardboards, plastics, and common clothing fabrics but with a wavelength short enough to allow a submillimeter imaging resolution. X-rays have a similar ability to penetrate materials; however, the low energy of THz photons means that THz radiation does not pose the same ionizing risk to biological tissues.

Perhaps the most compelling feature of THz waves is the fact that a large number of atoms and molecules have rotational, translational, and vibrational responses which uniquely occur in the THz band [3]. Gases, in particular, have strong and narrow absorption lines. These responses allow substances to be distinguished and identified based on their THz spectral signature or fingerprint. Some examples of this include detecting biological agents [4, 5] and explosives [6–8] and chemical recognition of gas mixtures [9, 10]. While surface plasmon-based sensors are typically sensitive over a narrow frequency range, they can be designed to target specific narrowband spectral features, allowing a specific substance to be detected.

One area where THz surface plasmon sensors may play an important role is in the field of biosensing. THz radiation allows for the direct probing of the molecular dynamics of many biomolecules. For large biomolecules, the THz response is primarily determined by the low-frequency vibrational modes and side chain rotations [11, 12]. The high density of state of these modes means that individual absorption lines cannot be easily resolved leading to a broad THz absorption spectrum that is relatively featureless; however, this spectrum tends to be different between biomolecules allowing them to be distinguished [11–13].

THz waves have been shown to be highly sensitive to hydration, binding, and conformational change of proteins which may prove useful for sensing applications [12]. Nagel et al. demonstrated the label-free analysis of DNA molecules using THz radiation [14]. Combining surface plasmon sensors with microfluidic chips offers the most promising approach to realizing sensitive THz biochips [15, 16].

8.2 THz PLASMONICS

The metals commonly employed in the visible part of the spectrum for supporting SPPs, such as gold and silver, have significantly different properties in the THz regime which make them less suitable for sustaining SPPs at a flat interface. Far below their plasma frequency, these materials have significantly larger negative real and positive imaginary parts of the permittivity, more closely resembling a perfect electric conductor (PEC). Higher conductivity means that an SPP's electric field only slightly penetrates the metal causing it to be loosely bound to the surface and exhibit poor confinement, losing many of its advantages.

There are two common approaches for overcoming this lack of confinement. The first is to replace the metal with a semiconductor whose lower free-carrrier density causes its plasma frequency to fall in the THz band. This approach allows many plasmonic devices operating in the visible part of the spectrum to be simply scaled to operate in the THz band and has the additional advantage that the material dielectric properties can be tuned through optical, thermal, or electronic means, providing a way to actively control the SPPs propagation. The second approach is to use metals whose surface has been textured so that the electric field effectively penetrates further into the metal side of the interface. These engineered metal surfaces support SPP-like modes which are called *designer* or *spoof surface plasmon polaritons* (SSPPs) because their dispersion properties closely resemble real SPPs. The theory and properties of SPPs and SSPPs as well as their sensing applications will be discussed in Sections 8.3 and 8.4, respectively.

8.3 SPPs ON SEMICONDUCTOR SURFACES

The complex dielectric function of many semiconductors in the THz regime can be described by the Drude model, which we reprint here for convenience:

$$\epsilon(\omega) = \epsilon_\infty - \frac{\omega_p^2}{\omega^2 + i\gamma\omega}, \tag{8.1}$$

where ϵ_∞ is the high-frequency permittivity and $\gamma = e/m^*\mu$ is the characteristic collision frequency with μ being the carrier mobility. The plasma frequency is given by $\omega_p = \sqrt{ne^2/\epsilon_0 m^*}$, where n is the free-carrier density, e the electron charge, m^* the effective charge mass, and ϵ_0 the permittivity of free space.

Since the plasma frequency is primarily determined by the free-carrier density, choosing an intrinsic or doped semiconductor with an appropriate carrier density allows the plasma frequency to be shifted to a desired frequency in the THz band. Ideally, the semiconductor should have a high carrier mobility to reduce the carrier collision frequency and hence loss due to absorption. Materials that suit these requirements include InSb, GaAs, and InAs. Another commonly used material is p- or n-doped silicon despite its comparatively lower carrier mobility.

Intrinsic InSb has one of the highest carrier mobilities for a bulk semiconductor and is often the material of choice for THz plasmonics [17]. With a carrier density of $n \approx 1.6 \times 10^{16}$ cm^{-3}, it has a plasma frequency of $\omega_p = 9.60$ THz and collision frequency of $\gamma = 1.79$ THz [18]. Its permittivity properties are plotted in Figure 8.1a and the dispersion characteristics for an SPP propagating at air–InSb and silicon–InSb interfaces are shown in Figure 8.1b. It can be seen that a modest amount of confinement is possible when the frequency approaches the surface plasmon frequency, limited primarily by the loss in the InSb.

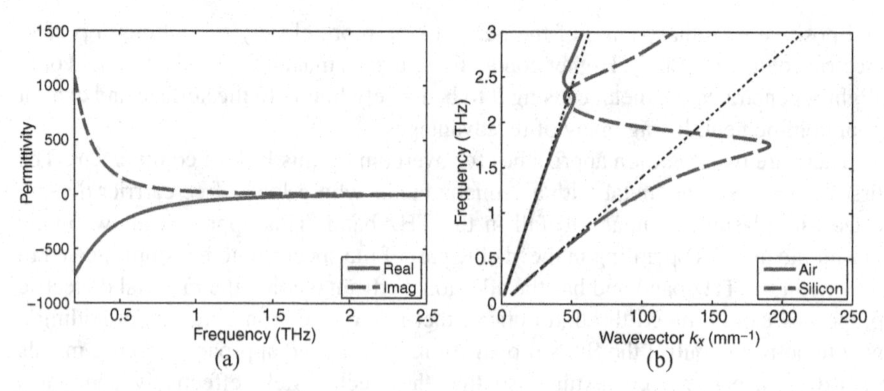

FIGURE 8.1 (a) Complex permittivity of InSb. (b) Dispersion properties of an SPP propagating on air–InSb and silicon–InSb interfaces.

Gómez-Rivas et al. [19] and Kuttge et al. [20] have demonstrated the propagation of SPPs on gratings made from n- and p-doped silicon wafers. The grating structuring of the surface was shown to provide additional degrees of freedom for tailoring the SPP propagation characteristics allowing the group velocity to be controlled as well as the creation of stop gaps in the spectral response. It was shown that resonant scattering at the edge of the stop gap reduced the group velocity by a factor of two which is useful for broadband spectroscopy and nonlinear applications [19].

Isaac et al. demonstrated the much greater sensitivity offered by an SPP propagating on the surface of InSb compared to a planar gold surface in Reference 21. In their experimental configuration, depicted in Figure 8.2, they used razor blades as scatterers to couple free-space propagating THz waves to and from an SPP traveling on an InSb or gold surface. They showed that the enhanced confinement of the SPP on the InSb surface leads to a stronger light–matter interaction than for the gold case and was capable of sensing thin polymer layers which were a thousand times thinner than the THz free-space wavelength.

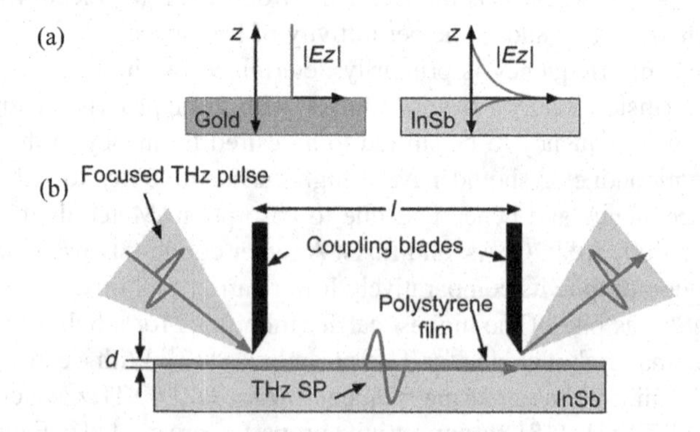

FIGURE 8.2 (a) Improved confinement of the electric field on an InSb surface compared with a gold surface. (b) Diagram of the experimental configuration showing how the THz energy is coupled into and out of the propagating SPP mode. Reprinted with permission from Reference 21. Copyright 2008 American Institute of Physics.

8.3.1 Active Control of Semiconductor Plasmonics

The nature of the electronic band structure of semiconductors allows the tuning of their material properties, which in turn creates the possibility of controlling the characteristics of SPPs excited on their surface. The existence of a band gap and unfilled electron energy states in the conduction band of semiconductors means that electrons can be promoted from the valence band to the conduction band, varying the free-carrier density which modifies the Drude dielectric response of the material. This can be achieved through photo-excitation of electrons or by varying the material's temperature. Alternatively, an applied electric field can be used to create a local depletion region or free carriers can be directly injected to increase the local density.

Examining the Drude model given by Equation 8.1, it can be seen that the square of the plasma frequency is proportional to the free-carrier density. The collision frequency is also affected by an increase in free-carrier density through the accompanying reduction in carrier mobility. This means that as a semiconductor's temperature is increased or is illuminated with light containing photons with energy greater than the semiconductor's band gap, the material properties become more metallic-like with the skin depth and SPP confinement being reduced. Of all the methods for modifying the free-carrier density, photo-excitation holds the most promise since it offers the possibility of controlling plasmons on picosecond timescales [22], opening the door to ultrafast active plasmonic components in the THz range.

A significant amount of work on active THz plasmonics has focused on modifying the transmission properties of THz radiation through gratings of subwavelength apertures [17,22–26]. The sensitivity of these structures to the dielectric properties of the underlying semiconductor makes them ideal for active control. The phenomena of ET will be discussed later in Section 8.6.

8.4 SSPP ON STRUCTURED METAL SURFACES

Surface waves propagating on metal–dielectric interfaces have been studied for more than a century, beginning with the seminal work of Sommerfeld investigating propagation on cylindrical surfaces (Sommerfeld waves) in 1899 [27] and Zenneck examining propagation on flat surfaces (Zenneck waves) in 1907 [28]. The surface waves studied were highly delocalized with the field extending a significant distance into the dielectric and closely resemble a plane wave propagating at a grazing angle to the surface. In 1950, Goubau [29] demonstrated that structuring the metal surface or the addition of a thin dielectric coating to the interface allowed the field extension to be controlled and confinement improved.

More recently, Pendry et al. [30] showed that a highly conductive metal surface perforated with subwavelength holes would support a surface wave and pointed out its connection with SPPs in the optical regime. As a result, these waves have since been termed spoof surface plasmon polaritons. The addition of the holes to the metal surface allows the field to effectively penetrate further into the metal, improving confinement of the wave. The use of general texturing of the metal surface allows us to engineer a Drude-like response similar to that of a metal close to the plasma frequency and gives us control over the propagation characteristics and plasmon confinement [30–32]. This ability to tailor the surface characteristics means that we can replicate the plasmon sensing undertaken in the visible part of the spectrum in the THz region while taking advantage of the unique properties of THz radiation. Some potential applications of SSPPs are subwavelength waveguiding [33–36], superfocusing [37, 38], near-field imaging, and biosensing with improved sensitivity.

A number of researchers have examined THz SSPPs propagating on corrugated or grooved surfaces [39, 40], hole arrays [40–44], and gratings [45]. More exotic spoof plasmon waveguides considered include wedges [33], V-channels [34], dominos [35], and slanted grooves [36]. As an illustrative example, we consider the relatively simple case of an SSPP propagating on a 1D grooved highly conductive metal surface. The dispersion relationship for this geometry can be derived by matching the field inside the grooves with the field above in vacuum. This is done by expanding the field inside the grooves in terms of the forward and backward propagating fundamental TE mode. The field above the surface is expressed in terms of an incident p-polarized free-space wave and its reflection from the surface. Matching the fields and applying the relevant boundary condition leads to the dispersion equation [32,40]

$$\frac{\sqrt{k_x^2 - k_0^2}}{k_0} = \frac{a}{d}\tan(k_0 h), \tag{8.2}$$

where k_0 is the free-space wave vector, d the groove spacing, a the groove width, and h the groove height. Plotting this dispersion relationship in Figure 8.3 for the case $a = 0.2d$ and $h = d$, we can see that despite the highly conductive nature of the metal, the wave vector k_x bends away from the light line and strong confinement is achieved. It can be observed that by simply varying the parameters defining the geometry (d, a, and h) the surface wave properties can be controlled and this is demonstrated for several values of d in Figure 8.3.

We next consider the case of an SSPP propagating on the surface formed by a 2D array of holes in a highly conductive metal which are filled with a dielectric material of permittivity ϵ. The holes are assumed to be square with dimension $a \times a$ and have an array period of d in both dimensions of the interface. Using a similar approach to the grooved surface case, we look for divergences in the reflection coefficient of the incident field due to coupling to a surface mode. From this the dispersion relationship can then be shown to be [40]

$$\frac{\sqrt{k_x^2 - k_0^2}}{k_0} = \frac{S_0^2 k_0}{\sqrt{(\pi/a)^2 - \epsilon k_0{}^2}}, \tag{8.3}$$

where

$$S_0 = \frac{2\sqrt{2}a}{\pi d}\frac{\sin(k_x a/2)}{k_x a/2}. \tag{8.4}$$

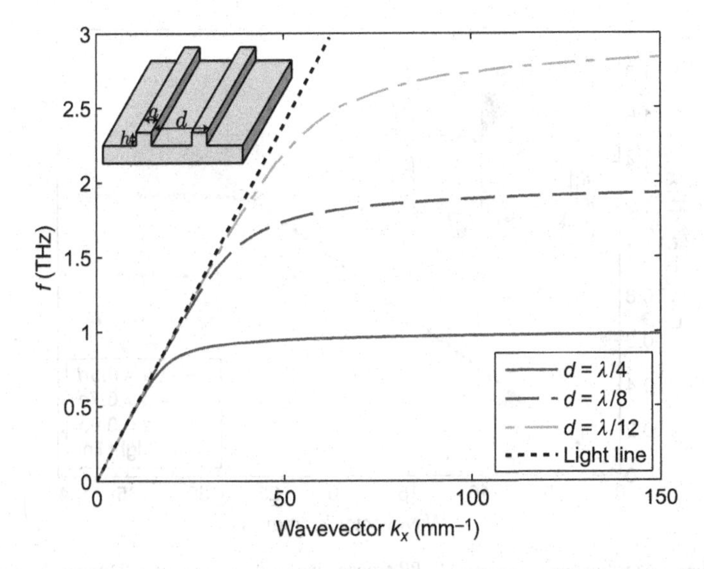

FIGURE 8.3 Dispersion diagram for an SSPP propagating on a highly conductive grooved surface. The plot was generated using the parameters $a = 0.2d$ and $h = d$, with several different groove spacings d as indicated in the plot legend as a fraction of the free-space wavelength λ at 1 THz.

Figure 8.4 plots this dispersion characteristic when $d = 75$ μm for the three cases: $a = 0.5d$, $a = 0.7d$, and $a = 0.9d$. It can be again seen that by structuring the surface, in this case with appropriately sized and spaced holes, the dispersion relationship can be engineered to achieve a tight confinement at any desired frequency in the THz band.

Metal wires can also be used to efficiently guide THz radiation [37,46–50]. Maier et al. demonstrated how surface corrugations on wires can be exploited to confine and focus an SSPP to make a THz probe in Reference 37. Figure 8.5 illustrates how the surface geometry of the corrugated wire was used to greatly reduce the radial extent of the SSPP electric field and achieve subwavelength concentration of the energy at the tip. This probe could be used as a flexible endoscope for spectroscopy or imaging well below the diffraction limit [48].

8.5 THz PLASMONIC ANTENNAS

Antennas provide an efficient means of transferring freely propagating electromagnetic waves from the far field to the near field, wherein the energy can be localized and concentrated into a subwavelength volume. The advantage of using an antenna is that it allows the diffraction limit to be overcome and the field intensity to be enhanced by several orders of magnitude at a defined location. It is for these reasons that optical nanoantennas have been found useful for a range of applications [51,52] and it is therefore useful to extend the concept to the THz region. Sensing applications of THz

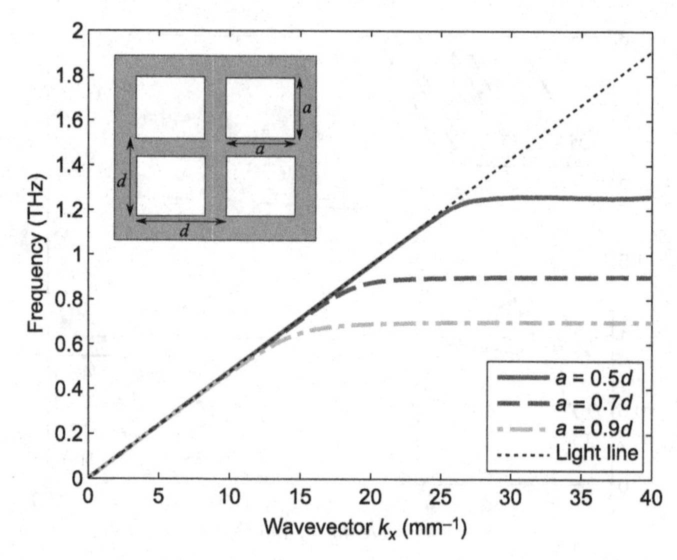

FIGURE 8.4 Dispersion diagram for an SSPP propagating on the surface of a 2D square hole array in a highly conductive metal filled with a dielectric of permittivity $\epsilon = 10$. The hole period is $d = 75$ μm with several different hole dimensions $a \times a$ as indicated in the plot legend.

antennas will likely be found in technological areas such as spectroscopy, biosensing, and security.

It is possible to adapt conventional (typically metal-based) antenna designs commonly used in the radiowave and microwave regimes for use in the THz band since metals remain reasonably conductive. However, it has been shown by Giannini et al. [53] that the adoption of plasmonic antennas allows greater enhancement of THz radiation. For example, in Figure 8.6 the field enhancement of a rectangular dimer antenna made from gold is compared with one made from InSb. It can be seen that the localized plasmonic resonance in the InSb antenna achieves a field enhancement of the order of 10^2–10^3 in the gap between dimers, which is greater than that for the gold antenna. Bow-tie antennas which are formed from two triangular dimers with their

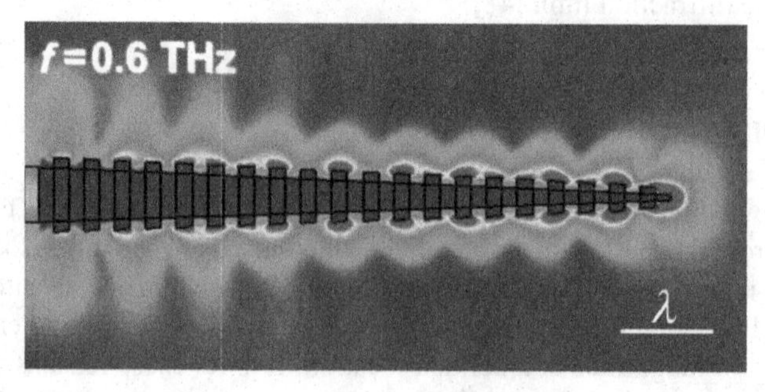

FIGURE 8.5 Focusing and guiding of an SSPP on a corrugated PEC cone surface. Reprinted with permission from Reference 49. Copyright 2008 IEEE.

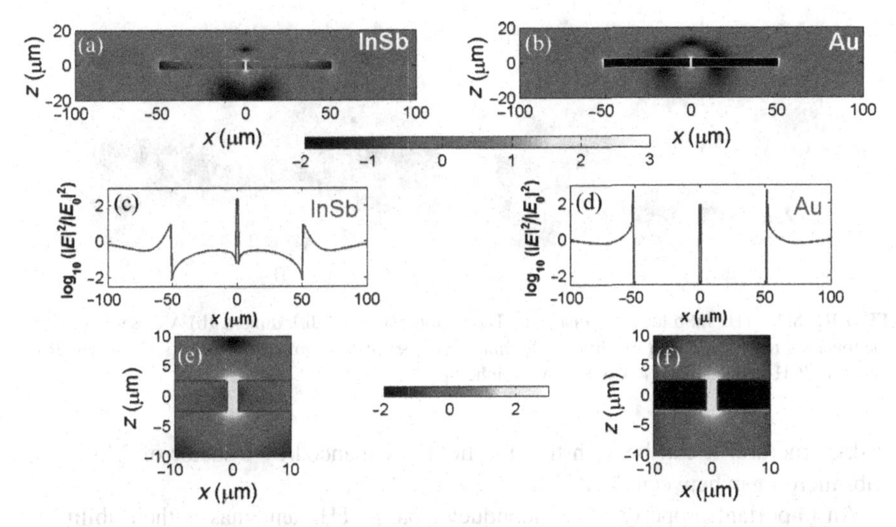

FIGURE 8.6 (a) Near-field intensity distribution at the resonance of an InSb dimer antenna with a gap $\Delta = 2$ μm, length $L = 50$ μm, height $h = 5$ μm, and $f = 1.62$ THz. (b) Gold antenna with the same dimension as (a) but at $f = 1.73$ THz. (c) Horizontal cut of (a) at $z = 0$. (d) Horizontal cut of (b) at $z = 0$. (e) Close-up view of the gap of (a). (f) Close-up view of the gap of (b). Reprinted with permission from Reference 53. Copyright 2010 Optical Society of America.

apex facing each other and a subwavelength gap in-between have been demonstrated to have even higher field enhancement than rectangular dimer antennas, partly due to the lightning rod effect which occurs at the triangle corners [53,54].

Plasmonic antennas that have a broadband response can be designed through the use of transformation optics, which has been demonstrated as a technique for transferring the broad bandwidth behavior of plasmonic waveguides to nano- and micro-antennas using geometric singularities [55]. A broad bandwidth response is important for sensing applications where the object being sensed has a broad dielectric response such as in the case of large biomolecules rather than a narrowband resonance.

Hanham et al. [18] investigated a periodic array of THz InSb disk dimers which were only just touching, a geometry shown in Figure 8.7 which has been predicted by transformation optics to efficiently harvest light over a broad bandwidth [55]. He demonstrated that the plasmonic interaction between the disks led to spectral broadening of the absorption cross section of the structure due to the excitation of multiple plasmonic modes. When an incident wave's polarization aligns with the dimer axis then single dipole and double dipole longitudinal modes are excited on the structure. Their charge distribution shown in Figure 8.8a reveals the associated buildup of charge in the disk gap region which leads to a strong field enhancement in the gap. In contrast, the transverse dipole mode which is excited when illuminated with orthogonally polarized light does not display the same field enhancement, underlining the importance of the plasmonic interaction between the disks for field enhancement. Figure 8.8b plots the field enhancement seen around the outside of the disk dimers relative to the incoming wave amplitude for several modes supported by the touching

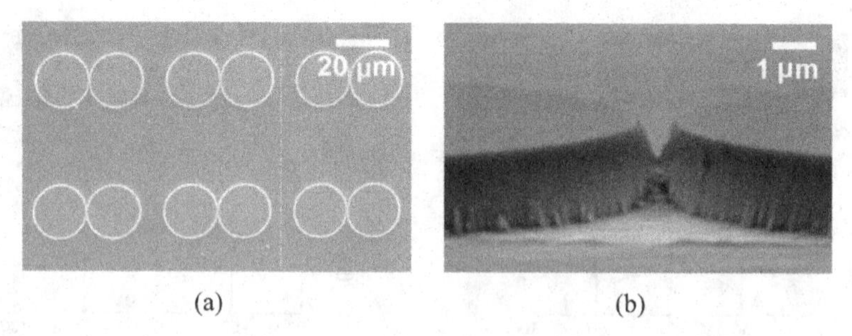

FIGURE 8.7 THz InSb touching disks. (a) The periodic array of disk dimers. (b) A close-up view of the touching region of one of the dimers. Reprinted with permission from Reference 18. Copyright 2012 WILEY-VCH Verlag GmbH & Co. KGaA, Weinheim.

disks structure. It can be seen that the field is enhanced by a factor of 270 in the sub-micron gap between the disks.

An important property of semiconductor-based THz antennas is their ability to be controlled and tuned. As discussed in Section 8.3.1, a semiconductor's dielectric properties depend on free-carrier concentration and mobility. By modifying the free-carrier concentration, the far-field and near-field properties of an antenna can be controlled. A potential use of this could be modifying the resonant frequency of an antenna to scan in frequency for a spectroscopy-type application. Berrier et al. demonstrated this capability in Reference 54 where the resonant frequency of a bow-tie antenna was blue-shifted and its extinction spectra increased by 40% by increasing the illumination fluence of the silicon antenna by a frequency-doubled 800 nm pulsed laser.

Berrier et al. later demonstrated the use of doped silicon bow-tie antennas for enhanced sensing of thin dielectric layers [56]. By exciting LSPR modes on individual antennas and monitoring the resonant frequency shift due to the presence

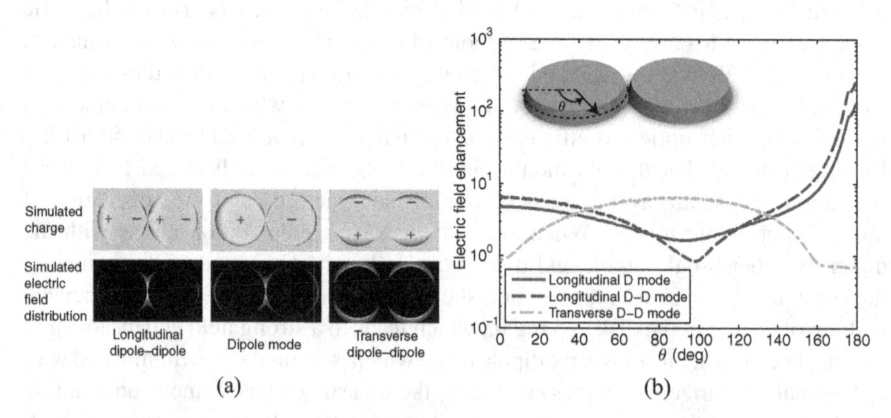

FIGURE 8.8 (a) Charge and electric field distribution for the modes supported by the touching disk dimers. (b) Simulated electric field amplitude enhancement for the different modes. Reprinted with permission from Reference 18. Copyright 2012 WILEY-VCH Verlag GmbH & Co. KGaA, Weinheim.

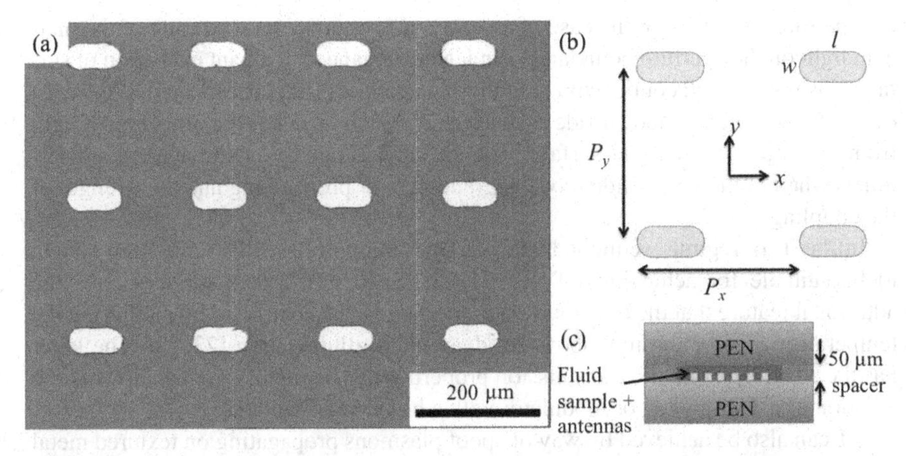

FIGURE 8.9 (a) Microscope image of the rectangular antenna array. (b) Schematic diagram of THz antenna array with antennas of length $l = 85$ μm, width $w = 38$ μm, and periods $P_x = 200$ μm, and $P_y = 220$ μm in the x and y directions, respectively. (c) Schematic diagram of the fluid chamber assembly used for transmission measurements. Reprinted with permission from Reference 57. Copyright 2011 Optical Society of America.

of an inorganic film they were able to detect films 3750 times thinner than the free-space wavelength. This is a 10 times improvement in sensitivity compared to the approach using an SPP propagating on a flat plasmonic surface [21, 56]. This technique is ideally suited for applications involving small sensing volumes such as cells, microorganisms, and explosives detection.

An interesting approach to sensing is the use of the collective response of an antenna array. In Reference 57, a 2D array of gold rectangular antennas are printed on a 100 μm thick polyethylene naphthalate substrate, shown in Figure 8.9. The array is designed such that a diffractive mode of a normal incident THz wave couples to the individual dipole resonances of each antenna, creating a combined or lattice resonance of the array. This resonance can be used for sensing since the addition of a thin dielectric layer will modify the phase-matching condition required for the coupling between antennas resulting in a frequency shift of the resonance, which corresponds to a minimum in the extinction spectra of the array. This approach was shown to be useful for sensing thin films of liquids and gases and has a relatively high Q-factor of 20.

8.6 EXTRAORDINARY TRANSMISSION

The concept of *extraordinary transmission* (ET) where light incident on an aperture has a transmission efficiency greater than unity when normalized by the aperture area can be extended to the THz regime. This phenomenon has been explained by the tunneling of surface plasmons propagating on one or both sides of the surface containing the aperture [58–60]. For the case of an array of apertures, the scattering of the incident light on the aperture boundaries can lead to enhanced resonant excitation of surface plasmons which couple with neighboring apertures [61]. The excited SPP waves excite the waveguide mode inside each aperture which mediates the tunneling of light from the entrant to the exit surface. The size and geometry of the apertures determines whether the waveguide mode is evanescent or propagating and the strength of the coupling.

In the THz regime, semiconductor surfaces supporting SPPs have been shown to be suitable for achieving ET [17, 21, 23, 25, 26, 62]. These surfaces have the additional feature that the transmission properties can be controlled by modifying the temperature [26] or through above-bandgap photo-illumination [23, 25]. The latter has the advantage that the transmission properties can be rapidly varied allowing the creation of a THz switch or modulator with a broad bandwidth [23].

ET can also be achieved by way of spoof plasmons propagating on textured metal surfaces [63–66]. In this case, the requisite texturing of the metal surface to support the SSPP can be provided by the array of apertures in the surface. The periodicity of the apertures helps convert the normal incident light into SSPPs by providing the additional momentum required for their excitation [58]. The excited SSPPs propagate along the surface and couple to neighboring apertures aiding the excitation of the aperture waveguide mode.

The resonant nature of ET makes it extremely sensitive to the local dielectric environment near the apertures, which is ideal for the detection of small quantities of samples located in the vicinity. Miyamaru et al. have demonstrated the use of metal hole arrays for sensing thin dielectric films [66, 67]. By monitoring the frequency shift of the peak in transmission through the hole array, they were able to detect a 5 μm thick dielectric layer. They explained that the relatively long localization of the SSPP in the vicinity of the apertures leads to improved sensitivity over conventional THz spectroscopy approaches [67].

An interesting demonstration of the near-field imaging capabilities of a metal hole array is shown in Figure 8.10. Miyamaru et al. [65] acquired a THz image of a fingerprint deposited on a dielectric film using a metal plate containing a triangular

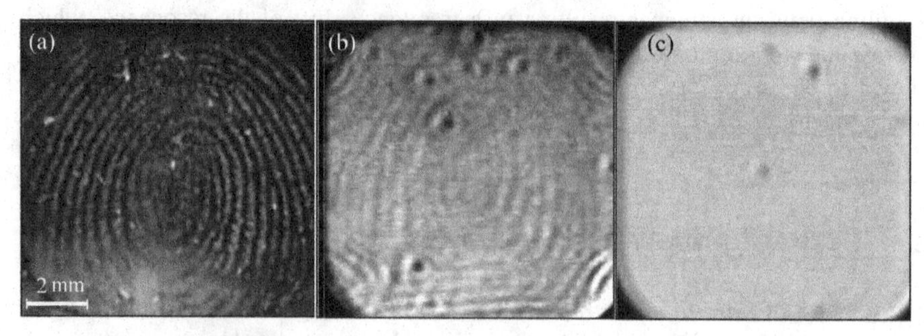

FIGURE 8.10 (a) Photograph of a fingerprint deposited on a polypropylene film. THz image of the fingerprint created (b) with and (c) without the metal hole array. Reprinted with permission from Reference 65. Copyright 2007 Optical Society of America.

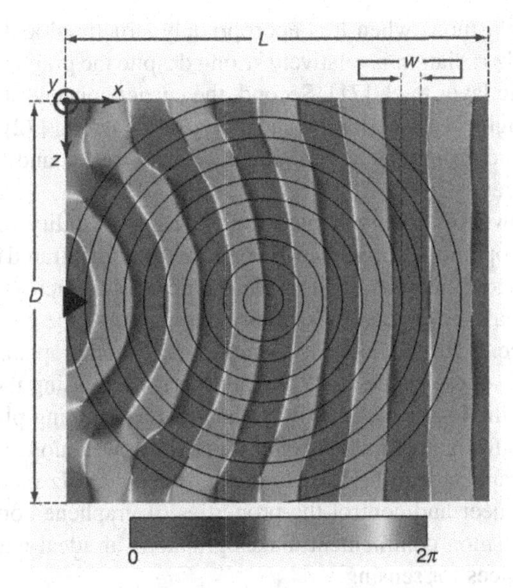

FIGURE 8.11 The collimation of an SPP by a Luneburg lens made from graphene. The plot shows the phase of the y-component of the electric field at $f = 30$ THz. In the plot $D = 1.5$ mm, $w = 75$ nm, $L = 1.6$ mm. Reprinted from Reference 72 with permission from AAAS.

lattice of holes and compared it with an image taken the conventional way. The metal hole array was placed in contact with the dielectric film to take advantage of the enhanced field strength around the holes and the subwavelength spatial sensitivity. The significantly improved resolution and sensitivity can be clearly seen in the images obtained.

Other applications of aperture arrays include the creation of THz filters for probing specific frequency windows for possible spectral features, which may prove useful for applications in chemistry and biology. The advantage of using a hole array is that the transmitted frequency can be varied simply by adjusting the array period.

Seo et al. [68] showed that THz radiation could pass through a nano-slit in a metal plate with a gap width of only 70 nm, which is below the metal's skin depth. The transmission of the THz radiation was accompanied by a two orders of magnitude increase in the electric field intensity inside the gap. It is thought that this approach could find application in near-field imaging and allow focusing of THz waves beyond the skin depth toward the Thomas–Fermi length scale.

8.7 THz PLASMONS ON GRAPHENE

Graphene is a promising material for the development of high-speed electronic devices due to its unique electronic structure which leads to useful phenomena such as massless charge carriers and ballistic transport [69, 70]. Graphene has several characteristics which make it extremely attractive for THz plasmonics. First, graphene will support THz plasmons when it is appropriately structured and the coupling of the plasmons to THz radiation is relatively strong despite the graphene surface being only a single atomic layer thick [71]. Second, the carrier concentration can be readily controlled through chemical doping and electrostatic gating [72]. This capability allows the plasmonic properties to be dynamically controlled and tailored across a single graphene sheet.

Engheta and coworkers demonstrated the extreme flexibility of graphene as a plasmonic material operating above the THz band in the mid-infrared in Reference 72. They were able to show theoretically how plasmonic devices such as waveguides and Luneburg lenses could be created by spatially manipulating the conductivity across the graphene surface. Figure 8.11 shows a simulation of their graphene Luneburg lens collimating a point source-excited SPP. By appropriately biasing the graphene they were able to grade the lenses' conductivity such that a diverging plasmonic surface wave is refracted to form a collimated beam. Ju et al. showed along similar lines that microribbons of graphene can be used to create tuneable THz metamaterials [71]. This ability to engineer and control the properties of graphene combined with the extreme surface plasmon confinement makes graphene an ideal material for future THz plasmonic devices for sensing.

REFERENCES

1. Siegel PH (2002) Terahertz technology. *IEEE Trans. Microw. Theory Tech.* 50(3): 910–928.
2. Siegel PH (2004) Terahertz technology in biology and medicine. *IEEE Trans. Microw. Theory Tech.* 52(10): 2438–2447.
3. Mittleman DM, Jacobsen RH, Nuss MC (1996) T-ray imaging. *IEEE J. Sel. Top. Quant.* 2(3): 679–692.
4. Brown ER, Woolard DL, Samuels AC, Globus T, Gelmont B (2002) Remote detection of bioparticles in the THz region. *IEEE MTT-S Digest*. Vol. 3. pp 1591–1594.
5. Wang S, Ferguson B, Abbott D, Zhang XC (2003) T-ray imaging and tomography. *J. Biol. Phys.* 29(2): 247–256.
6. Tribe WR, Newnham DA, Taday PF, Kemp MC (2004) Hidden object detection: security applications of terahertz technology. *Proc. SPIE* 5354: 168-176.
7. Huang F, Schulkin B, Altan H, Federici JF, Gary D, Barat R, Zimdars D, Chen M, Tanner DB (2004) Terahertz study of 1, 3, 5-trinitro-s-triazine by time-domain and Fourier transform infrared spectroscopy. *Appl. Phys. Lett.* 85(23): 5535–5537.
8. Federici JF, Schulkin B, Huang F, Gary D, Barat R, Oliveira F, Zimdars D (2005) THz imaging and sensing for security applications–explosives, weapons and drugs. *Semicond. Sci. Technol.* 20: S266-S280.
9. Jacobsen RH, Mittleman DM, Nuss MC (1996) Chemical recognition of gases and gas mixtures with terahertz waves. *Opt. Lett.* 21(24): 2011–2013.
10. Hassani A, Skorobogatiy M (2008) Surface plasmon resonance-like integrated sensor at terahertz frequencies for gaseous analytes. *Opt. Express* 16(25): 20206–20214.

11. Smye SW, Chamberlain JM, Fitzgerald AJ, Berry E (2001) Topical review: the interaction between terahertz radiation and biological tissue. *Phys. Med. Biol.* 46: R101–R112.
12. Markelz AG (2008) Terahertz dielectric sensitivity to biomolecular structure and function. *IEEE J. Sel. Top. Quant.* 14(1): 180–190.
13. Globus T, Woolard D, Bykhovskaia M, Gelmont B, Werbos L, Samuels A (2003) THz-frequency spectroscopic sensing of DNA and related biological materials. *Int. J. High Speed Electron. Syst.* 13(4): 903–936.
14. Nagel M, Richter F, Haring-Bolivar P, Kurz H (2003) A functionalized THz sensor for marker-free DNA analysis. *Phys. Med. Biol.* 48: 3625-3636.
15. Laurette S, Treizebre A, Bocquet B (2011) Co-integrated microfluidic and THz functions for biochip devices. *J. Micromech. Microeng.* 21: 065029.
16. George PA, Hui W, Rana F, Hawkins BG, Smith AE, Kirby BJ (2008) Microfluidic devices for terahertz spectroscopy of biomolecules. *Opt. Express* 16(3): 1577–1582.
17. Gómez-Rivas J, Janke C, Bolivar P, Kurz H (2005) Transmission of THz radiation through InSb gratings of subwavelength apertures. *Opt. Express* 13(3): 847–859.
18. Hanham SM, Fernández-Domínguez AI, Teng JH, Ang SS, Lim KP, Yoon SF, Ngo CY, Klein N, Pendry JB, Maier SA (2012) Broadband terahertz plasmonic response of touching InSb disks. *Adv. Mater.* 24(35): OP226–OP230.
19. Gómez-Rivas J, Kuttge M, Bolivar PH, Kurz H, Sánchez-Gil JA (2004) Propagation of surface plasmon polaritons on semiconductor gratings. *Phys. Rev. Lett.* 93(25): 256804.
20. Kuttge M, Kurz H, Rivas JG, Sánchez-Gil JA, Bolivar PH (2007) Analysis of the propagation of terahertz surface plasmon polaritons on semiconductor groove gratings. *J. Appl. Phys.* 101(2): 023707.
21. Isaac TH, Barnes WL, Hendry E (2008) Determining the terahertz optical properties of subwavelength films using semiconductor surface plasmons. *Appl. Phys. Lett.* 93(24): 241115–241115.
22. Gómez-Rivas J, Kuttge M, Kurz H, Bolivar PH, Sánchez-Gil JA (2006) Low-frequency active surface plasmon optics on semiconductors. *Appl. Phys. Lett.* 88(8): 082106.
23. Janke C, Rivas JG, Bolivar PH, Kurz H (2005) All-optical switching of the transmission of electromagnetic radiation through subwavelength apertures. *Opt. Lett.* 30(18): 2357–2359.
24. Hendry E, Lockyear MJ, Rivas JG, Kuipers L, Bonn M (2007) Ultrafast optical switching of the THz transmission through metallic subwavelength hole arrays. *Phys. Rev. B* 75(23): 235305.
25. Hendry E, García-Vidal FJ, Martín-Moreno L, Rivas JG, Bonn M, Hibbins AP, Lockyear MJ (2008) Optical control over surface-plasmon-polariton-assisted THz transmission through a slit aperture. *Phys. Rev. Lett.* 100(12): 123901.
26. Gómez-Rivas J, Bolivar PH, Kurz H (2004) Thermal switching of the enhanced transmission of terahertz radiation through subwavelength apertures. *Opt. Lett.* 29(14): 1680–1682.
27. Sommerfeld A (1899) Ueber die fortpanzung elektrodynamischer wellen längs eines drahtes. *Ann. Phys-Berlin* 303(2): 233–290.
28. Zenneck J (1907) Über die fortpanzung ebener elektromagnetischer wellen längs einer ebenen leiterfläche und ihre beziehung zur drahtlosen telegraphie. *Ann. Phys-Berlin* 328(10): 846–866.

29. Goubau G (1950) Surface waves and their application to transmission lines. *J. Appl. Phys.* 21(11): 1119–1128.

30. Pendry JB, Martín-Moreno L, García-Vidal FJ (2004) Mimicking surface plasmons with structured surfaces. *Science* 305(5685): 847–848.

31. Raether H (1988) *Surface Plasmons on Smooth and Rough Surfaces and on Gratings.* Vol. 111. Berlin: Springer.

32. Fernández-Domínguez AI, Martín-Moreno L, García-Vidal FJ (2011) In: Surface Electromagnetic Waves on Structured Perfectly Conducting Surfaces. Maradudin AA, editor. *Structured Surfaces as Optical Metamaterials.* Cambridge University Press. Chapter 7.

33. Fernández-Domínguez AI, Moreno E, Martín-Moreno L, García-Vidal FJ (2009) Terahertz wedge plasmon polaritons. *Opt. Lett.* 34(13): 2063–2065.

34. Fernández-Domínguez AI, Moreno E, Martín-Moreno L, García-Vidal FJ (2009) Guiding terahertz waves along subwavelength channels. *Phys. Rev. B* 79(23): 233104.

35. Martín-Cano D, Nesterov ML, Fernández-Domínguez AI, García-Vidal FJ, Martín-Moreno L, Moreno E (2010) Domino plasmons for subwavelength terahertz circuitry. *Opt. Express* 18(2): 754–764.

36. Wood JJ, Tomlinson LA, Hess O, Maier SA, Fernández-Domínguez AI (2012) Spoof plasmon polaritons in slanted geometries. *Phys. Rev. B* 85(7): 075441.

37. Maier SA, Andrews SR, Martín-Moreno L, García-Vidal FJ (2006) Terahertz surface plasmon-polariton propagation and focusing on periodically corrugated metal wires. *Phys. Rev. Lett.* 97(17): 176805.

38. Zhan H, Mendis R, Mittleman DM (2010) Superfocusing terahertz waves below $\lambda/250$ using plasmonic parallel-plate waveguides. *Opt. Express* 18(9): 9643–9650.

39. Mukina LS, Nazarov MM, Shkurinov AP (2006) Propagation of THz plasmon pulse on corrugated and at metal surface. *Surf. Sci.* 600(20): 4771–4776.

40. García-Vidal FJ, Martín-Moreno L, Pendry JB (2005) Surfaces with holes in them: new plasmonic metamaterials. *J. Opt. A Pure Appl. Opt.* 7: S97–S101.

41. Ulrich R, Tacke M (1973) Submillimeter waveguiding on periodic metal structure. *Appl. Phys. Lett.* 22(5): 251–253.

42. Williams CR, Andrews SR, Maier SA, Fernández-Domínguez AI, Martín-Moreno L, García-Vidal FJ (2008) Highly confined guiding of terahertz surface plasmon polaritons on structured metal surfaces. *Nat. Photonics* 2(3): 175–179.

43. Williams C, Misra M, Andrews S, Maier S, Carretero-Palacios S, Rodrigo S, García-Vidal F, Martín-Moreno L (2010) Dual band terahertz waveguiding on a planar metal surface patterned with annular holes. *Appl. Phys. Lett.* 96: 011101.

44. Maier SA, Andrews SR (2006) Terahertz pulse propagation using plasmon-polariton-like surface modes on structured conductive surfaces. *Appl. Phys. Lett.* 88(25): 251120.

45. Mills DL, Maradudin AA (1989) Surface corrugation and surface-polariton binding in the infrared frequency range. *Phys. Rev. B* 39(3): 1569.

46. King M, Wiltse J (1962) Surface-wave propagation on coated or uncoated metal wires at millimeter wavelengths. *IRE Trans. Ant. Prop.* 10(3): 246–254.

47. Wächter M, Nagel M, Kurz H (2005) Frequency-dependent characterization of THz Sommerfeld wave propagation on single-wires. *Opt. Express* 13(26): 10815–10822.

48. Wang K, Mittleman DM (2004) Metal wires for terahertz wave guiding. *Nature* 432(7015): 376–379.

49. Fernández-Domínguez AI, Martín-Moreno L, García-Vidal FJ, Andrews SR, Maier SA (2008) Spoof surface plasmon polariton modes propagating along periodically corrugated wires. *IEEE J. Sel. Top. Quant.* 14(6): 1515–152.

50. Fernández-Domínguez AI, Williams CR, García-Vidal FJ, Martín-Moreno L, Andrews SR, Maier SA (2008) Terahertz surface plasmon polaritons on a helically grooved wire. *Appl. Phys. Lett.* 93(14): 141109.

51. Giannini V, Fernández-Domínguez AI, Heck SC, Maier SA (2011) Plasmonic nanoantennas: fundamentals and their use in controlling the radiative properties of nanoemitters. *Chem. Rev.* 111: 3888–3912.

52. Novotny L, van Hulst N (2011) Antennas for light. *Nat. Photonics* 5(2): 83–90.

53. Giannini V, Berrier A, Maier SA, Sánchez-Gil JA, Rivas JG (2010) Scattering efficiency and near field enhancement of active semiconductor plasmonic antennas at terahertz frequencies. *Opt. Express* 18(3): 2797–2807.

54. Berrier A, Ulbricht R, Bonn M, Rivas JG (2010) Ultrafast active control of localized surface plasmon resonances in silicon bowtie antennas. *Opt. Express* 18(22): 23226–23235.

55. Aubry A, Lei DY, Fernández-Domínguez AI, Sonnefraud Y, Maier SA, Pendry JB (2010) Plasmonic light-harvesting devices over the whole visible spectrum. *Nano Lett.* 10(7): 2574–2579.

56. Berrier A, Albella P, Poyli MA, Ulbricht R, Bonn M, Aizpurua J, Gómez-Rivas J (2012) Detection of deep-subwavelength dielectric layers at terahertz frequencies using semiconductor plasmonic resonators. *Opt. Express* 20(5): 5052–5060.

57. Ng B, Hanham SM, Giannini V, Chen ZC, Tang M, Liew YF, Klein N, Hong MH, Maier SA (2011) Lattice resonances in antenna arrays for liquid sensing in the terahertz regime. *Opt. Express* 19(15): 14653–14661.

58. Ebbesen TW, Lezec HJ, Ghaemi HF, Thio T, Wolff PA (1998) Extraordinary optical transmission through sub-wavelength hole arrays. *Nature* 391(6668): 667–669.

59. Martín-Moreno L, García-Vidal FJ, Lezec HJ, Pellerin KM, Thio T, Pendry JB, Ebbesen TW (2001) Theory of extraordinary optical transmission through subwavelength hole arrays. *Phys. Rev. Lett.* 86(6): 1114–1117.

60. Genet C, Ebbesen TW (2007) Light in tiny holes. *Nature* 445(7123): 39–46.

61. Popov E, Bonod N (2011) Physics of extraordinary transmission through hole arrays. In: Maradudin AA, editor. *Structured Surfaces as Optical Metamaterials*. Cambridge University Press. Chapter 1.

62. Gómez-Rivas J, Schotsch C, Bolivar PH, Kurz H (2003) Enhanced transmission of THz radiation through subwavelength holes. *Phys. Rev. B* 68(20): 201306.

63. Qu D, Grischkowsky D, Zhang W (2004) Terahertz transmission properties of thin, subwavelength metallic hole arrays. *Opt. Lett.* 29(8): 896–898.

64. Cao H, Nahata A (2004) Resonantly enhanced transmission of terahertz radiation through a periodic array of subwavelength apertures. *Opt. Express* 12(6): 1004–1010.

65. Miyamaru F, Takeda MW, Suzuki T, Otani C (2007) Highly sensitive surface plasmon terahertz imaging with planar plasmonic crystals. *Opt. Express* 15(22): 14804–14809.

66. Miyamaru F, Sasagawa Y, Takeda MW (2010) Effect of dielectric thin films on reflection properties of metal hole arrays. *Appl. Phys. Lett.* 96(2): 021106.
67. Miyamaru F, Hayashi S, Otani C, Kawase K, Ogawa Y, Yoshida H, Kato E (2006) Terahertz surface-wave resonant sensor with a metal hole array. *Opt. Lett.* 31(8): 1118–1120.
68. Seo MA, Park HR, Koo SM, Park DJ, Kang JH, Suwal OK, Choi SS, Planken PCM, Park GS, Park NK, Park HQ, Kim SD (2009) Terahertz field enhancement by a metallic nano slit operating beyond the skin-depth limit. *Nat. Photonics* 3(3): 152–156.
69. Novoselov KS, Geim AK, Morozov SV, Jiang D, Katsnelson MI, Grigorieva IV, Dubonos SV, Firsov AA (2005) Two-dimensional gas of massless Dirac fermions in graphene. *Nature* 438(7065): 197–200.
70. Castro Neto AH, Guinea F, Peres NMR, Novoselov KS, Geim AK (2007) The electronic properties of graphene. *Rev. Mod. Phys.* 81(1): 109–162.
71. Ju L, Geng B, Horng J, Girit C, Martin M, Hao Z, Bechtel HA, Liang X, Zettl A, Shen YR, Wang F (2011) Graphene plasmonics for tuneable terahertz metamaterials. *Nat. Nanotechnol.* 6(10): 630–634.
72. Vakil A, Engheta N (2011) Transformation optics using graphene. *Science* 332(6035): 1291.

9

Subwavelength Imaging by Extremely Anisotropic Media

PAVEL A. BELOV
Queen Mary University of London, London, UK
National Research University ITMO, St. Petersburg, Russia

9.1 INTRODUCTION TO CANALIZATION REGIME OF SUBWAVELENGTH IMAGING

Lens-based imaging devices such as conventional microscopes cannot provide resolution better than $\lambda/2$, where λ is the wavelength of radiation. This restriction is known as the diffraction limit, and holds irrespectively of the frequency of operation—from microwave frequencies up to the visible range. Conventional lenses operate only with the far field of the source, which is formed by propagating spatial harmonics. The information related to the near field of the source is transported by evanescent spatial harmonics, which exhibit exponential decay in all natural materials as well as in free space. For this reason, the subwavelength details of the near field are not directly accessible with conventional imaging systems. In practice, the near field can be scanned, point by point, using small near-field probes. However, scanning near-field microscopy is a rather slow process, and thus lenses that may enable the simultaneous imaging of the whole region of interest with super-resolution would be extremely useful.

The diffraction limit can be surpassed only if one adopts a conceptually different imaging technique. In order to obtain imaging with subwavelength resolution, it may be necessary to overstep beyond the frontiers of naturally available materials and to use specially engineered artificial media with electromagnetic properties not readily available in nature. In recent years, several techniques have been put forward aiming at imaging with super-resolution in different ranges of the electromagnetic spectrum. These proposals include perfect lenses [1], silver superlenses [2], hyperlenses [3–7],

Active Plasmonics and Tuneable Plasmonic Metamaterials, First Edition. Edited by Anatoly V. Zayats and Stefan A. Maier.

and stimulated emission depletion (STED) fluorescence microscopes [8]. Another option, which is the topic of this chapter, is based on the use of materials with extreme optical anisotropy.

Let us consider an uniaxial dielectric medium with permittivity dyadic of the form

$$\overline{\overline{\varepsilon}} = \begin{pmatrix} \varepsilon_{xx} & 0 & 0 \\ 0 & \varepsilon & 0 \\ 0 & 0 & \varepsilon \end{pmatrix}. \tag{9.1}$$

The dispersion equation for extraordinary waves in such medium reads

$$\frac{k_x^2}{\varepsilon} + \frac{k_y^2 + k_z^2}{\varepsilon_{xx}} = \frac{\omega^2}{c^2}, \tag{9.2}$$

where k_x, k_y, k_z are the components of wave vector $\mathbf{k}$; ω is the frequency of operation; and c is the speed of light in vacuum.

If the material is characterized by extreme optical anisotropy such that $|\epsilon_{xx}| \gg \epsilon$, then Equation 9.2 has the following solution:

$$k_x = \pm\sqrt{\varepsilon}\frac{\omega}{c} \text{ and } k_y, k_z \text{ are arbitrary.} \tag{9.3}$$

Therefore, all extraordinary waves in a medium with extreme optical anisotropy are dispersionless: The electromagnetic waves travel along the optical axis of the material with a fixed phase velocity independently of the transverse phase variations. Note that ϵ_{xx} does not have to be necessarily a real number. It may be a complex number with a large absolute value. The imaginary part of ϵ_{xx} representing losses has very little influence on the wave propagation.

To a good approximation, the class of dielectrics with extreme optical anisotropy can be described by a dielectric function of the form

$$\overline{\overline{\varepsilon}} = \begin{pmatrix} \infty & 0 & 0 \\ 0 & \varepsilon & 0 \\ 0 & 0 & \varepsilon \end{pmatrix}. \tag{9.4}$$

The infinite value appearing in Equation 9.4 should not be treated as something exceptional or unusual. The corresponding material relations read as $E_x = 0$, $D_y = \varepsilon E_y$, and $D_z = \varepsilon E_z$. This actually means that the medium is perfectly conducting along the x-direction.

The latter fact gives a hint of how media with extreme optical anisotropy can be created artificially. An array of parallel metallic wires, also known as "wire medium" [9], may provide the necessary properties at frequencies up to the infrared band [10] (see Sections 9.2, 9.3, and 9.4 for more details). In the visible range, one of the options may be a layered metal–dielectric structure [11] which has an effective permittivity consistent with Equation 9.4 (see Section 9.6 for more details).

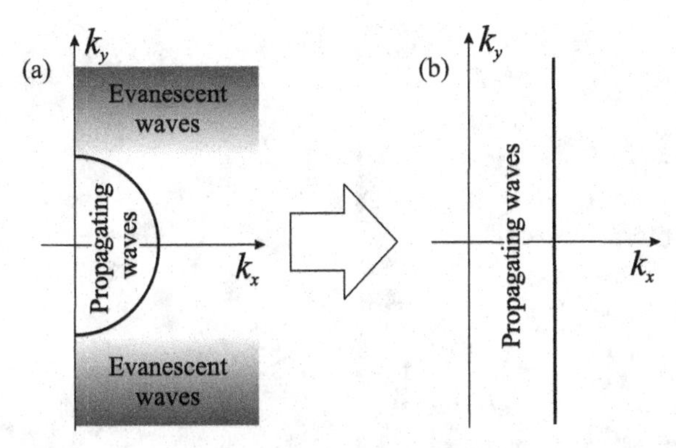

FIGURE 9.1 (a) Isofrequency contours in an isotropic material (e.g., free space). For $|k_y| > \omega/c$ the spatial harmonics correspond to evanescent waves (shaded region). (b) Isofrequency contours in a material with extreme optical anisotropy. All the spatial harmonics are propagating waves.

Materials with extreme anisotropy have remarkable properties: All the extraordinary waves supported by the media are propagating waves (see Fig. 9.1). Thus, theoretically, it is possible to transport an arbitrary field distribution with suitable polarization through such material with no loss of resolution. Lenses formed by slabs of such materials provide a unique opportunity to transmit the near field with super-resolution. At the front interface with free space, the evanescent waves are transformed into propagating waves, which prevents their decay and preserves the subwavelength information. The propagating waves travel through the slab and reproduce the image at the back interface of the lens. This phenomenon is known as *canalization* [2]. This regime is especially appropriate for the transmission of images with super-resolution over significant distances in terms of the wavelength. This effect cannot be achieved using other available imaging techniques.

Basically, the canalization regime is very similar to the fiber-optic effect which is observed in the "TV rock" (see Fig. 9.2). This mineral consists of parallel tiny optical fibers which are capable of transporting images pixel by pixel. The significant difference is that the canalization regime provides super-resolution, whereas the resolution provided by the TV rock is determined by dimensions of the fibers which are greater than the wavelength. In order to get super-resolution one has to use an array of subwavelength waveguides such as an array of metallic wires.

An alternative and widely known opportunity of subwavelength imaging is the perfect lens concept [1]. It is based on the amplification of evanescent harmonics by a metamaterial, rather than on their conversion into propagating waves. This may enable imaging objects which are placed at some significant distance from the lens interface, like buried objects. However, the perfect lens relies on the resonant excitation of surface plasmon polaritons at the interfaces of a metamaterial slab, and hence it is very sensitive to losses [13]. The amplification of the evanescent waves requires such a low level of losses in the metamaterial that it may be extremely difficult to achieve in practice. Even if the current race to create the metamaterial required for the implementation of the perfect lens in the optical range [14–17] is

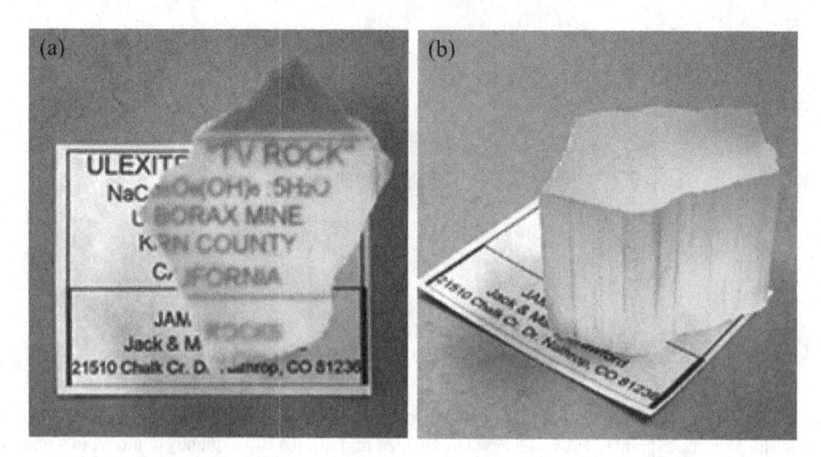

FIGURE 9.2 The fiber-optic effect by "TV rock," a mineral ulexite ($NaCaB_5O_9{\cdot}8H_2O$, hydrated sodium calcium borate hydroxide). (a) Top view. (b) Side view.

successfully accomplished then such material will be inevitably lossy and its use as a subwavelength imaging device may be of limited interest.

The effect of losses on the canalization regime is much less restrictive. Losses cause the decay of propagating waves inside of the material, but the decaying factor is the same for the whole spatial spectrum. This means that losses cause only a degradation of intensity but do not corrupt the form of the image [18]. This makes lenses operating in the canalization regime easier to fabricate as compared to perfect lenses, regardless of the frequency range of operation.

9.2 WIRE MEDIUM LENS AT THE MICROWAVE FREQUENCIES

In principle, materials with extreme optical anisotropy are not directly available in nature. However, these materials may be synthesized as microstructured materials. In particular, an array of long thin parallel metallic wires (Fig. 9.3) is characterized by extreme anisotropy in the long wavelength limit. The emergence of extreme anisotropy in this structure is due to the strong electric polarizability of the rods along the direction of the axes. The wire medium supports electromagnetic modes that are effectively described by a dielectric function of the form (9.4) with ε equal to the permittivity of the host matrix [9]. Such "transmission line" modes travel along the wires with the speed of light in the host medium and can have an arbitrary transverse wave vector.

It is important to mention that the description of the wire medium using the permittivity model (9.4) is only a rough approximation. In reality, the wire medium is a spatially dispersive material and thus its electrodynamics is more complicated than suggested by Equation 9.4 [9]. However, for very long wavelengths the spatial dispersion effects are to some extent of second order [19], and the dielectric function (9.4) may describe fairly well the response of the material.

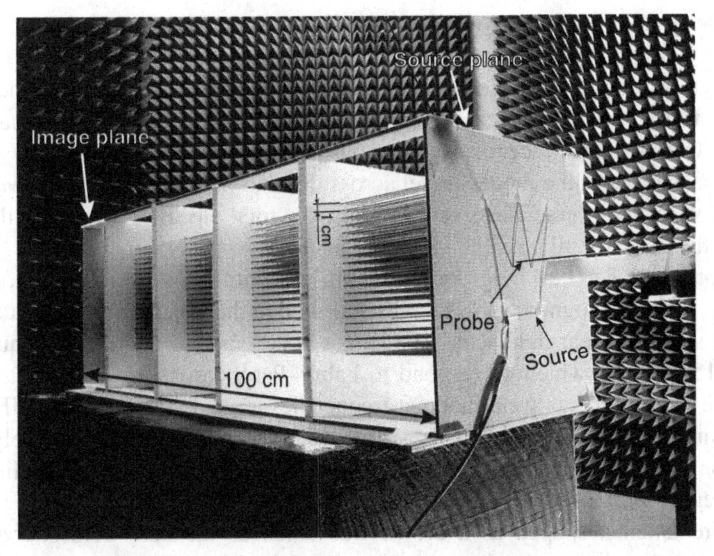

FIGURE 9.3 Photograph of the wire medium lens used in experiment [20]: 21×21 array of parallel 1 m long aluminum wires. The period of the array is 1 cm. The radius of the wires is 1 mm. The source in the form of a crown is adjacent to the front interface of the lens. The source is fed by a coaxial cable connected to a network analyzer. The small electric probe used for near-field scan is shown together with the supporting arm. Reprinted with permission from Reference 20. Copyright 2008 American Physical Society.

A planar wire medium slab (Fig. 9.3) enables the transformation of the near field produced by a source placed in the vicinity of the slab into propagating transmission line modes. These modes propagate along the direction normal to the interfaces and transfer the distribution of electric field from the front interface of the lens to the back interface without distortion, closely mimicking the behavior of a material with extreme anisotropy.

In a different but equivalent perspective, the slab of a wire medium may also be regarded as a bundle of very subwavelength waveguides performing pixel-to-pixel imaging. Each wire plays the role of a waveguide delivering information from one side of the lens to the other. The electric field distribution transported by a specific wire is approximately radial, and due to this reason the waveguides are nearly decoupled. This enables the propagation of waves along neighboring wires with a negligibly small interaction. The imaging resolution is approximately two periods of the wire medium [19], and, in principle, can be made as small as required by a given application.

In contrast to conventional lenses, near field lenses have to be placed in the close vicinity of the source in order to capture evanescent waves. Thus, it is important to ensure that the lens does not distort the source field distribution. An ideal near field lens should not produce any reflection. The problem of possible harmful reflections from a slab of a material with extreme anisotropy can be eliminated by properly choosing its thickness so that it operates as a Fabry–Perot resonator. The thickness should be equal to an integer number of half-wavelengths inside of the host material. In the case of an ideal material described by the dielectric function (9.4), the Fabry–Perot condition is verified simultaneously for all angles of incidence, including evanescent waves. This collective resonance makes the lens virtually transparent to any propagating or evanescent wave. This outstanding property is justified by the fact that all transmission line modes travel across the slab with the same phase velocity, irrespective of transverse wave vector. Thus, the total electrical length of the slab remains the same for all possible incidence angles.

A prototype of a 1 m long wire medium lens with 1 cm period is reported in Figure 9.3. The photograph shows the lens located in the anechoic chamber together with the setup for near field scan. The lens was designed to operate at frequencies around $150n$ MHz, which correspond to Fabry–Perot resonances of the nth order [20]. The measurements were made at frequencies around 900 and 1050 MHz corresponding to Fabry–Perot resonances of the 6th and 7th orders, respectively. This prototype is about seven times longer than the 15 cm long lens used for the preliminary experiments reported in References 21 and 22.

A wire antenna shaped as a crown was taken as the near field source. The source was placed near the front interface of the wire medium lens as shown in Figure 9.2, and the near field was scanned using an automatic mechanical planar near field scanner at a distance of 2 mm from the front and back interfaces of the lens (source and image planes). In order to confirm that the lens does not perturb or distort the near field of the source, the collected data were compared with the reference near field distributions obtained by scanning the near field created by the source in free space. The results of the near field scan at 898 and 1038 MHz are reported in Figures 9.4 and 9.5. A near field distribution reproducing the crown shape is clearly visible at the image plane. It is practically indistinguishable from the distributions in the source plane both with and without the presence of the wire medium lens. Despite the distance in-between the source and image planes at 898 and 1038 MHz being as large as 3 and 3.5 wavelengths, respectively, the imaging resolution is about two periods of the wire medium (2 cm), which in terms of the wavelength corresponds to $\lambda/15$. To the best of our knowledge, this is up to date the only demonstration of subwavelength imaging with such fine resolution over a distance much larger than the wavelength. In all other subwavelength imaging experiments reported in the literature, the distance between the source and image planes was only a small fraction of the wavelength, whereas the imaging resolution was much worse.

The near-field scan results reported in Figures 9.4 and 9.5 enable to clarify how the resolution of the wire medium lens depends on the frequency of operation. As mentioned before, the lens thickness is supposed to verify the Fabry–Perot condition. Actually, as it was shown in Reference 19, the best imaging is achieved at a frequency slightly below the Fabry–Perot condition. At higher frequencies subwavelength imaging is still observed, but with worse resolution. This can be clearly seen in Figure 9.4: The widths of the lines at the image plane become wider above the Fabry–Perot resonance. The amount of reflection provided by the wire medium slab at these frequencies is quite small: In Figure 9.4 the difference between the source plane distributions with and without the presence of the lens is not significant.

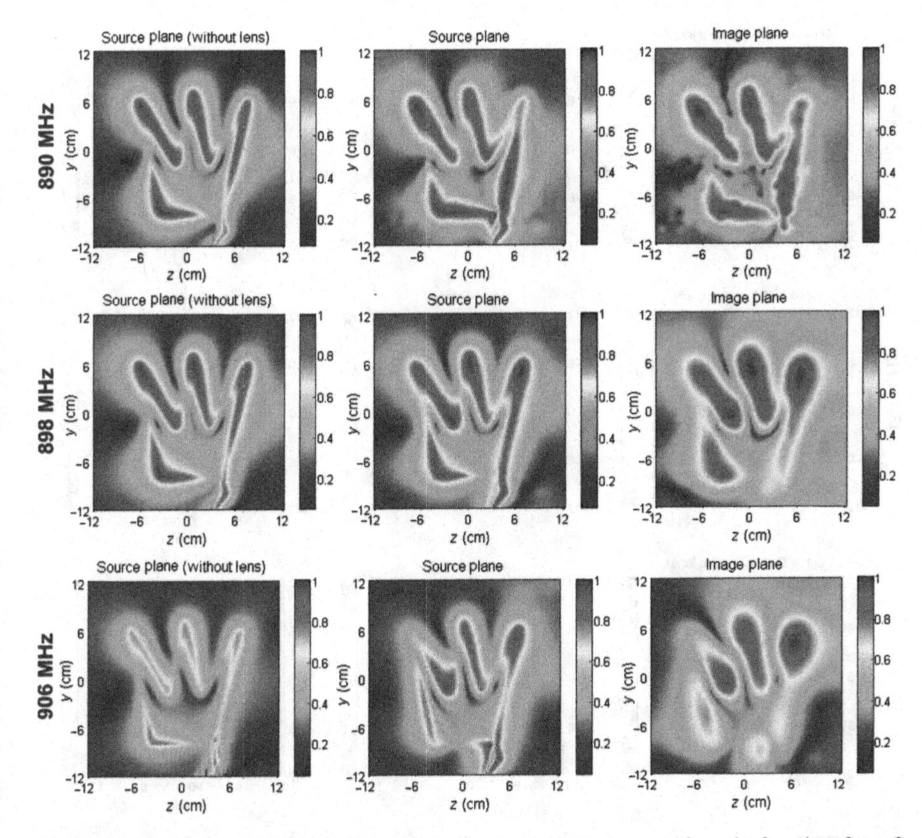

FIGURE 9.4 Results of the near field scan at the source plane (2 mm away from the front interface of the lens) without (left column) and with (central column) the wire medium lens and (right column) at the image plane (2 mm away from the back interface of the lens) at 890, 898, and 906 MHz.

These properties are typical of wire medium lenses [19,22]. At frequencies below the Fabry–Perot resonance the imaging is disturbed by the excitation of surface waves at both the front and back interfaces of the lens. This effect can be clearly seen in Figure 9.5 where the image and the source plane field distributions in the presence of the lens at frequencies below Fabry–Perot resonance are completely distorted by the ripples caused by the excitation of surface wave modes [19,22].

Despite the described dependency of the resolution on frequency, the resolving capabilities of wire medium lenses are reasonably wide-band. The actual bandwidth of operation depends on the complexity of the near field source and on its sensitivity to external fields [22], but it cannot be smaller than the fundamental theoretical limit formulated in Reference 19. In practice, a bandwidth of the order of a few percents is observed: 4.5% in Reference 22 for a half-wavelength thick lens with a complex meander-like source, and 2% in experiment with a 3 wavelength thick lens with a crown-shaped source [20]. For less complex sources the same lenses may provide much better bandwidth. For example, in Reference 21 the bandwidth was 18%, for a half-wavelength thick lens with a source shaped in the form of letter "P."

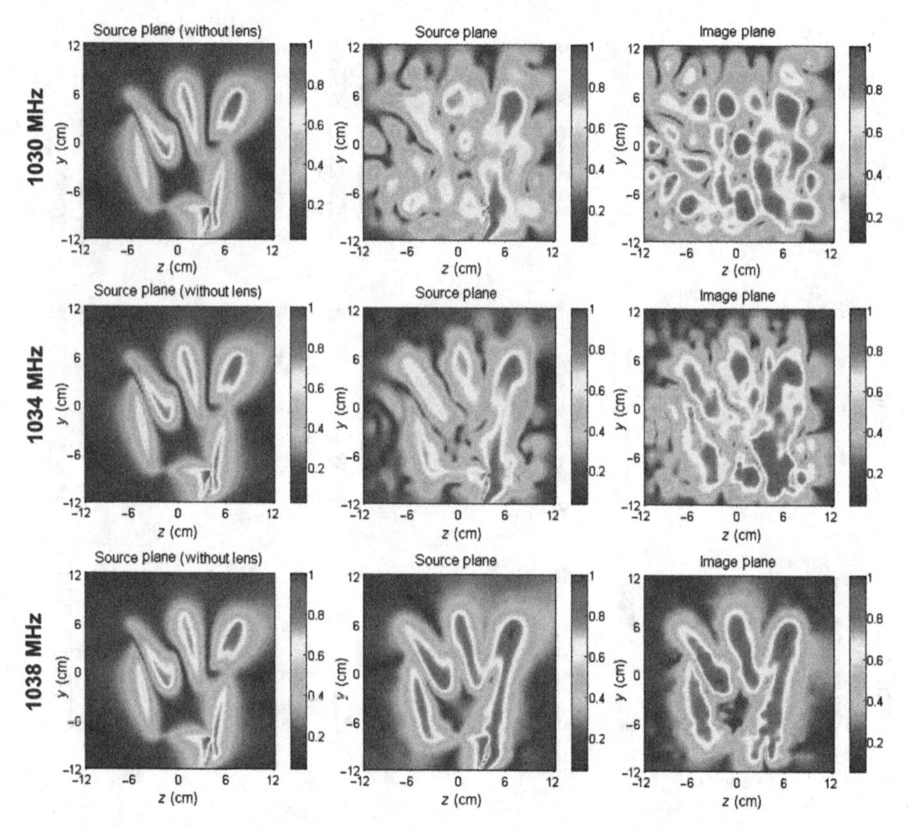

FIGURE 9.5 Results of the near field scan at the source plane (2 mm away from the front interface of the lens) without (left column) and with (central column) the wire medium lens and (right column) at the image plane (2 mm away from the back interface of the lens) at 1030, 1034 and 1038 MHz.

In principle, the resolution of the wire medium lens can be made as fine as required by a specific application by just using materials with smaller periods. For example, instead of a lens with 1 cm period and wires with 1 mm radius (as in Fig. 9.3) providing resolution $\lambda/15$, one can manufacture a structure with 1 mm period and 100 μm radius of wires which would provide a resolution of $\lambda/150$ and so on.[1] Basically, the minimum achievable resolution is limited only by the manufacturing capabilities and by the skin depth of the metal. In order to operate the wire medium in the canalization regime it is necessary that the radius of the rods is greater than the skin depth of the metal [23]. At microwaves, the skin depth of good metals is of the order of a few micrometers, and is thus negligible. The ultimate limit of resolution of wire medium lenses in different frequency regimes is discussed in Section 9.4.

[1]However, one should keep in mind that to take advantage of a lens with a better resolution the source may need to be placed closer to the front interface.

9.3 MAGNIFYING AND DEMAGNIFYING LENSES WITH SUPER-RESOLUTION

Processing of subwavelength images, such as magnification or demagnification, can be performed by tapering the array of wires. Thus, by gradually increasing (decreasing) the spacing between wires from the front interface to the back interface, it is possible to magnify (demagnify) the subwavelength details of an image.

A numerical simulation of the tapered version of the 1 m lens described in Section 9.2 was performed using a commercial electromagnetic simulator based on the method of moments (MoM) [24]. The geometry of the structure is presented in Figure 9.6a. The spacing between the wires at the front interface is 1 cm (as in Fig. 9.3) and it gradually increases up to 3 cm at the back interface. The front and back interfaces of the lens are spherical surfaces with 50 and 150 cm radii, respectively. The lens is excited by a planar center-fed wire antenna in the form of letter "M," which is placed in the close vicinity of the front interface. The frequency of operation

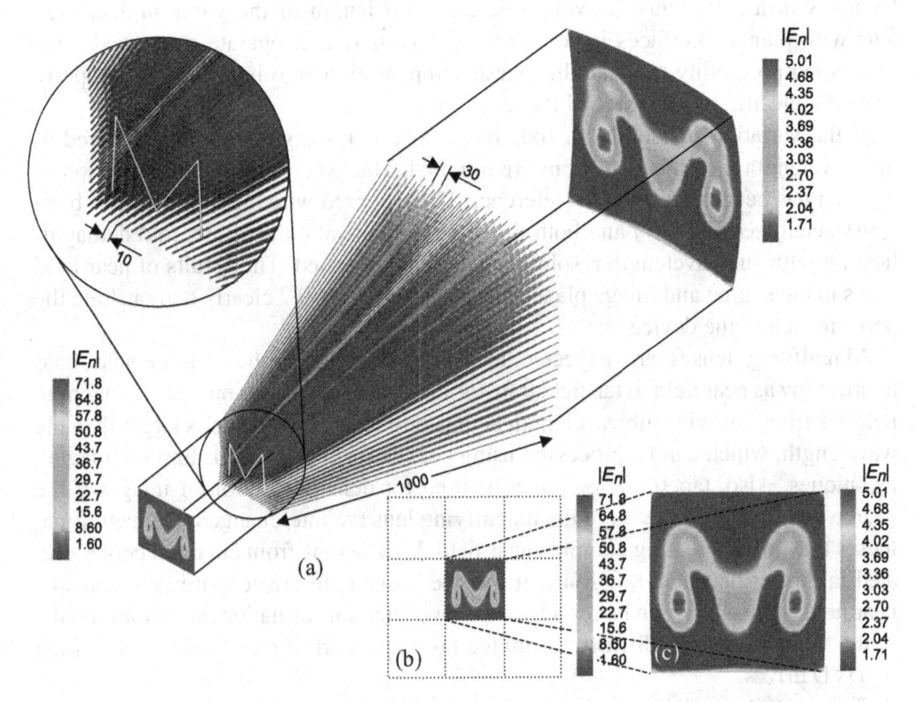

FIGURE 9.6 (a) Geometry of the magnifying wire medium lens excited by a planar near field source shaped as the letter "M." The distance between the source and the center of the front interface is 13 mm. All the length dimensions in the figure are given in millimeters. Calculated distributions of the electric field normal component: (b) on a planar surface located (13+8) mm away from the lens input interface (the reference plane is displaced 8 mm from the source) and (c) on a spherical surface displaced 15 mm from the output interface. The frequency of operation is 910 MHz. Reprinted with permission from Reference 20. Copyright 2008 American Physical Society.

(910 MHz) is chosen slightly higher than the Fabry–Perot resonance of the 6th order (900 MHz) in order to achieve the best quality of imaging [25].

This structure enables a threefold magnification of the source. This fact is clearly seen in Figures 9.6b and 9.6c where the distributions of the normal (with respect to the interface) component of electric field at 8 mm distance from the source and 15 mm distance from the back interface are plotted. The distance in-between the source and the image is about 3 wavelengths. In both figures a sharp near field distribution in the form of the letter M is visible, but remarkably the image is magnified three times as compared to the original. A larger magnification is possible with a similar structure, but it requires the use of a thicker lens.

The proposed tapered lens (Fig. 9.6a) has spherical input and output interfaces, but some applications may require the use of planar interfaces. The difficulty of making a tapered lens with planar interfaces is that it inevitably requires the use of wires with different physical lengths. This may destroy the collective Fabry–Perot resonance and affect the performance of the lens. However, returning to the explanation that each wire of the lens operates as a separate subwavelength waveguide, one can conclude that the physical length of the wires does not really matter, provided its electrical length is invariant. Thus, provided the electrical length of the wires in a tapered lens with planar interfaces is constant, the lens may still operate as desired. One interesting possibility to make this phase compensation is to incorporate a properly shaped dielectric block inside of the lens [26].

If the variation of lengths of rods in the lens is not significant as compared to the wavelength then the distortions are minimal. The experimental demonstration of this fact has been performed in Reference 27. A tapered wire medium lens has been constructed (see Fig. 9.7) and both threefold magnification and threefold demagnification with subwavelength resolution were demonstrated. The results of near field scans in the source and image planes presented in Figure 9.7 clearly demonstrate the performance of the device.

Magnifying lenses are expected to find immediate application in near field microscopy as near field to far-field transformers [4–7] since they may allow mapping field distributions with subwavelength details into images with details larger than the wavelength, which can be processed using conventional diffraction-limited imaging techniques. Also, tapered lenses may be used for demagnification of images if the source and image interfaces of the magnifying lens are interchanged. Demagnifying lenses may allow creating complex near field distributions from enlarged copies created in the far field. On this route, it may be possible to create extremely compact near field spots which can be used for dense writing of information on various media, in the same manner as diffraction-limited laser spots are currently used for writing on DVD drives.

Tapered wire medium lenses, especially in the terahertz range, can be regarded as multipixel endoscopes capable of performing direct and inverse manipulations with the electromagnetic fields in the subwavelength spatial scale: capturing an electromagnetic field profile created by deeply subwavelength objects at the endoscope's tip and magnifying it for observation, or projecting a large mask at the endoscope's base onto a much smaller image at the tip [28].

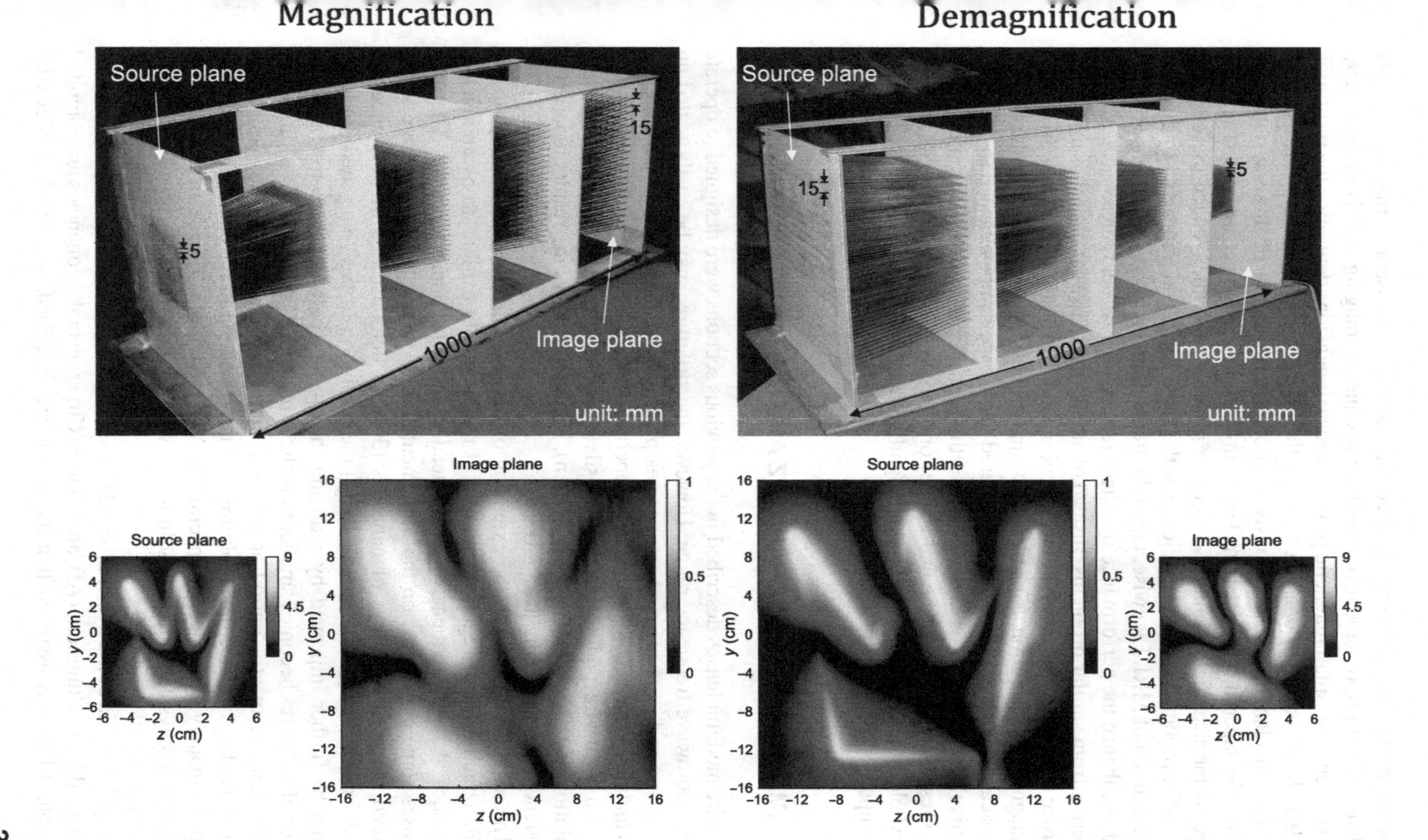

FIGURE 9.7 Magnification and demagnification experiments with tapered array of brass wires. The geometries of the lenses and the near field distributions in the source and image planes are shown. The frequency of operation for magnification regime is 1047 MHz and for demagnification 455 MHz. Reprinted with permission from Reference 27. Copyright 2010 American Institute of Physics.

In the microwave range the tapered wire medium endoscopes may be possibly applied for the improvement of magnetic resonance imaging (MRI) systems [29]. Such systems use small magnetic probes, whose dimensions determine the resolution of MRI. To ensure a reasonable efficiency the probes cannot be made much smaller than the wavelength. This obviously restricts the resolution of these systems. By placing the tapered wire medium lenses in-between the object under test and the probe of the near field scanner, it may be possible to further reduce the effective size of the probe keeping its efficiency at the same level, and improving in this way the resolution of MRI. Likewise, tapered wire medium endoscopes and lenses can be used to enhance the performance of other mechanical near field microwave scanners suffering from similar problems. Instead of attempting to use small probes in order to improve resolution one can magnify the near field distribution under test and detect the magnified distribution using a conventional probe. It is envisioned that the tapered lenses can be used in two ways: The tapered endoscope can be attached to the probe effectively reducing its size (in this case the scan is performed along the surface under test using the tip of the endoscope) or a large tapered lens can be attached to the whole sample under test and the magnified near field distribution is scanned at the output interface of the lens.

9.4 IMAGING AT THE TERAHERTZ AND INFRARED FREQUENCIES

The wire medium lenses described in the previous sections were designed to operate at the microwave frequency range. However, the structures may be easily scaled up to terahertz and even infrared frequencies provided the metal used to fabricate the rods maintains reasonable conductive properties. For example, in Reference 30 it was reported that a wire medium lens made of silver (Ag) nanorods enables subwavelength resolution of $\lambda/10$ at 30 THz ($\lambda = 10$ μm). Figure 9.8 shows the results reported in Reference 30 for imaging at 31 THz.

The main difficulty with the extension of the wire medium lens concept to the terahertz and infrared regimes is that most metals do not exhibit strong conductive properties at these frequencies, but instead have a plasmonic response. This fact imposes some constraints on the implementation of the canalization regime at the infrared range using arrays of nanorods [30]. As demonstrated in Reference 23, for real metals (with finite conductivity) the resolution of the "wire medium lens" cannot be further improved by reducing the spacing between the wires after the radius of the wires becomes smaller than the skin depth of the metal. In addition, for geometrical reasons the period of the array should be at least two times larger than the radius of rods. Since the resolution of a lens operated in the canalization regime is approximately two times the spacing between the wires [19], we can estimate that the ultimate physical limit of resolution of wire medium lenses is about four times the skin depth of the metal.

To give an idea of the possibilities, the limits of resolution yielded by rods of silver (Ag), gold (Au), aluminum (Al), and copper (Cu) at several frequencies are compiled in Table 9.1. It was assumed that the metals are described by a Drude model with

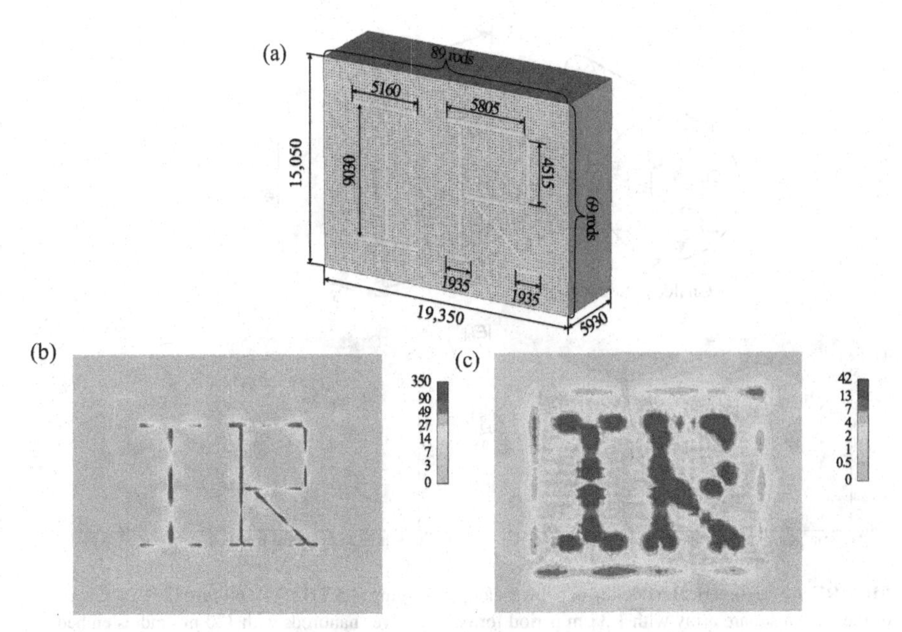

FIGURE 9.8 Numerical modeling of subwavelength imaging at 31 THz ($\lambda = 10$ μm) [30]. (a) Geometry of the lens: a square array with 215 nm period formed by silver nanorods with 21.5 nm radius embedded into a block of chalcogenide glass (relative permittivity $\varepsilon = 2.2$). All dimensions in the figure are given in nanometers. The lens is excited by a planar antenna shaped in the form of the letters "IR," and located 107.5 nm apart from the front interface of the lens. Calculated distributions of the normal component of electric field: (b) at the front interface (source plane) and (c) at the back interface (image plane). Reprinted with permission from Reference 30. Copyright 2007 American Physical Society.

the plasma and damping frequencies as in References 31 and 32. It is seen that at microwave frequencies the ultimate limit of resolution is several orders of magnitude larger than the classical diffraction limit, and that the best results are achieved with silver. Even at terahertz frequencies the resolution may be of the order of $\lambda/300$. In particular, one may conclude that at the terahertz range, arrays of parallel rods made of noble metals and their tapered versions may enable the same manipulations in

TABLE 9.1 Ultimate Limit of Resolution of Wire Medium Slabs for Several Metals and Frequencies

	100 MHz	1 GHz	10 GHz	100 GHz	1 THz	10 THz
Ag	$\lambda/116{,}000$ (26 μm)	$\lambda/37{,}000$ (8.2 μm)	$\lambda/11{,}600$ (2.6 μm)	$\lambda/3700$ (0.8 μm)	$\lambda/1265$ (237 nm)	$\lambda/320$ (94 nm)
Au	$\lambda/95{,}000$ (32 μm)	$\lambda/30{,}000$ (10 μm)	$\lambda/9500$ (3.2 μm)	$\lambda/3020$ (1.0 μm)	$\lambda/1010$ (297 nm)	$\lambda/300$ (100 nm)
Al	$\lambda/90{,}000$ (33 μm)	$\lambda/28{,}000$ (11 μm)	$\lambda/9000$ (3.3 μm)	$\lambda/2850$ (1.1 μm)	$\lambda/923$ (325 nm)	$\lambda/324$ (93 nm)
Cu	$\lambda/75{,}000$ (41 μm)	$\lambda/23{,}000$ (13 μm)	$\lambda/7500$ (4.1 μm)	$\lambda/2340$ (1.3 μm)	$\lambda/776$ (387 nm)	$\lambda/248$ (121 nm)

The Metals were Characterized using the experimental data from Reference 31.

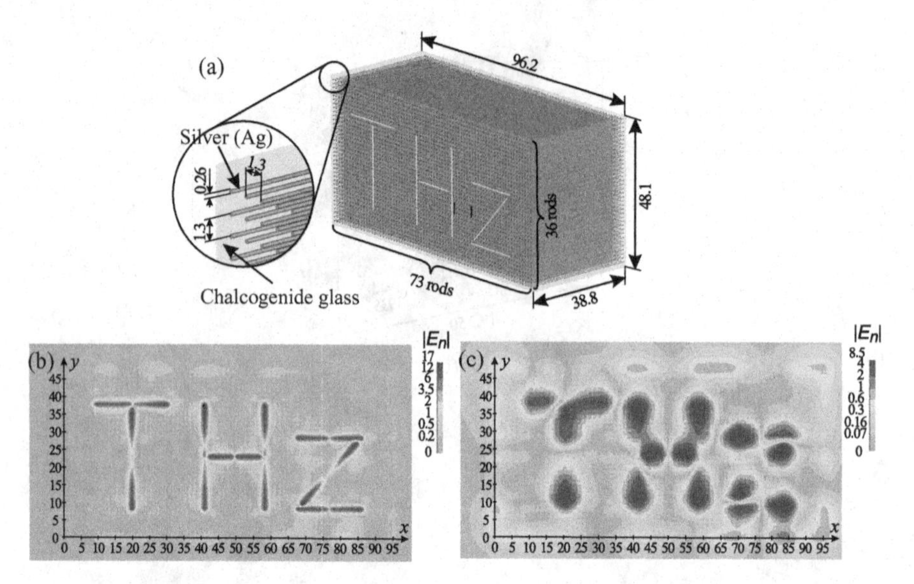

FIGURE 9.9 Numerical modeling of subwavelength imaging at 5 THz ($\lambda = 60$ μm) [23]. (a) Geometry of the lens: a square array with 1.3 μm period formed by silver nanorods with 130 nm radius embedded into a block of chalcogenide glass (relative permittivity $\varepsilon = 2.2$). All dimensions in the figure are given in micrometers. The lens is excited by a planar antenna shaped in the form of the letters "THz" and located 650 nm apart from the front interface of the lens. Calculated distributions of the normal component of electric field: (b) at the front interface (source plane) and (c) at the back interface (image plane).

the subwavelength spatial scale as their microwave counterparts (Figs. 9.3 and 9.6a). The only difference is that the structures for the terahertz range are more difficult to fabricate.

To illustrate the potentials of wire medium lenses at terahertz frequencies, we consider the structure depicted in Figure 9.9a. The lens is formed by an array of silver nanorods supported by a block of chalcogenide glass, which has low losses at terahertz frequencies. The source antenna is shaped as the letters "THz" in order to emphasize the frequency of operation: 5 THz. The performance of the lens was simulated using the commercial electromagnetic solver [33]. The calculated near field distributions at the front and back interfaces are presented in Figures 9.9b and 9.9c, respectively. The letters "THz" are clearly visible in both distributions. The effect of realistic losses in silver at 5 THz [31, 32] was fully taken into account in the simulation. In this example the radius of the rods R was chosen so that $R = 4.8\delta$ (δ is the skin depth of silver), which enables very good imaging. The resolution in this example is $\lambda/23$ (2.6 μm) and the image is formed at a distance 0.65λ (38.8 μm) from the source.

Interestingly, high losses in the rods material may not affect the imaging performance provided the absolute value of permittivity is large enough. This happens because the field canalized through the lens is mainly concentrated in the host material in-between the rods. It is extruded from the rods because of their high permittivity. This means that extremely lossy materials, for example, materials operated near absorption bands, may also be good candidates for the realization of wire media with extreme anisotropy.

Regarding fabrication issues, it has to be noted that the realization of wire medium lenses does not need to be very precise. The lens performance is not sensitive to the exact locations of the rods in the matrix, variations of the radii of the rods, or the precise shape of the rods cross section. The only parameter that it is necessary to control is the length of the rods. It should be roughly the same for all rods, and the rods need to be continuous. Broken rods or rods with conductivity defects may significantly affect the imaging quality.

An experimental demonstration of subwavelength imaging by array of metallic nanorods in the infrared range has been performed in Reference 34. A manufactured nanolens consisted of aligned gold nanowires, with 12 nm diameters and lattice spacing of 25 nm, embedded in 10 μm thick porous alumina template matrix. Figure 9.10 shows the near field optical microscope scans in the source and image planes at 1550 nm wavelength. It is clear that the subwavelength details of the source were

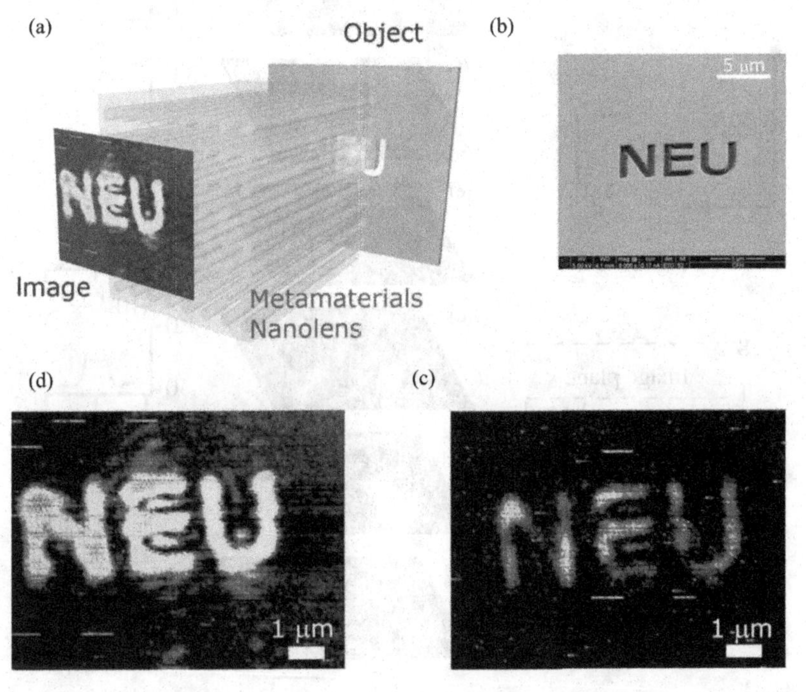

FIGURE 9.10 (a) Imaging with subwavelength resolution by the metamaterial nanolens at 1550 nm. The nanolens consists of high aspect ratio metallic nanowires which are embedded in a host dielectric medium. This bulk metamaterial transports subwavelength details of an object at a significant distance of more than six times the wavelength (λ). (b) Scanning electron microscope (SEM) image of the NEU letters (acronym for Northeastern University) milled in 100 nm thick gold metallic film. The letters have 600 nm wide arms (0.4λ). (c) near field scanning optical microscope (NSOM) scan of the source object in the near field at 1550 nm wavelength. (d) NSOM scan of the corresponding image by the metamaterial nanolens above the nanolens surface. Reprinted with permission from Reference 34. Copyright 2010 American Institute of Physics.

transmitted to the image plane with $\lambda/4$ resolution over 5λ distance. The manufacturing of nanolenses using porous alumina templates is a very prospective technique for mass production since it uses chemical processes and self-organization mechanisms at the nanoscale which are quite cheap and well established.

9.5 NANOLENSES FORMED BY NANOROD ARRAYS FOR THE VISIBLE FREQUENCY RANGE

In the visible frequency range it may also be possible to achieve subwavelength imaging using metallic nanorods [35]; see Figure 9.11. This possibility is based on the excitation of surface plasmon-polaritons in the metallic rods. However, this regime is highly sensitive to losses and may introduce important phase and amplitude

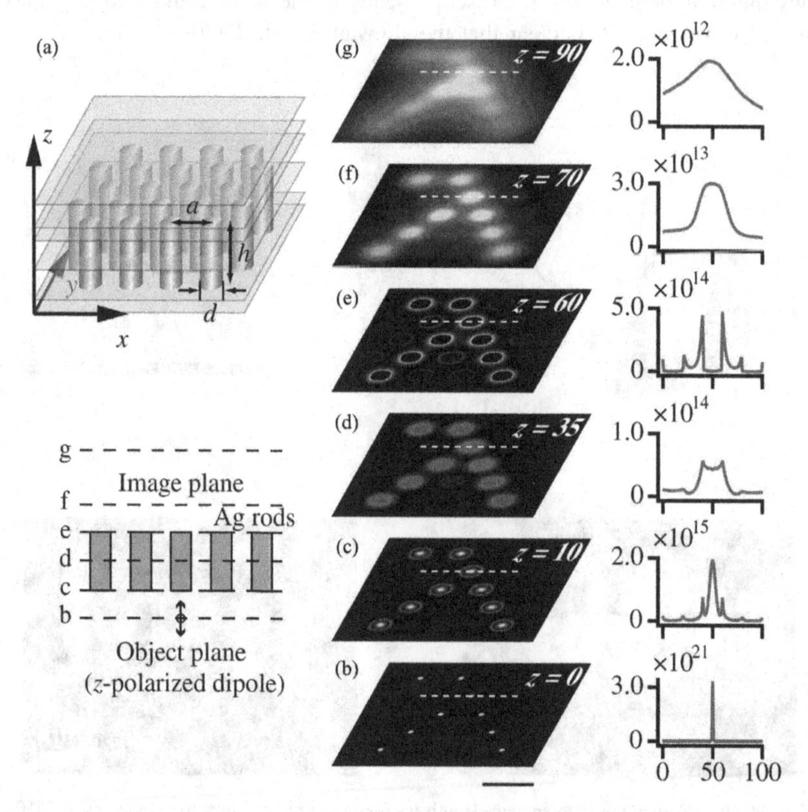

FIGURE 9.11 Subwavelength plasmonic image transfer of a character pattern "λ" with a metallic nanorod array device [35]. (a) The structural model of the device, which is constructed by hexagonally arranged silver nanorods of 20 nm diameter, 50 nm height, and 40 nm pitch, respectively. (b–g) Field propagation process in the image transfer obtained at each longitudinal position by the FDTD simulation. The character pattern is composed of an array of z-polarized dipoles. Left-side images show the cross-sectional intensity distributions in the x–y plane. The plots on the right side show the cross-sectional line profiles of dashed lines in the left images. The object plane and imaging plane are defined as $z = 0$ and $z = 70$ nm. The operation wavelength is 488 nm. The size of the scale bar is 50 nm. Reprinted with permission from Reference 35. Copyright 2005 American Physical Society.

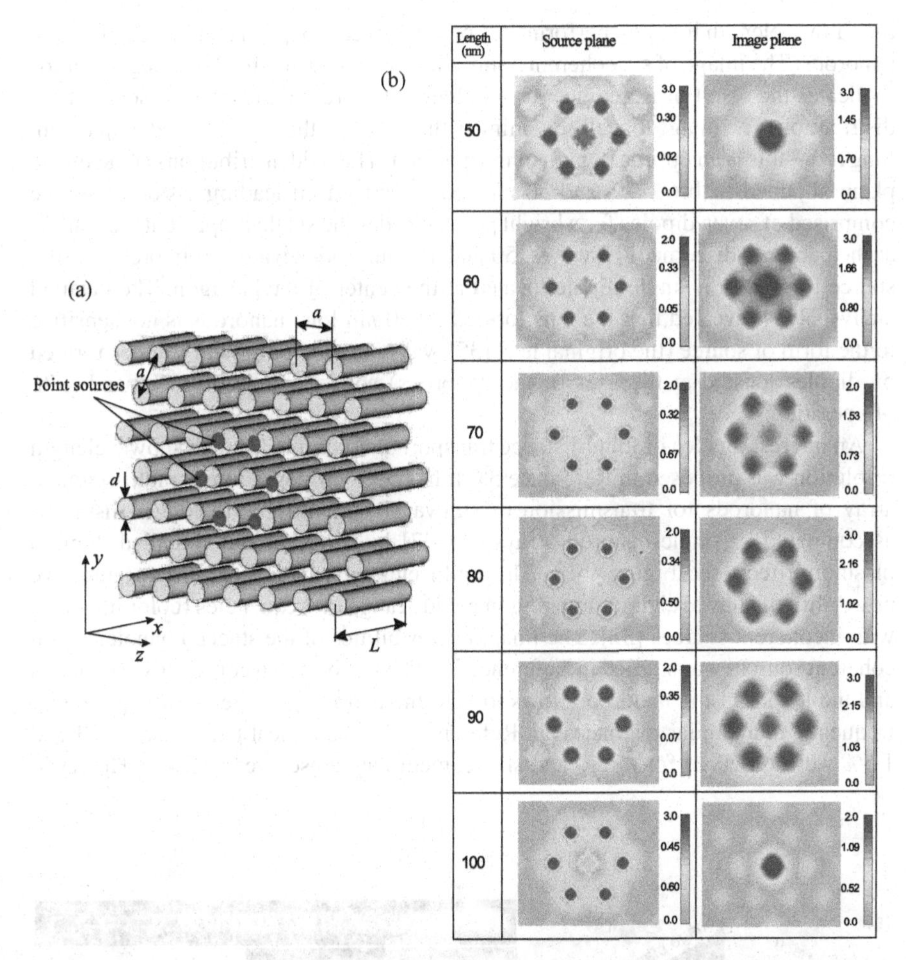

FIGURE 9.12 (a) Geometry of the finite array of silver nanorods with triangular-lattice arrangement. The point coherent sources are aligned with the axis of the nanorod and are placed at a distance 10 nm away from array interface. The diameter of the nanorods is $d = 20$ nm, the period of the structure is $a = 40$ nm. The length of nanorods L is varied. (b) Source and image plane field distribution for various lengths of the nanorod at 488 nm wavelength. The normal component of electric field is sampled at the front interface of the nanolens and across a plane 10 nm away from back interface, respectively. Reprinted with permission from Reference 36. Copyright 2010 American Physical Society.

distortions in the image due to the harmful interaction among surface plasmons traveling along adjacent rods. In particular, this may result in the impossibility of resolving certain source field distributions at the image plane.

The nanorod array proposed in Reference 35 operates with incoherent sources and does not image properly coherent sources with certain phase distributions. This happens because the length of nanorods is not properly chosen for imaging at the wavelength of interest. It has been shown in Reference 36 that the proper choice of the nanorod length (so that the length obeys Fabry–Perot resonance condition and thus, the lens operates in canalization regime) makes possible to image arbitrary coherent sources with the same geometry of nanorod array. Figure 9.12 shows how the subwavelength imaging performance of the nanolens depends on the length of the nanorods. The image of six coherent point sources obtained with 50 nm long nanorods replicates the result of Reference 35 and does not represent the original source field distribution. The image that closely mimics the source is the one obtained with 80 nm long rod—this is an optimal length of the nanorod. The field distributions at the image plane obtained with 60, 70, and 90 nm rods give us a misleading vision of source comprised of seven dipoles for a bright point, besides the six that represent the source, at the center is discerned. However, 50 and 100 nm rods give us an impression of a source formed of a single dipole located at the center of the hexagon. The detailed analysis [37] revealed that the lens formed by 80 nm long nanorods is not sensitive to the form of source (the original lens [35] was operating only with sources formed by dipoles located at the axes of the nanorods) and offers about 10% bandwidth of operation.

An interesting idea of long-distance transport of color images with subwavelength resolution was proposed in Reference 38. It has been proposed to use multi-segment array of nanorods for transmission of subwavelength images to longer distances as compared to single-nanorod arrays [35–37] by virtue of the so-called domino plasmon effect; see Figure 9.13. The multisegment nanolens offers an attractive possibility of subwavelength imaging in a wide range of frequencies (color imaging) with incoherent sources [38]. The imaging capabilities of the stacked nanolens with coherent sources was studied in Reference 39. The gap between consecutive segments and the length of the nanorod allows to tune the nanolens to operate at a particular frequency in a similar manner as in Reference 36. The typical bandwidths of 3 and 1.5% were observed for three- and six-segment nanolenses, respectively. However,

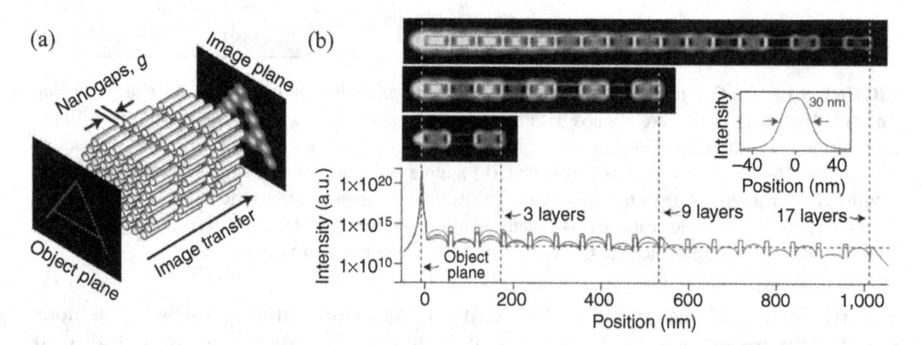

FIGURE 9.13 Long-distance image transfer realization through a stacked arrangement of nanorod arrays. (a) The basic concept of the stacked arrangement of nanorod arrays for long-distance image transfer. In this example, a three-layered stacked arrangement is presented. The image of the letter A is obtained from simulations based on 3D-FDTD calculations. (b) Intensity profiles across 3-, 9-, and 17-layered stacked arrays, showing that neither intensity nor the point spread function has any noticeable dependence on distance, at least up to micrometer-scale distances. The inset shows the line width of the point spread function at the exit of 17 layers. Reprinted with permission from Reference 38. Copyright 2008 Macmillan Publishers Limited.

despite all the positive traits for the performance of stacked nanolenses, the color imaging of coherent sources remains elusive.

9.6 SUPERLENSES AND HYPERLENSES FORMED BY MULTILAYERED METAL–DIELECTRIC NANOSTRUCTURES

In the visible frequency range there is an interesting alternative opportunity to create extremely anisotropic material (9.4) without employing metallic nanorods. This can be achieved by using multilayered metal–dielectric nanostructures [40]. Let us consider a layered structure presented in Figure 9.14.

Such a metamaterial can be described using permittivity tensor of the form

$$\overline{\overline{\varepsilon}} = \begin{pmatrix} \varepsilon_\perp & 0 & 0 \\ 0 & \varepsilon_\parallel & 0 \\ 0 & 0 & \varepsilon_\parallel \end{pmatrix} \tag{9.5}$$

where

$$\varepsilon_\parallel = \frac{\varepsilon_1 d_1 + \varepsilon_2 d_2}{d_1 + d_2}, \quad \varepsilon_\perp = \left[\frac{\varepsilon_1^{-1} d_1 + \varepsilon_2^{-1} d_2}{d_1 + d_2}\right]^{-1}.$$

In order to get $\varepsilon_\parallel = 1$ and $\varepsilon_\perp = \infty$ and obtain a material with permittivity tensor of the form (9.4), required for implementation of the canalization regime, it is necessary to choose parameters of the layered material so that $\varepsilon_1/\varepsilon_2 = -d_1/d_2$ and $\varepsilon_1 + \varepsilon_2 = 1$. From the first equation it is clear that one of the layers should have negative permittivity and thus the structure has to be formed by one dielectric layer and one metallic layer. For example, one can choose $\varepsilon_1 = 2$, $\varepsilon_2 = -1$, and $d_1/d_2 = 2$, or $\varepsilon_1 = 15$, $\varepsilon_2 = -14$, and $d_1/d_2 = 15/14$.

Note that no layered structure required for canalization regime can be formed using equally thick layers $d_1 = d_2$. The layered metal–dielectric structures considered in References 41–43 have completely different properties. As it is noted in Reference 43, the structures with $d_1 = d_2$ and $\epsilon_1 = -\epsilon_2$ correspond to $\epsilon_\perp = 0$ and $\varepsilon_\parallel = \infty$ and operate in tunneling regime as an array of wires embedded into the medium with zero permittivity. The absence of matching causes strong reflections

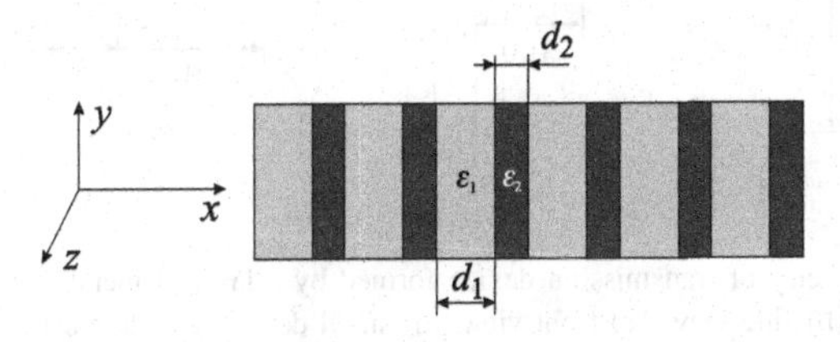

FIGURE 9.14 Geometry of layered metal–dielectric metamaterial.

and restricts slab thickness to be thin. More details about differences between canalization and tunneling regimes in multilayered metal–dielectric structures are available in References 44 and 45.

The canalization regime implementation using the suggested metal–dielectric layered structure has been demonstrated in Reference 11 numerically using CST Microwave Studio package [33]. A subwavelength source (a loop of the current in the form of letter P) is placed at 20 nm distance from a 300 nm thick multilayer slab composed of 10 and 5 nm thick layers with $\varepsilon_1 = 2$ and $\varepsilon_2 = -1$, respectively. The detailed geometry of the structure is presented in Figure 9.15. The wavelength of operation λ is 600 nm. The metal with $\epsilon = -1$ at 600 nm wavelength can be

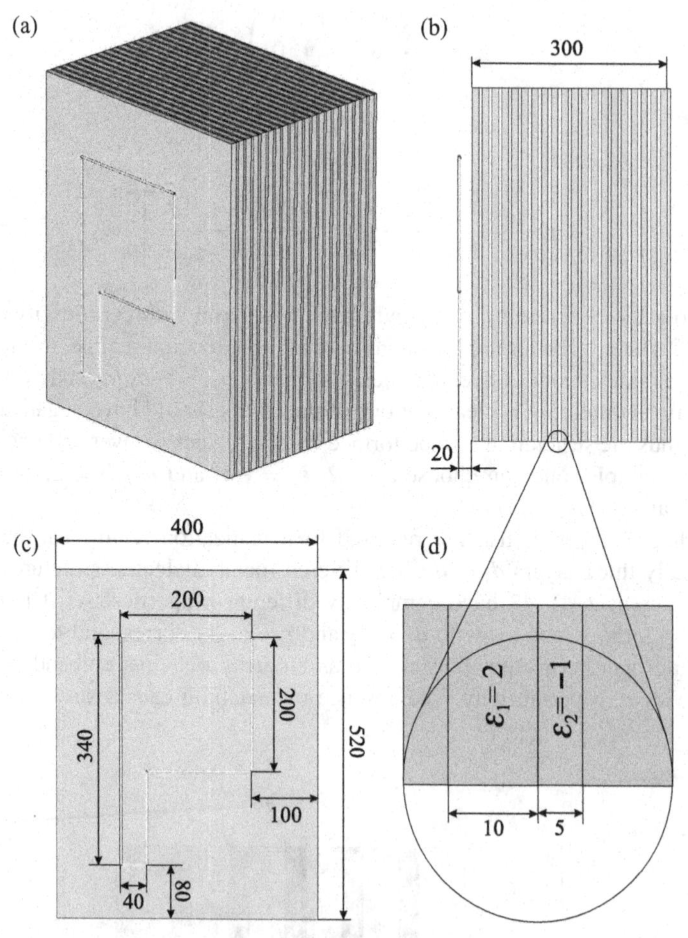

FIGURE 9.15 Geometry of transmission device formed by a layered metal–dielectric metamaterial: (a) perspective view, (b) side view, (c) front view, (d) small details. All dimensions are in nanometers. Reprinted with permission from Reference 11. Copyright 2006 American Physical Society.

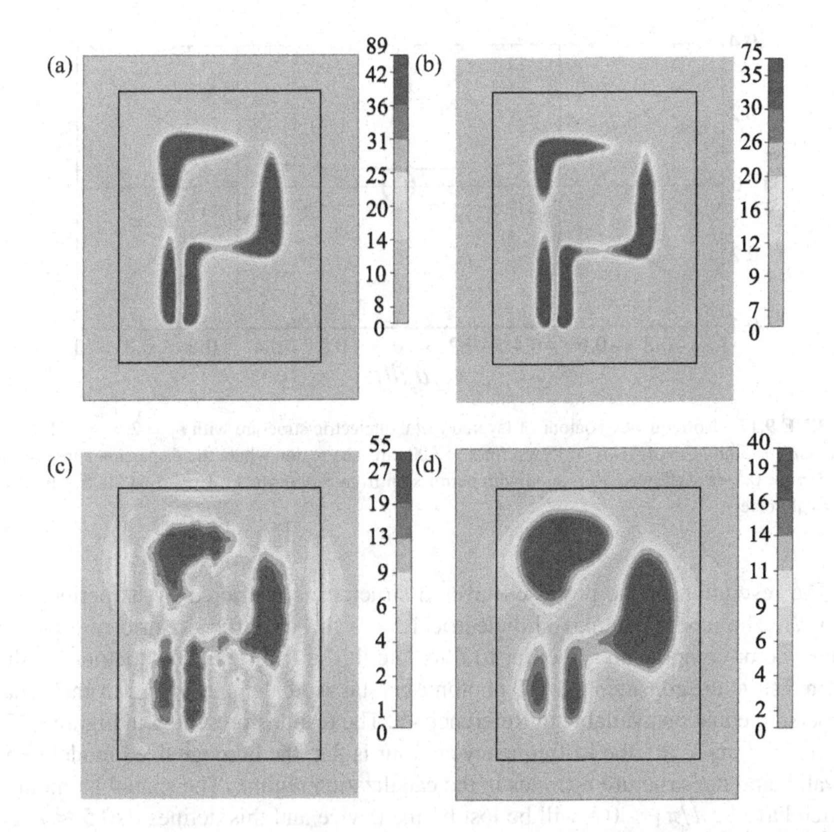

FIGURE 9.16 Distributions of absolute value normal to the interface component of electrical field (arbitrary units): (a) in free space, 20 nm from source; (b) at the front interface; (c) at the back interface; (d) in free space, 20 nm from back interface.

created by doping some lossless dielectric by a small concentration of silver which has $\epsilon = -15$ at such frequencies, in a similar manner to the ideas of works [46,47].

The field distributions in the planes parallel to the interface of the structure plotted in Figure 9.16 clearly demonstrate imaging with 30 nm resolution ($\lambda/20$). Figure 9.16a shows the field produced by the source in free space at 20 nm distance. It is practically identical to the field observed at the front interface of the structure (Fig. 9.16b). It confirms that the reflections from the front interface are negligibly small. Actually, the main contribution to the reflected field comes from diffraction at corners and wedges of the device. Figure 9.16c shows the field at the back interface of the structure. The image is clearly visible, but it is a little bit distorted by plasmon-polariton modes excited at the back interface. Being diffracted into the free space this distribution forms an image without distortions produced by plasmon-polariton modes; see Figure 9.16d presenting field distribution at 20 nm distance from the back interface.

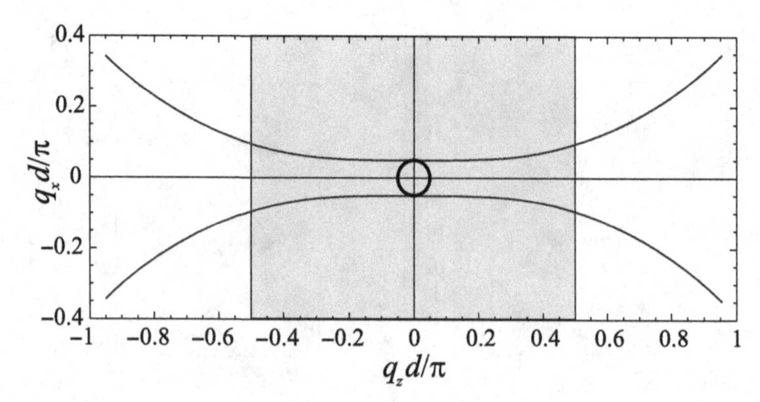

FIGURE 9.17 Isofrequency contour of layered metal–dielectric structure with $\varepsilon_1 = 2$, $\varepsilon_2 = -1$, $d_1 = 10$ nm, $d_2 = 5$ nm, $d = d_1 + d_2 = 15$ nm for $\lambda = 600$ nm. The region where the dispersion curve is flat ($|q_z d/\pi| < 0.5$) is shadowed. Reprinted with permission from Reference 11. Copyright 2006 American Physical Society.

The resolution of the proposed layered structure is restricted by its period $d = d_1 + d_2$. The model of uniaxial dielectric 1.5 is valid only for a restricted range of wave-vector components. In order to illustrate this an isofrequency contour for the layered structure considered as 1D photonic crystal was calculated using an analytical dispersion equation available in Reference 48. The result is presented in Figure 9.17. While $|q_z d/\pi| < 0.5$ the isofrequency contour is flat, the homogenized model (9.5) is valid, and the structure operates in the canalization regime. The spatial harmonics which have $|q_z d/\pi| > 0.5$ will be lost by the device and this defines $d/0.5 \approx \lambda/20$ resolution. The calculation of isofrequency contour for the case of $\varepsilon_1 = 15$, $\varepsilon_2 = -14$, $d_1 = 7.76$ nm, and $d_2 = 7.24$ nm for the same wavelength of 600 nm revealed its flatness for $|q_x d/\pi| > 1.5$. This allows us to predict $d/1.5 \approx \lambda/60$ resolution for this case. The last structure can be constructed using silicon as dielectric and silver as metal, but very accurate fabrication with error no more than 0.05 nm will be required

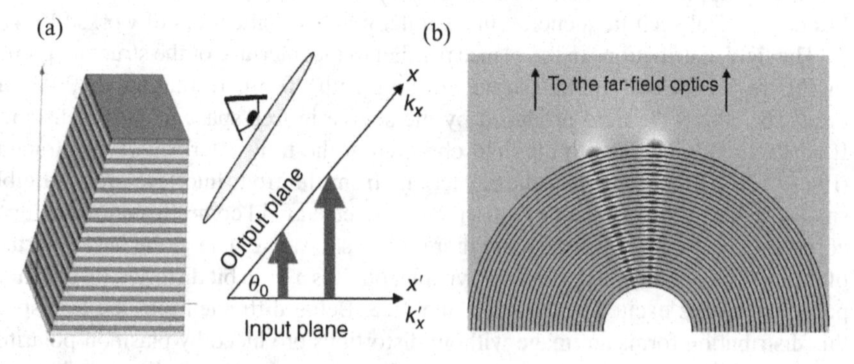

FIGURE 9.18 Hyperlenses. (a) Prism geometry. (b) Cylindrical geometry. Reprinted with permission from Reference 4. Copyright 2006 American Physical Society.

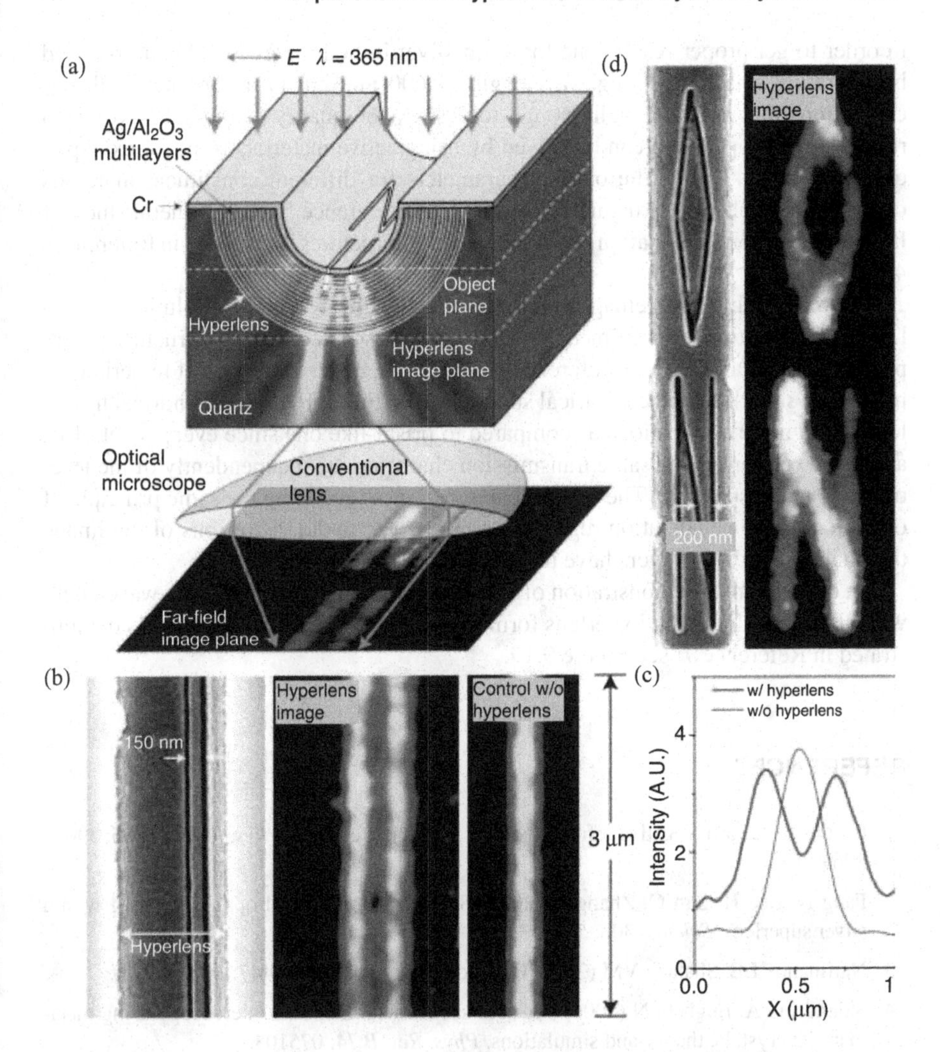

FIGURE 9.19 Magnifying optical hyperlens. (a) Schematic of hyperlens and numerical simulation of imaging of subdiffraction-limited objects. (b) Hyperlens imaging of line pair object with line width of 35 nm and spacing of 150 nm. From left to right, scanning electron microscope image of the line pair object fabricated near the inner side of the hyperlens, magnified hyperlens image showing that the 150-nm-spaced line pair object can be clearly resolved, and the resulting diffraction-limited image from a control experiment without the hyperlens. (c) The averaged cross section of hyperlens image of the line pair object with 150 nm spacing (red), whereas a diffraction-limited image obtained in the control experiment (green). A.U., arbitrary units. (d) An arbitrary object ON imaged with subdiffraction resolution. Line width of the object is about 40 nm. The hyperlens is made of 16 layers of Ag/Al_2O_3. Reprinted with permission from Reference 7. Copyright 2007 American Association for the Advancement of Science.

in order to get proper result. The losses in silver in both cases are already reduced by operating at a rather long wavelength of 600 nm, but in accordance with our estimations they are still high enough to destroy the quality of the subwavelength resolution. This problem can be solved by using active materials [49], for example, doped silicon [43, 46]. The optimal parameters for different constituent materials of the metal–dielectric structure are available in Reference 50. The general study of limitations on homogenization of periodic layered structures is provided in References 48 and 51.

The magnifying and demagnifying lenses with subwavelength resolution (hyperlenses, see Section 9.3) formed by layered metal–dielectric nanostructures were proposed simultaneously in References 4 and 5. There are two types of hyperlenses: prism-like superlens and cylindrical superlens (see Fig. 9.18). The cylindrical hyperlens found more applications as compared to prism-like one since every point of its aperture experiences the same transmission characteristic independently of the level of losses in the structure. The cylindrical hyperlenses are using the same principle of operation as the canalization regime, but in order to avoid distortions of the image the thicknesses of the layers have to be tuned accordingly [52].

An experimental demonstration of subwavelength imaging at 365 nm wavelength with 40 nm resolution by hyperlens formed by silver and alumina layers was demonstrated in Reference 7; see Figure 9.19.

REFERENCES

1. Pendry JB (2000) Negative refraction index makes perfect lens. *Phys. Rev. Lett.* 85: 3966–3969.
2. Fang N, Lee H, Sun C, Zhang X (2005) Subdiffraction-limited optical imaging with a silver superlens. *Science* 308: 534–537.
3. Narimanov EE, Shalaev VM (2007) Beyond diffraction. *Nature* 447: 266–267.
4. Salandrino A, Engheta N (2006) Far-field subdiffraction optical microscopy using metamaterial crystals: theory and simulations. *Phys. Rev. B* 74: 075103.
5. Jacob Z, Alekseyev LV, Narimanov E (2006) Optical hyperlens: far-field imaging beyond the diffraction limit. *Opt. Express* 14: 8247–8256.
6. Smolyaninov II, Hung Y-J, Davis CC (2007) Magnifying superlens in the visible frequency range. *Science* 315: 1699–1701.
7. Liu Z, Lee H, Xiong Y, Sun C, Zhang X (2007) Far-field optical hyperlens magnifying sub-diffraction-limited objects. *Science* 315: 1686.
8. Westphal V, Hell SW (2005) Nanoscale resolution in the focal plane of an optical microscope. *Phys. Rev. Lett.* 94: 143903.
9. Belov PA, Marques R, Maslovski SI, Nefedov IS, Silverinha M, Simovski CR, Tretyakov SA (2003) Strong spatial dispersion in wire media in the very large wavelength limit. *Phys. Rev. B* 67: 113103.

10. Silveirinha M (2006) Nonlocal homogenization model for a periodic array of epsilon-negative rods. *Phys. Rev. E* 73: 046612.

11. Belov PA, Hao Y (2006) Subwavelength imaging at optical frequencies using a transmission device formed by a periodic layered metal-dielectric structure operating in the canalization regime. *Phys. Rev. B* 73: 113110.

12. Belov PA, Simovski CR, Ikonen P (2005) Canalization of sub-wavelength images by electromagnetic crystals. *Phys. Rev. B.* 71: 193105.

13. Podolskiy VA, Narimanov EE (2005) Near-sighted superlens. *Opt. Lett.* 30: 75–77.

14. Yen TJ, Padilla WJ, Fang N, Vier DC, Smith DR, Pendry JB, Basov DN, Zhang Z (2004) Tetrahertz magnetic response from artificial materials. *Science* 303: 1494–1496.

15. Linden S, Enkrich C, Wegener M, Zhou J, Kochny T, Soukoulis C (2004) Magnetic response of metamaterials at 100 terahertz. *Science* 306: 1351–1353.

16. Zhang S, Fan W, Panoiu NC, Malloy KJ, Osgood RM, Brueck SRJ (2005) Experimental demonstration of near-infrared negative-index metamaterials. *Phys. Rev. Lett.* 97: 137404.

17. Shalaev VM (2007) Optical negative-index metamaterials. *Nat. Photonics* 1: 41–48.

18. Ikonen P, Belov PA, Simovski CR, Maslovski SI (2006) Experimental demonstration of subwavelength field channeling at microwave frequencies using a capacitively loaded wire medium. *Phys. Rev. B* 73: 073102.

19. Belov PA, Silveirinha MG (2006) Resolution of sub-wavelength lenses formed by a wire medium. *Phys. Rev. E* 73: 056607.

20. Belov PA, Zhao Y, Tse S, Ikonen P, Silveirinha MG, Simovski CR, Tretyakov SA, Hao Y, Parini C (2008) Transmission of images with subwavelength resolution to distances of several wavelengths in the microwave range. *Phys. Rev. B* 77: 193108.

21. Belov PA, Hao Y, Sudhakaran S (2006) Subwavelength microwave imaging using an array of parallel conducting wires as a lens. *Phys. Rev. B* 73: 033108.

22. Belov PA, Zhao Y, Sudhakaran S, Alomainy A, Hao Y (2006) Experimental study of the sub-wavelength imaging by a wire medium slab. *Appl. Phys. Lett.* 89: 262109.

23. Silveirinha MG, Belov PA, Simovski CR (2008) Ultimate limit of resolution of subwavelength imaging devices formed by metallic rods. *Opt. Lett.* 33: 1726–1728.

24. FEKO Suite 5.2, EM SOftware & Systems - SA (Pty) Ltd, http://www.feko.info.

25. Ikonen P, Simovski C, Tretyakov S, Belov P, Hao Y (2007) Magnification of subwavelength field distributions at microwave frequencies using a wire medium slab operating in the canalization regime. *Appl. Phys. Lett.* 91: 104102.

26. Zhao Y, Palikaras G, Belov PA, Dubrovka RF, Simovski CR, Hao Y, Parini CG (2010)t Magnification of subwavelength field distributions using a tapered array of metallic wires with planar interfaces and an embedded dielectric phase compensator. *New J. Phys.* 12: 103045.

27. Belov PA, Palikaras GK, Zhao Y, Rahman A, Simovski CR, Hao Y, Parini C (2010) Experimental demonstration of multiwire endoscopes capable of manipulating near fields with subwavelength resolution. *Appl. Phys. Lett.* 97: 191905.

28. Shvets G, Trendafilov S, Pendry JB, Sarychev A (2007) Guiding, focusing, and sensing on the subwavelength scale using metallic wire arrays. *Phys. Rev. Lett.* 99: 053903.
29. Radu X, Dardenne X, Craeye C (2007) Experimental results and discussion of imaging with a wire medium for mri imaging applications. In: *Proceedings of IEEE Antennas and Propagation Society International Symposium*. pp 5499–5502.
30. Silveirinha MG, Belov PA, Simovski CR (2007) Sub-wavelength imaging at infrared frequencies using an array of metallic nanorods. *Phys. Rev. B* 75: 035108.
31. Ordal MA, Bell RJ, Alexander RW, Long LL, Querry MR (1985) Optical properties of fourteen metals in the infrared and far infrared: Al, Co, Cu, Au, Fe, Pb, Mo, Ni, Pd, Pt, Ag, Ti, V, and W. *Appl. Opt.* 24: 4493–4499.
32. El-Kady I, Sigalas MM, Biswas R, Ho KM, Soukoulis CM (2000) Metallic photonic crystals at optical wavelengths. *Phys. Rev. B* 62: 15299–15302.
33. CST Microwave Studio 2006, CST GmbH, http://www.cst.com.
34. Casse BDF, Lu WT, Huang YJ, Gultepe E, Menon L, Sridhar S (2010) Super-resolution imaging using a three-dimensional metamaterials nanolens. *Appl. Phys. Lett.* 96: 023114.
35. Ono A, Kato J, Kawata S (2005) Subwavelength optical imaging through a metallic nanorod array. *Phys. Rev. Lett.* 95: 267407.
36. Rahman A, Belov PA, Hao Y (2010) Tailoring silver nanorod arrays for subwavelength imaging of arbitrary coherent sources. *Phys. Rev. B* 82: 113408.
37. Rahman A, Kosulnikov SYu, Hao Y, Parini C, Belov PA (2011) Subwavelength optical imaging with an array of silver nanorods. *J. Nanophotonics* 5: 051601.
38. Kawata S, Ono A, Verma P (2008) Subwavelength colour imaging with a metallic nanolens. *Nat. Photonics* 2: 438–442.
39. Voroshilov PM, Rahman A, Kivshar YS, Belov PA (2011) Efficiency of subwavelength imaging with multisegment nanolens. *J. Nanophotonics* 5: 053516.
40. Shen JT, Catrysse PB, Fan S (2005) Mechanism for designing metallic metamaterials with a high index of refraction. *Phys. Rev. Lett.* 94: 197401.
41. Shamonina E, Kalinin VA, Ringhofer KH, Solymar L (2001) Imaging, compression and Poynting vector streamlines for negative permittivity materials. *Electron. Lett.* 37: 1243–1244.
42. Anantha Ramakrishna S, Pendry JB, Wiltshire MCK, Stewart WJ (2003) Imaging the near field. *J. Mod. Opt.* 50: 1419–1430.
43. Anantha Ramakrishna S, Pendry JB (2003) Removal of absorption and increase in resolution in a near field lens via optical gain. *Phys. Rev. B* 67: 201101.
44. Li X, He S, Jin Y (2007) Subwavelength focusing with a multilayered Fabry-Perot structure at optical frequencies. *Phys. Rev. B* 75: 045103.
45. Kotynski R, Stefaniuk T (2009) Comparison of imaging with sub-wavelength resolution in the canalization and resonant tunnelling regimes. *J. Opt. A Pure Appl. Opt.* 11: 015001.
46. Garcia de Abajo FJ, Gomez-Santos G, Blanco LA, Borisov AG, Shabanov SV (2005) Tunneling mechanism of light transmission through metallic films. *Phys. Rev. Lett.* 95: 067403.

47. Cai W, Genov DA, Shalaev VM (2005) Superlens based on metal-dielectric composite. *Phys. Rev. B* 72: 193101.

48. Orlov AA, Voroshilov PM, Belov PA, Kivshar YuS (2011) Engineered optical nonlocality in nanostructured metamaterials. *Phys. Rev. B* 84: 045424.

49. Vincenti MA, de Ceglia D, Rondinone V, Ladisa A, D'Orazio A, Bloemer MJ, Scalora M (2009) Loss compensation in metal-dielectric structures in negative-refraction and super-resolving regimes. *Phys. Rev. A* 80: 053807.

50. Pastuszczak A, Kotynski R (2011) Optimized low-loss multilayers for imaging with sub-wavelength resolution in the visible wavelength range. *J. Appl. Phys.* 109(8): 084302.

51. Chebykin AV, Orlov AA, Vozianova AV, Maslovski SI, Kivshar YuS, Belov PA (2011) Nonlocal effective medium model for multilayered metal-dielectric metamaterials. *Phys. Rev. B* 84: 115438.

52. Kildishev AV, Narimanov EE (2007) Impedance-matched hyperlens. *Opt. Lett.* 32: 3432–3434.

10

Active and Tuneable Metallic Nanoslit Lenses

SATOSHI ISHII, XINGJIE NI, VLADIMIR P. DRACHEV*, MARK D. THORESON, VLADIMIR M. SHALAEV, AND ALEXANDER V. KILDISHEV

Birck Nanotechnology Center and School of Electrical and Computer Engineering, Purdue University, West Lafayette, Indiana, USA

**Department of Physics, University of North Texas, Denton, Texas, USA*

10.1 INTRODUCTION

In diffractive optics, typical feature sizes are on the order of the wavelength, which is around 1 μm for visible and near-infrared light. Although nanometer-scale features do not diffract optical waves, nanosized structures could enable us to design novel diffractive optical devices. Plasmonics and metamaterials are the two ways to couple light into nanoscale. Plasmonics allow the coupling of light with a free-electron gas, thus generating oscillation modes known as plasmons. Metamaterials can be used to create artificial electromagnetic spaces characterized by effective permittivities and permeabilities. In this chapter, we focus our discussion on diffraction lenses made of arrays of nanoslits milled in metallic films. We show that diffraction lenses could go beyond classical optics with plasmonics and metamaterials.

One way to take advantage of plasmonic materials is to make waveguides out of them. In plasmonic waveguides, light propagates as a surface plasmon polariton (SPP) which is nanoscale in size and oscillates at a few hundreds of terahertz. Plasmonic waveguides have the potential to help solve the problem of ever-increasing demand for smaller and faster communication devices. In the design of our nanoslit lenses, we utilize the unique dispersion features of plasmonic waveguides.

Among optical metamaterials [1], a unique class of metal–dielectric lamellar structures called hyperbolic metamaterials (HMMs) have received much recent attention.

Active Plasmonics and Tuneable Plasmonic Metamaterials, First Edition. Edited by Anatoly V. Zayats and Stefan A. Maier.

An HMM has strong uniaxial anisotropy where the principal elements of the permittivity tensor have opposite signs. Hence, the isofrequency curve of hyperbolic media is a hyperbola; this is different from an anisotropic dielectric whose dispersion is ellipsoidal. As a consequence of their hyperbolic dispersion, HMMs theoretically could possess many unique properties such as unbounded effective refractive index and super-resolution capabilities. When subwavelength slits are combined with HMMs, the diffractive nature becomes drastically different from diffraction in free space. For example, inside an HMM, the propagation of diffracted light from a nanoslit is highly directional. By utilizing this unique feature, focusing is possible with a subwavelength spot size [2].

The initial devices based on plasmonics and metamaterials were passive, and thus their functionalities were determined by the initial designs and the materials. As these research fields continue to mature, the demands for the devices to be active and tuneable became much stronger for practical and engineering applications. Therefore, we present our results on active nanoslit lenses and tuneable nanoslit lenses in the following sections. Since plasmonic and metamaterial devices use metals whose losses are not negligible, it is of critical interest to develop active devices. In our chapter, we study the possibility of loss compensation of the HMMs by including a gain medium in the design. When modeling the dispersions of HMMs, we tested four different methods that are commonly used in the community.

We have organized the chapter in the following way. We begin by presenting our recent work on planar plasmonic diffractive focusing devices which consist of arrays of nanoslits milled in gold films [3]. Our planar nanoslit metal lenses work for both TM and TE polarizations and can emulate both convex and concave conventional dielectric lenses. We give a brief overview on the experimental characterization and full-wave numerical simulations of light propagation in the fabricated prototypes. Next we show that the focusing properties of nanoslit lenses can be externally tuned by filling the slits with index-changing materials such as a liquid crystal (LC) material. Using LCs, we can control the focal intensity [4] of the device. With a different index-changing material, in this case a nonlinear medium, we can shift the focal length [5] of the nanoslit lens. Finally, we discuss our work on HMM-assisted nanoslit devices for subwavelength focusing. A case study on active HMM dispersions is provided and discussed [6], followed by the numerical simulations of subwavelength diffraction with active HMMs [7] and the experimental demonstration of subwavelength diffraction with a passive HMM [8].

10.2 POLARIZATION-SELECTIVE GOLD NANOSLIT LENSES

The performance of diffraction optics can be improved by using noble metals in the visible spectral range. When light propagates through metallic apertures (e.g., slits, holes), SPP modes can be generated and used in diffraction lenses. The wavevectors of the SPP mode depend on the dimensions of the apertures. This allows for very flexible control over the phase of the outgoing light, and thus makes it possible to design a focusing device by adjusting the size and the position of the apertures.

Recently, a planar metallic lens made of parallel nanoslits milled in a thick metal film has been proposed [9, 10] and then demonstrated experimentally [11]. This nanoslit lens can focus an incoming plane wave, similar to the operation of a traditional convex dielectric lens.

In contrast to convex dielectric lenses, however, nanoslit lenses are flat, are scalable to the order of the wavelength, and can be easily integrated for on-chip applications. In addition, it is possible to design a nanoslit lens whose depth of focus is as short as a few micrometers, which is difficult to achieve with a dielectric lens. Due to these attractive features, planar metallic lenses have been studied extensively [12], and some other planar metal lenses based on similar principles have been proposed. However, those devices, including, for example, an array of nanoholes [13, 14], do not modulate the phase front of the transmitted light by varying the dimensions of each aperture individually. This is a significant difference between our lens design and these previous metal-based lenses.

10.2.1 Design Concept of Gold Nanoslit Lenses

In order to design and analyze our nanoslit lenses, we considered light propagating in a slit milled through a metal film to act as a wave propagating in a metal–insulator–metal (MIM) waveguide (see Fig. 10.1a), where two different modes with distinct dispersion dependencies are possible. The first mode is a long-range SPP mode excited only with TM polarization, as only TM-polarized light can couple to the plasmonic mode. The other mode is the lowest one in the TE polarization (a photonic mode).

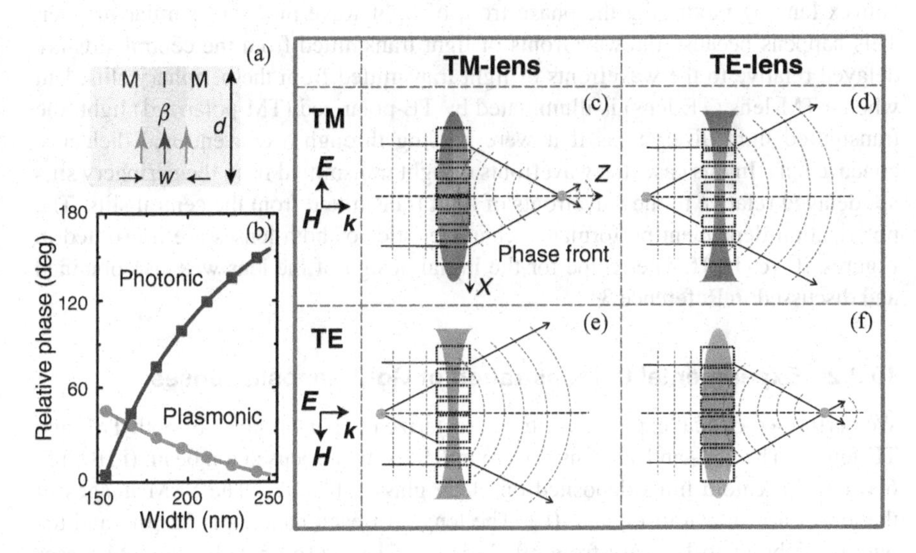

FIGURE 10.1 (a) Schematic image of a MIM waveguide. (b) Relative phases plotted against the width of the gold parallel-plate waveguide for 531 nm light. The length of the waveguide, d, is 600 nm. (c–f) Schematic images of TM- and TE-lenses in TM- and TE-polarized light.

Equations 10.1 and 10.2 show the dispersion relations calculated from a boundary value problem analysis [15] for the plasmonic and photonic modes, respectively:

$$\tanh\left(\tfrac{1}{2}k_1 w\right) = -\varepsilon_\mathrm{d} k_2 / (\varepsilon_\mathrm{m} k_1),\ k_1 = \sqrt{\beta^2 - \varepsilon_\mathrm{d} k_0^2}, \tag{10.1}$$

$$\tan\left(\tfrac{1}{2}k_1 w\right) = -k_2/k_1,\ k_1 = \sqrt{\varepsilon_\mathrm{d} k_0^2 - \beta^2}. \tag{10.2}$$

Note that $k_2 = \sqrt{\beta^2 - \varepsilon_\mathrm{m} k_0^2}$ for both Equations 10.1 and 10.2. Also, w is the width of the dielectric core between the two parallel plates, β is the propagation constant, ε_m and ε_d are the permittivities of the cladding metal and the dielectric core, respectively, and $k_0 = 2\pi/\lambda$ is the wavenumber for the free-space wavelength λ. After a wave propagates through a parallel-plate waveguide for a finite distance d, the relative phase of the outgoing light is $\phi = \arg(\beta d)$, where ϕ also depends on w. In Figure 10.1b, the relative phases of the outgoing light are plotted after propagating through a 600 nm long gold–polymer–gold waveguide, where the permittivity of gold is taken from Reference 16 and the polymer is polyvinylpyrrolidone (PVP) whose refractive index is 1.5. The phase of the plasmonic and photonic modes has opposite dependencies as we increase the width of the MIM waveguide so that we can design devices that will focus TM-polarized light (TM-lenses) or TE-polarized light (TE-lenses).

In a planar TM-lens, the narrowest slit is milled at the center of the lens, while wider slits are milled farther away from the center. A planar TE-lens has an opposite design: The narrowest slits are milled at the periphery of the lens, while the widest slit is milled at the center. Both lenses emulate the behavior of a conventional dielectric convex lens by modifying the phase front of light wave in a very similar manner. This happens because the wavefronts of light transmitted from the central slits are delayed relative to the wavefronts of light transmitted from the periphery slits. But when a TM-lens (TE-lens) is illuminated by TE-polarized (TM-polarized) light, the transmitted light diverges as if it were passing through a conventional dielectric concave lens. In this case, the wavefronts of light transmitted from the periphery slits are delayed relative to the wavefronts of light transmitted from the central slits. The polarization-dependent performance characteristics of both lenses are illustrated in Figures 10.1c–10.1f. The recipe for the initial design of the lens was also obtained and discussed in Reference 3.

10.2.2 Experimental Demonstration of Gold Nanoslit Lenses

We performed optical experiments to characterize the performance of both TM- and TE-lenses. The TM- and TE-lenses were milled using a focused ion beam (FIB) into 600 nm thick gold films deposited on clean glass substrates. The SEM images of the lenses are shown in Figure 10.2. The length of each nanoslit is 22 μm, and the widths of the nanoslits vary from 69 (215) to 154 nm (163 nm) for the TM-lenses (TE-lenses). Then, a thin film of PVP was spin-coated onto the samples, filling the nanoslits to increase the contrast of the refractive indexes in the MIM slit waveguides.

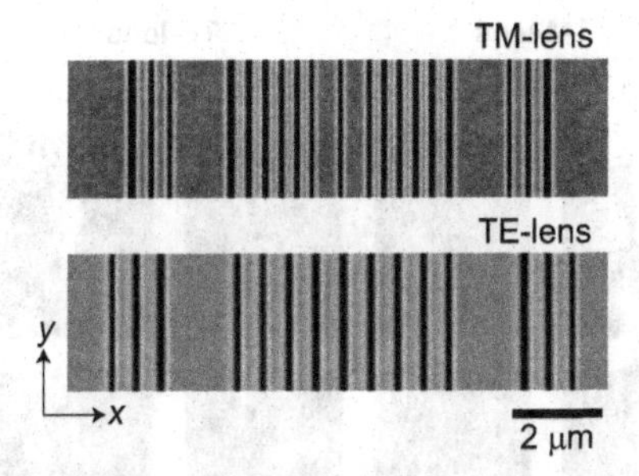

FIGURE 10.2 SEM images for the TM-lens (top) and the TE-lens (bottom).

The transmission properties of the samples were measured using a conventional optical microscope with a 100X objective lens (NA = 0.75, W.D. = 0.98 mm). A schematic setup is shown in Figure 10.3. The stage resolution along the vertical direction (z-axis) and the depth of focus for the 100X objective lens are less than 1 μm, which were sufficient for our measurements. The samples were mounted on the microscope stage with the film side up and were illuminated from the substrate side with a linearly polarized, CW laser at 531 nm. The transmission images from the samples were recorded by a CCD camera by moving the height stage in increments of 1 μm. Note that the focal point of the objective lens and the surface of the sample are coincident at $z = 0$, and $x = 0$ is the center of the sample. From the obtained images, 2D maps of the transmitted light were created.

Figures 10.4a–10.4d show the transmitted light irradiance from both the TM-lens and TE-lens for both TM and TE polarizations. The TM-lens (TE-lens) focuses TM-polarized (TE-polarized) light, but diverges TE (TM) incident light. These results

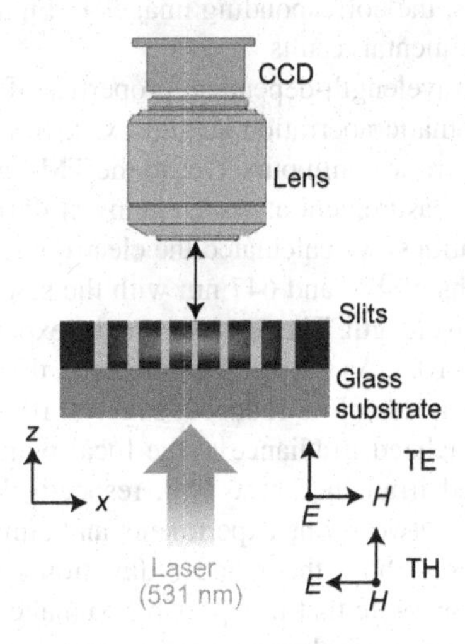

FIGURE 10.3 Schematic of the measurement setup. The distance between the lens and the sample was changed to record the irradiance at each Z position.

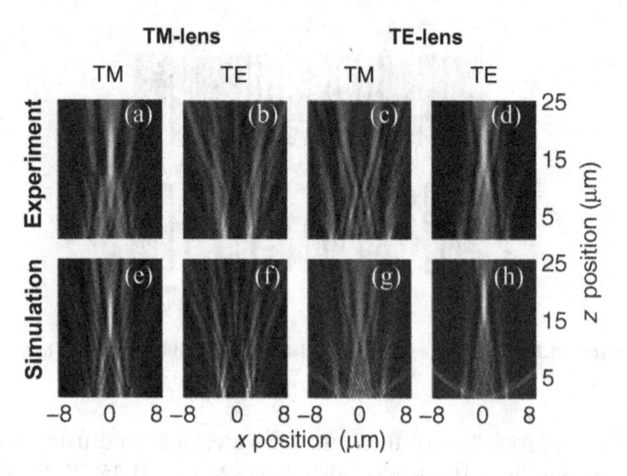

FIGURE 10.4 Irradiance of the transmitted light through the nanoslit lenses. (a and b) Measurement results for the TM-lens; (c and d) measurement results for the TE-lens; (e and f) simulations for the TM-lens; (g and h) simulations for the TE-lens. Incident illumination is TM-polarized for (a), (c), (e), and (g) and TE-polarized for (b), (d), (f), and (h). In (a–h), the color scale is normalized to its maximum.

are a clear indication that the lenses, which were designed based on the dispersion relations above, do indeed operate as shown in Figure 10.1. The results for both lenses obtained from full-wave numerical simulations using commercial finite-element method (FEM) software (COMSOL Multiphysics) are shown in Figures 10.4e–10.4h. Considering that (i) the recorded images show the average intensity over the depth of focus of the objective lens and (ii) the two outgoing beams in Figures 10.4g and 10.4h are not observed experimentally because of the finite field of view of the objective lens, the corresponding images in Figure 10.4 are in excellent agreement with the experimental results.

Next, we studied the wavelength-dependent properties of nanoslit lenses. Gold is dispersive, and thus chromatic aberration should exist. Since the TM-lens and the TE-lens work similarly, here we limit ourselves to the TM-lens case. In experiments, we repeated the optical measurement at wavelengths of 476 and 647 nm with TM-polarized light. In simulations, we calculated the electromagnetic field intensities in the device for wavelengths of 476 and 647 nm with the same conditions as 531 nm except for the incident wavelength. The results from the experiments and simulations are summarized in Figure 10.5. As shown in Figure 10.5, the focus distances become shorter as the incident wavelength is reduced. Figures 10.5g and 10.5h depict the cross section of the normalized irradiance at the focal plane heights and the cross section of the normalized irradiance at $x = 0$, respectively, proving that a good quantitative match exists between our experiments and simulations. The widths of the focal spots are wider in both the x and z directions, which could be due to imperfection of fabrication. Note that it is possible to make use of this wavelength-dependent property (i.e., chromatic aberration) as a micro spectrometer to measure the incident wavelength.

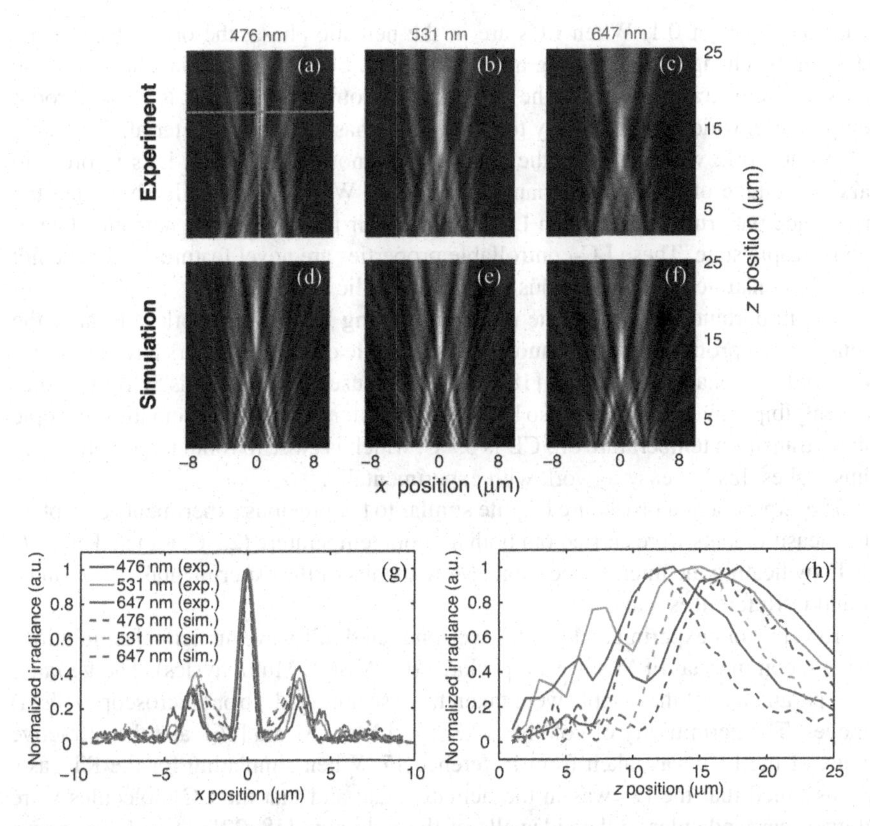

FIGURE 10.5 Wavelength-dependent irradiance of the transmitted light through the TM-lens with TM-polarized light at 467, 531, and 647 nm. (a)–(c) are from the experiments and (d)–(f) are from the FEM simulations. (g) Cross section at the focus distances for the TM-lens at each wavelength. (h) Cross sections at $x = 0$ μm for the TM-lens at each wavelength. In (g) and (h), each curve is normalized to its maximum; the legend for (g) applies to (h) as well.

10.3 METALLIC NANOSLIT LENSES WITH FOCAL-INTENSITY TUNEABILITY AND FOCAL LENGTH SHIFTING

The nanoslit lenses shown above are passive devices. By incorporating index-changing materials into the nanoslits, we can achieve external control over the focusing properties of the lenses. We chose LCs and a nonlinear semiconductor material as our index-tuneable materials. We then studied the focusing properties of the lenses whose slits were filled with these index-changing materials.

10.3.1 Liquid Crystal-Controlled Nanoslit Lenses

In optics, LCs are known to exhibit large refractive index changes when their orientation or phase changes [17]. Typically, the absolute value of a refractive index change is on the order of 0.1. When LCs are in the nematic phase, the orientation of the LCs can be changed by applying an electric field. Some LC materials have a phase transition temperature between the nematic and isotropic states that is close to room temperature, which makes it easy to control the phase of the LC material.

In our work, we have filled the slits of the nanoslit lenses with LCs in order to take advantage of their index-changing property. We experimentally show that the irradiance pattern changes when LCs change their phase from the nematic state to the isotropic state. These LC-controllable properties are novel features that have not been demonstrated in any previous work on metallic lenses [3, 11].

It is important to note that the purpose of using LCs in our work is to alter the transmission profiles from the nanoslit lenses, while other researchers have used LCs to shift the resonance conditions [18–23]. The LC used in our study is 5CB (4-cyano-4′-pentylbiphenyl), which has also been used in Reference 16] The nematic–isotropic phase transition temperature of 5CB is 35°C, which is close to room temperature and thus makes this LC easy to work with experimentally.

The experimental procedure is quite similar to the previous experiments except all the measurements were carried out both at room temperature (22°C, RT) and at 42°C (HT) by heating the microscope stage. More details of the experimental setup can be found in References 3 and 4.

To verify our experimental results, we conducted full-wave numerical simulations using a commercial FEM software package (COMSOL Multiphysics). The geometrical parameters of the sample were taken from scanning electron microscope (SEM) images. The permittivity of gold is –12.8 + 1.12i at 531 nm [16], and the refractive index of the LCs was taken from Reference 17. When simulating for the RT case, we assumed that the LC was in the nematic state and that the LC molecules were aligned perpendicular to the sidewalls of the gold slits [18, 22]. The LC refractive index tensor was defined as $n = \text{diag}(1.74, 1.55, 1.55)$. When simulating for the HT case, we assumed the LC to be isotropic and defined the refractive index tensor as $n = \text{diag}(1.6, 1.6, 1.6)$. The computational domain was truncated by a scattering boundary condition, and a monochromatic plane wave was incident from the sample/substrate boundary. The geometry of the model and the orientations of the LCs are shown in Figure 10.6.

Representative SEM images of the fabricated samples are shown in Figure 10.7. Each slit is 22 μm long for both the TM- and TE-lenses. Compared to the wavelength of the illuminating light, the slits are substantially long, and we therefore assume both lenses to be essentially two-dimensional. Both lenses are symmetric with respect to the centerline or center slit. We define the origin of the coordinate at the surface of the lens at center. The x-axis is perpendicular to the slits, the y-axis is along the slit, and the z-axis is vertical to the sample.

In Figure 10.8, we plot the irradiance of the transmitted light from the TM-lens. Figure 10.8a is at RT and Figure 10.8b is at HT. The color scales for both 2D plots are identical and normalized to the maximum value recorded at HT. It is clearly seen that the irradiance at HT is higher than the irradiance at RT. The observed change in irradiance can be attributed to the nematic–isotropic phase change of the LCs inside the slits. Figures 10.8c and 10.8d are the simulation results for TM-lens in the nematic

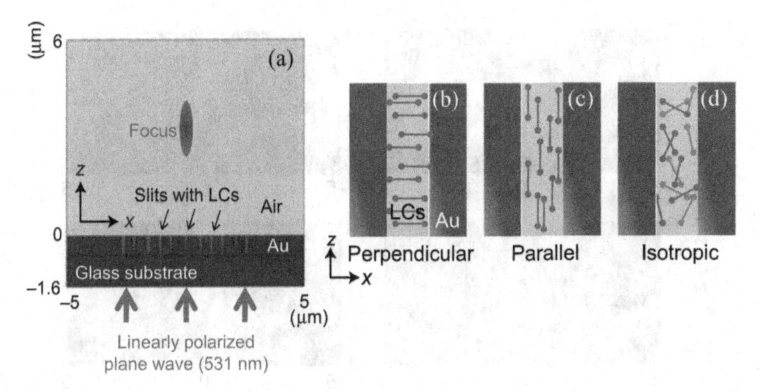

FIGURE 10.6 Simulation setting. (a) Geometry of the simulation (TM-lens) and (b–d) orientation of the LCs modeled in the simulations. The actual tapered shapes of the slits are modeled from the cross-section images of SEM.

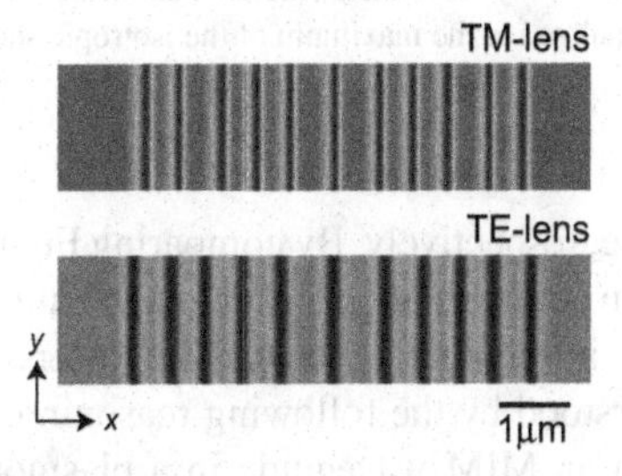

FIGURE 10.7 SEM images for the TM-lens (top) and the TE-lens (bottom).

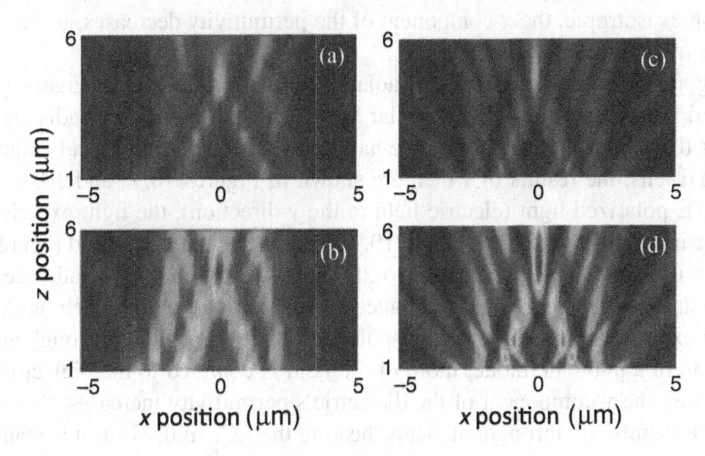

FIGURE 10.8 Irradiance of the transmitted light through the TM-lens. (a and b) 2D plot of the experimentally measured irradiance at RT (a) and at HT (b). The color scale is normalized to the maximum of HT. (c and d) 2D plot of the numerically calculated irradiance in the nematic state (c) and in the isotropic state (d). The color scale is normalized to the maximum of the isotropic state.

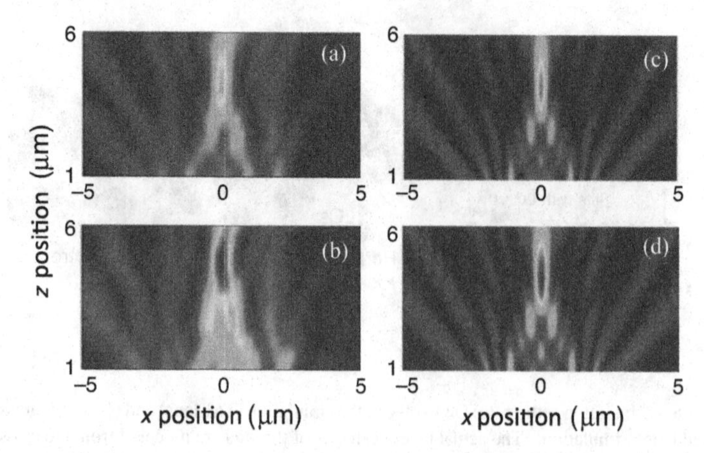

FIGURE 10.9 Irradiance of the transmitted light through the TE-lens. (a and b) 2D plot of the experimentally measured irradiance at RT (a) and at HT (b). The color scale is normalized to the maximum of HT. (c and d): 2D plot of the numerically calculated irradiance in the nematic state (c) and in the isotropic state (d). The color scale is normalized to the maximum of the isotropic state.

state and the isotropic state, respectively. By comparing Figures 10.8c and 10.8d, it is obvious that the simulation confirms our experimental results that irradiance is higher in the isotropic state than in the nematic state. The observed changes in irradiance can be qualitatively understood by the following reasoning. Consider TM-polarized light propagating through an MIM waveguide in a plasmonic mode. With a lower permittivity in the dielectric layer, less field penetrates into the metal, and thus losses are lower in the metal layer. In our case, when the temperature of the LC rises and the LC becomes isotropic, the *x*-component of the permittivity decreases, which results in higher transmission.

In Figure 10.9, we show similar irradiance plots for the TE-lens. Figure 10.9a is at RT and Figure 10.9b is at HT. Similar to the TM-lens case, the irradiance at HT is higher than the irradiance at RT. We have also carried out numerical simulations for the TE-lens, the results of which are shown in Figures 10.9c and 10.9d. In the case of TE-polarized light (electric field in the *y*-direction), the light experiences a refractive index change of 0.05. Figure 10.9c is for the nematic state and Figure 10.9d is for the isotropic state. From the two 2D plots, we see that the irradiance in the nematic state is higher than the irradiance in the isotropic state, which agrees well with our experimental results. For TE-polarized light propagating through an MIM waveguide in a photonic mode, more of the field is confined to the dielectric layer. Hence when the *y*-component of the dielectric's permittivity increases, the result is a higher transmission throughput. Thus, heating the LCs to the isotropic state again increases the transmission in this case. The difference in the transmitted irradiance of the TE-lens at the two temperatures is smaller compared to the case of the TM-lens. This is because the relevant refractive index change in the LCs in the TE-lens case is smaller than the relevant refractive index change in the LCs for the TE-lens.

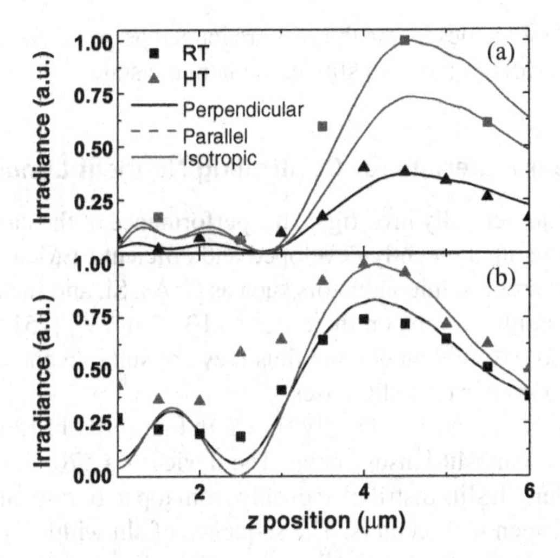

FIGURE 10.10 Irradiance along the z-axis for the TM-lens (a) and for the TE-lens (b).

To gain more quantitative insight into the operation of our nanoslit lenses, we have plotted the transmitted irradiances along the z-axis in Figures 10.10a and 10.10b for the TM-lens and the TE-lens, respectively. From the experimental results for the TM-lens, the focal lengths at RT and HT are 4.0 and 4.5 μm, respectively. At HT, the irradiance at the focus is 2.7 times higher than the irradiance at the focus at RT. From the simulation results, the focal lengths in the nematic state and the isotropic state are 4.500 and 4.565 μm, respectively. The simulation results also show 2.5 times higher irradiance at the focus for the isotropic state compared to the nematic state. In looking at the experimental results from the TE-lens, we see that the focal lengths at RT and HT are identical at 4.0 μm. The irradiance at focus for the HT case is 1.3 times higher than the irradiance at focus for RT. The simulation results show that the focal lengths in the nematic and isotropic states are both 4.129 μm, while the irradiance in the isotropic state at focus is 1.2 times higher than the irradiance in the nematic state. Considering the resolution of our measurement setup and the nonuniform distribution of the LCs inside the slits, the experimental results and the simulation results are in rather good agreement.

We also note that Sanda et al. have reported that the orientation of 5CB, which is the LCs studied in this paper, is parallel to the gold surface in their studies [24]. Our results can also be used to confirm the assumption of the orthogonal orientation of LC molecules relative to the wall (indicated as perpendicular in Fig. 10.6). In order to verify our assumption of the LC orientation, we also simulated a parallel LC orientation inside the slits (n = diag(1.55, 1.55, 1.74)). The results are included in Figure 10.10. In the case of the TM-lens, if the parallel orientation of the LCs is assumed for RT, the ratio of the irradiances at focus in the nematic state and the isotropic state is 1.35, which does not correspond to our experimental results. In the case of the TE-lens, the parallel orientation and the perpendicular orientation are equivalent because of the illumination polarization. Therefore, we conclude that the majority of the LC molecules inside the slits are indeed perpendicularly oriented with respect to the side walls of the gold slits in the nematic state.

10.3.2 Nonlinear Materials for Controlling Nanoslit Lenses

In this section, we numerically investigate the performance of the nanoslit lens with a nonlinear material using a recently developed and efficient FEM simulation scheme [5]. Some of the common semiconductors such as GaAs, Si, and InGaAsP are known to have high Kerr nonlinearities on the order of 10^{-18} m^2/V^2 [25]. These materials are widely used and easy to handle, and thus they are suitable choices for materials to be incorporated into our nanoslit lenses.

The simulated device is schematically shown in Figure 10.11 [26] and is conceptually similar to the nanoslit lenses above. The device is a 570 nm thick silver slab with five subwavelength slits distributed evenly from top to bottom and with identical 400 nm offsets between their centers. The sequence of slit widths is 100, 70, 60, 70, and 100 from one edge of the lens to the other. Note that the slit width sequence is symmetric about the center slit. The slits are filled with a nonlinear Kerr medium

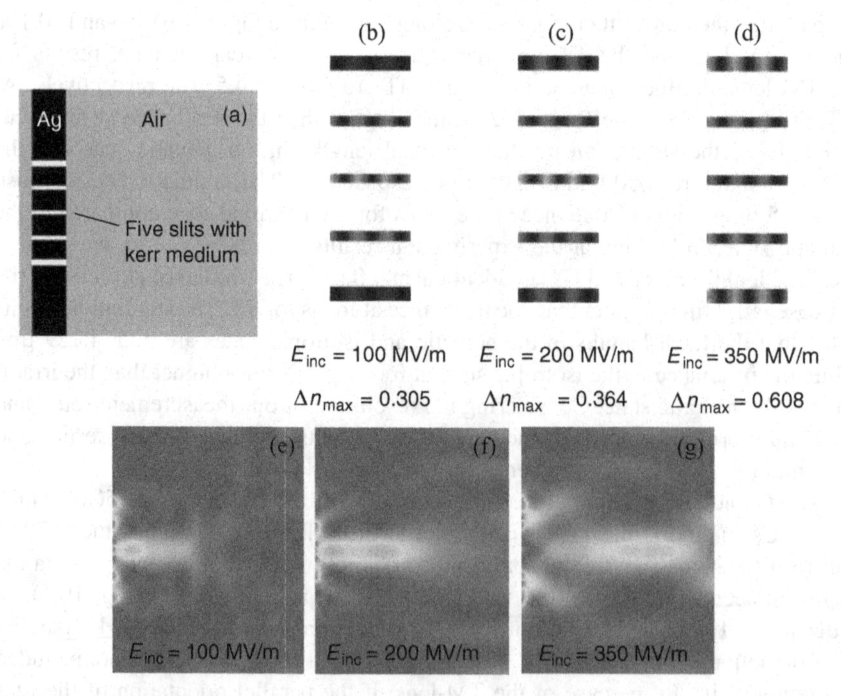

FIGURE 10.11 (a) Geometry of a nonlinear focusing device. A 570 nm thick silver film is illuminated by an 850 nm plane wave moving from left to right through five subwavelength slits (with a 100, 70, 60, 70, 100 nm slit width sequence) distributed evenly from top to bottom with 400 nm center-to-center offsets. Intensity-dependent refractive indexes at an incident field magnitude of 100 (b), 200 (c), and 350 MV/m (d). Intensity-tuneable focusing at the incident magnitude of 100 (e), 200 (f), and 350 MV/m (g).

(with $\varepsilon_1 = 2.25 + i0$ and $\chi^{(3)} = (10^{-18} + i0)\ m^2/V^2$), and the surround medium is air. The structure is illuminated by a plane wave with a wavelength of 850 nm moving from left to right (see Fig. 10.11) through the lens structure.

The intensity-tuneable focusing capability of this nanoslit lens enabled through the use of different SPP modes excited in the structure at different intensities of the incident light. Figure 10.11a depicts the intensity-tuneable refractive index in the slits. Figure 10.11b shows the distribution of the nonlinear part of the refractive index in the slits for an incident E-field magnitude of 100 MV/m. Here the maximum $\Delta n =$ 0.305 is obtained inside the central slit, and the widest slits (the 100 nm slits at the top and bottom) are not activated at this field level. Figure 10.11b shows the evolution of this process as the incident field magnitude approaches 200 MV/m. The maximum of $\Delta n = 0.305$ is still achieved in the central slit, but the widest 100 nm slits are already moderately engaged. Finally, as the incident magnitude reaches 350 MV/m, as shown in Figure 10.11d, the maximum $\Delta n = 0.608$ is now obtained in the fully activated widest slits, while the central slit is almost deactivated.

The above process results in an intensity-tuneable focusing performance sequentially shown in Figures 10.11e–10.11g as the intensity increases from 100 to 350 MV/m. The constructive interference of the different activated slits moves the focal point further away from the device as the incident intensity increases.

10.4 LAMELLAR STRUCTURES WITH HYPERBOLIC DISPERSION ENABLE SUBWAVELENGTH FOCUSING WITH METALLIC NANOSLITS

The nanoslit lenses considered so far focus incident light in the far zone in air; hence, their spot sizes are limited by the diffraction limit. In this section we show that nanoslit lenses can go beyond the diffraction limit with the use of HMMs. A number of researchers have already studied the focusing properties and near-field distributions or SPPs in order to achieve subwavelength focusing, including the use of surface plasmon lenses [27] and a near-field plate [28]. Our work differs from those methods in that we have dealt with diffracted fields propagating away from the apertures (slits, holes, etc.).

Two exemplary structures used to design HMMs are metal–dielectric lamellar composites and metallic nanowires. Between these two structures, planar multilayer structures are easier to fabricate in samples with sub-micrometer thicknesses, and hence we have chosen a metal–dielectric multilayer structure for our work. When the layer thicknesses of the metal and dielectric materials are much smaller than the free-space wavelength, the effective permittivity of the multilayer structures is calculated by effective medium theory (EMT) [29]. Depending on the signs of ε_x and ε_z, HMMs can be classified into two groups: transverse positive (TP, $\varepsilon_x > 0,\ \ \varepsilon_z < 0$) and transverse negative (TN, $\varepsilon_x < 0,\ \ \varepsilon_z > 0$). In the visible region, it is easy to find metals and dielectrics for a TN-HMM. Metals could be low-loss noble metals such as silver or gold, and the dielectric could be silica, alumina, or titanium dioxide. On the other hand, finding materials to obtain a TP-HMM in the optical region is not trivial as noble metals have a negative permittivity that is too large, while the indexes of dielectrics are on the order of unity. One combination could be titanium nitride (TiN) for the metal and aluminum nitride (AlN) for the dielectric [30]. Forming a planar multilayer structure with thick dielectric layers and ultrathin metal layers [31] is yet another option.

To begin our analysis, we consider a single nanoslit lens with an HMM-covered substrate, where diffracted large-wavenumber waves (high-k waves) are allowed to propagate inside the HMM slab [2]. The propagation direction of high-k waves in an HMM with negligible losses is approximated as

$$\tan\theta = S_x/S_z \approx \sqrt{-\mathrm{Re}\,[\varepsilon_x]/\mathrm{Re}\,[\varepsilon_z]}, \tag{10.3}$$

where θ is the angle from the normal axis, $\vec{S} = (S_x, S_z)$ is the Poynting vector, and $\bar{\varepsilon} = \mathrm{diag}(\varepsilon_x, \varepsilon_z)$ is the permittivity of the HMM. After careful optimization of a given double-slit design and choosing appropriate materials for the actual planar lamellar metal–dielectric composite comprising the HMM, the diffracted high-k waves could form a desired interference field pattern with subwavelength features [2]. When the design is adjusted such that the maximum field of the pattern occurs at the interface of the HMM and the substrate, the device could work as a subwavelength probe.

One significant drawback of a passive optical HMM is that the intrinsic loss in the metal always limits the overall functionality [32,33]. This is a fundamental challenge in the applications of HMMs, but it can be overcome by including active (gain) media in the system [34–36]. So far, gain media have been experimentally incorporated into various kinds of plasmonic and metamaterial systems including SPPs [37], localized SPPs [38], plasmonic waveguides [39], and negative index materials [35]. Incorporating gain within the multilayer system allows for the compensation of the metallic loss [40]. In fact, some experiments on metamaterials have even shown complete loss compensation or even overcompensation [35]. With gain, then, metal–dielectric HMM applications can be far more robust than with passive HMMs. This section starts with the modeling of a metal–dielectric HMM for the case when the dielectric is an active medium, and we study the dispersion relationship of such a system. Then, we numerically study the subwavelength diffraction properties of the double slits with the active HMM. As a preliminary demonstration, we show our experimental results on subwavelength diffraction patterns where the HMM is passive.

10.4.1 Active Lamellar Structures with Hyperbolic Dispersion

The cross section of a binary HMM, which consists of isotropic metal and dielectric layers, is schematically shown in Figure 10.12a. The permittivity and thickness of each layer are denoted respectively as ε_j and δ_j, where $j = 1$ for the metal layer and $j = 2$ for the dielectric layer, and the period of the structure is $\delta = \delta_1 + \delta_2$. We also define a linearly polarized plane wave with the following free-space parameters: wavelength λ, wavenumber $k_0 = 2\pi/\lambda$, and wavevector $\mathbf{k}_0 = k_0\,(\hat{\mathbf{x}}\cos\theta + \hat{\mathbf{y}}\sin\theta)$, where a given angle of incidence θ is aligned with the structure $(0 < \theta < \pi/2)$. As discussed above, HMMs made from thin, alternating layers of a metal and a dielectric

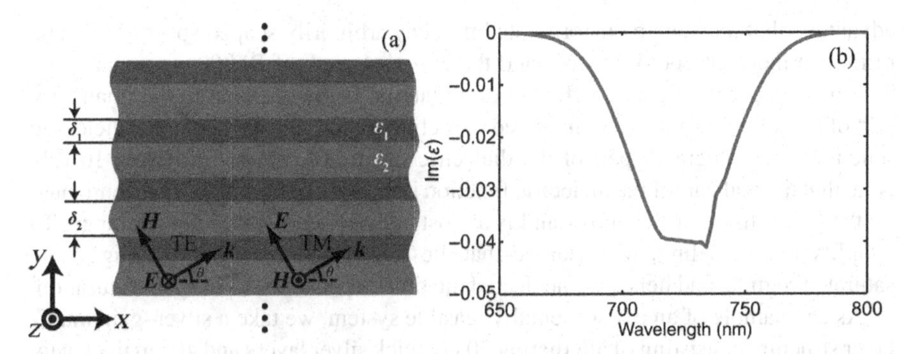

FIGURE 10.12 (a) Geometry of a metal–dielectric multilayer composite. The permittivities of the metal and dielectric layers are denoted, respectively, as ε_1 and ε_2, and the thicknesses are δ_1 and δ_2. All the layers are parallel to the x–z plane. (b) Extracted imaginary part of the dielectric function of dye-doped epoxy. Note that the real part of the dielectric function (not shown) is dominated by the permittivity of the epoxy, which is about 2.72.

can be considered to be metamaterials as long as their periodicity δ is significantly smaller than λ or, more precisely, when $k_0\delta \ll 1$.

We consider four different methods to analyze the dispersion relation of the HMM design above. When studying the dispersion relationships of multilayers, using T-matrices to form a set of nonlinear equations (NLE) is a rigorous method, and the resulting solutions are exact [41]. A simpler way is to use the standard EMT (denoted as EMT_1 here). The EMT homogenizes the multilayers, thus giving the effective dielectric constants of the medium as a whole. A more advanced EMT method (denoted EMT_2) [42] takes into account the nonlocal effects that are neglected in EMT_1. With regard to numerical methods, spatial harmonic analysis (SHA) is a robust method to study periodic systems [43,44]. All the details of the four methods discussed here are given in the addendum of Reference 6.

We have estimated the properties of metal–dielectric multilayers using the four methods mentioned above. We take silver as the material in all the metal layers. Note that thin silver films tend to form islands below 20 nm on a silica surface, which is not suitable for obtaining a multilayer structure. Recent studies have shown that a silver film evaporated on a 1 nm thick germanium layer is continuous and smooth even at a 5 nm Ag thickness [31]. In our calculations, we used realistic values of the optical properties of the materials; these values are taken from experimental results. We take the permittivity of silver from Reference 45, which uses a Drude–Lorentz model with three Lorentz terms. The parameters are fitted using measured data from Reference 16. The loss factor (a multiple of the collision rate in the Drude term) is set to 2 in order to take into account the size-dependent sliver loss [31]. Since annealing can reduce the loss of germanium to negligible levels in the red part of the spectrum, which is our region of interest, the germanium layer is not modeled in our simulations. For the active dielectric medium, we take the gain values extracted from experimentally measured data of an organic dye (Rh800) mixed in epoxy [35, 46]. Typically, organic dyes mixed with polymers have high gain coefficients and have the advantage that they can be incorporated into any arbitrarily shaped spaces [35]. The refractive index of epoxy is 1.65, and the emission peak of Rh800 occurs at about 720 nm. Because the gain coefficient of a material is proportional to the imaginary part of the refractive index, we modeled the active dielectric with a complex dielectric function. The imaginary part of the dielectric function is plotted in Figure 10.12b. Note that the real part of the dielectric function is not plotted here since it is dominated by the permittivity of the epoxy and is almost constant in this wavelength range. To simplify our modeling, we assumed that the active medium was operating in the saturated regime, and hence we neglected the time dependence of the gain saturation.

As an example of an experimentally feasible system, we take a silver–gain multilayer structure consisting of alternating 20 nm thick silver layers and 40 nm thick gain layers. In Figure 10.13, we plot the effective anisotropic permittivity of the structure calculated from the four methods (EMT_1, EMT_2, NLE, and SHA). The incidence angle (θ) is zero for the plot. Within the plotted wavelength range (650–800 nm), all four methods show that the multilayer has opposite signs for ε_x and ε_y, and the system therefore exhibits hyperbolic dispersion.

Figure 10.13 also shows that the effective permittivities resulting from NLE and SHA are perfectly matched in both the *x*- and *y*-directions. This is not surprising, as both methods are aimed at calculating the eigenvalues of the composite layer and differ only in their numerical realizations. The results from EMT_2 are very close to those values. However, there are significant discrepancies between the results from EMT_1 and those from the other methods. The relative errors are calculated using the results from SHA as the reference (exact value) using $|(\varepsilon_{\text{calc}} - \varepsilon_{\text{SHA}})/\varepsilon_{\text{SHA}}| \times 100\%$.

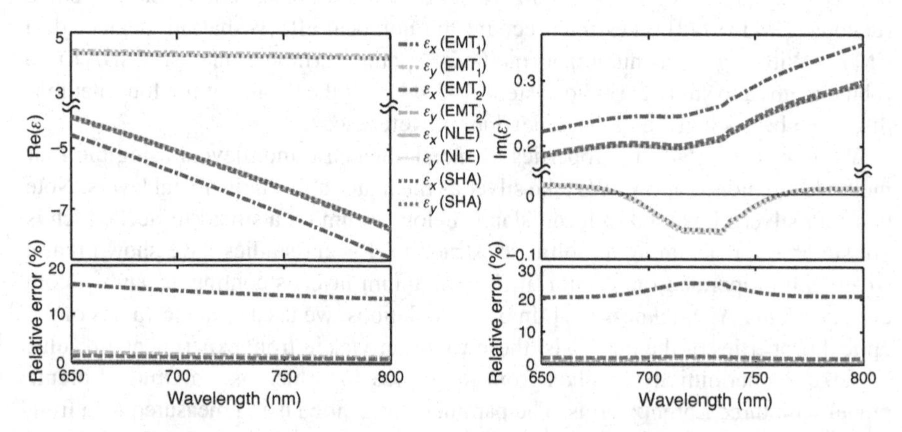

FIGURE 10.13 Wavelength dependence of the effective anisotropic permittivity for the silver–gain HMM at normal incidence ($\theta = 0$) calculated from EMT_1 (dashed-dotted line), EMT_2 (dashed line), NLE (solid line), and SHA (dotted line). Top left panel: the real part of the effective permittivity. Top right: the imaginary part of the effective permittivity. Bottom panels: relative errors in percent with respect to the SHA results. Note that the results from NLE and SHA are coincident, and their curves therefore overlap completely. The real part of permittivity in the *y*-direction from EMT_2 overlaps with the NLE and SHA curves (therefore it is obscured by the NLE and SHA curves and is not seen in the figure), and the rest of the EMT_2 curve partially overlaps with the NLE and SHA curves.

Comparing with EMT_1, EMT_2 is more accurate and has lower error over the calculated wavelength range (the EMT_2 error is less than 3% in this case). We also note that the relative errors of the imaginary parts of the permittivities from both EMT_1 and EMT_2 are higher in the active region, which can be observed from the relative error plots in Figure 10.13.

If we take a look at the fundamental mode for a TM-polarized wave propagating in the x direction in the structure, the real part of the modal index, which is $\sqrt{\varepsilon_y}$, is higher than the refractive index of the dielectric in the multilayer. This indicates that the dominant propagating mode is indeed a plasmonic mode. The imaginary part of the modal index, which has the same sign as $\mathrm{Im}(\varepsilon_y)$, is negative from 700 to 750 nm. Within this range, the propagating TM wave is loss-free. However, in the x direction, we can see that the imaginary part of ε_x stays positive, which means that the loss is not compensated for a TE-polarized wave inside the structure. It is important to point out that although the loss is not fully compensated in the x direction, it is still lower than it would be without gain. To achieve loss compensation in both directions in this type of bilayer metamaterial, we need an enormously large gain coefficient. It is also worth mentioning that the actual loss might be lower in real structures because the active medium within the metal–dielectric multilayer structure will give rise to an effective gain that is much higher than its bulk counterpart. The large value of gain is due to the local-field enhancement inherent in the plasmonic response of the structure [35].

In Figure 10.14, we plot the incidence-angle-dependent dispersion relationship of the silver–gain HMM. The incident wavelength is 720 nm. Here again we see perfect agreement between the NLE and SHA results. The results from EMT_1 do not show any dependence on the angle of incidence. The EMT_2 results show that this method can only handle angle-dependent problems in the y-direction, and it has

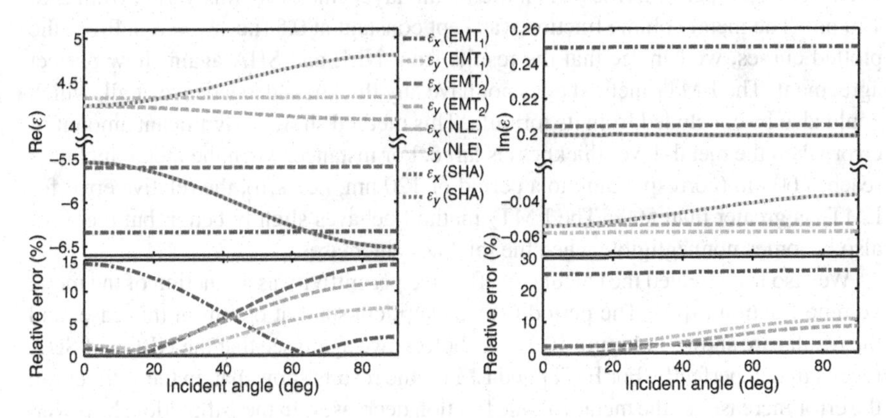

FIGURE 10.14 Incidence-angle dependence of the effective anisotropic permittivity for the silver–gain HMM at a wavelength of 720 nm calculated from EMT_1 (dashed-dotted line), EMT_2 (dashed line), NLE (solid line), and SHA (dotted line). Top left panel: real part of the effective permittivity. Top right: imaginary part of the effective permittivity. Bottom panels: relative errors in percent with respect to the SHA results. Note that the results from NLE and SHA are coincident, and their curves overlap.

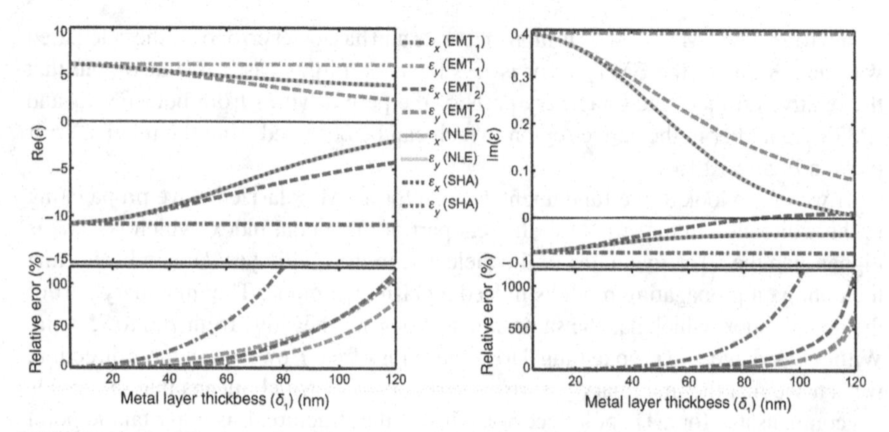

FIGURE 10.15 Effective anisotropic permittivity for the silver–gain HMM with different metal-layer thicknesses at a wavelength of 720 nm at normal incidence ($\theta = 0$) calculated from EMT_1 (dashed-dotted line), EMT_2 (dashed line), NLE (solid line), and SHA (dotted line). The volume fraction of the metal layer (δ_1/δ) is kept constant at 0.5. Top left panel: real part of the effective permittivity. Top right: imaginary part of the effective permittivity. Bottom panels: relative errors in percent with respect to the SHA results. Note that the results from NLE and SHA are coincident, and their curves overlap.

nonnegligible error compared to the NLE or SHA results. As shown in Figure 10.14, the permittivities in the x- and y-directions both show variations on the order of 1 in this case, which are rather strong changes. Thus, we should use either NLE or SHA to take into account the angular dependence of the permittivity if we are dealing with angles away from normal incidence.

Our studies included an analysis of how the effective permittivity changes when varying the thicknesses of the metal layers. Figure 10.15 shows the calculated effective anisotropic permittivities when the metal–layer thickness was varied from 5 to 120 nm. The metal volume fraction was kept constant at 0.5 (i.e., $\delta_1 = \delta_2$). From the plotted curves, we can see that the results from NLE and SHA again show perfect agreement. The EMT_1 method does not have any thickness dependence at all, which could also be concluded from its formula. This method shows a significant amount of error when the metal–layer thickness is large. For instance, when the metal thickness reaches 60 nm (corresponding to a period of 120 nm, i.e., $\lambda/6$), the relative error for EMT_1 is greater than 50%. The EMT_2 method behaves slightly better, but the error also becomes nonnegligible when the thickness increases.

We also investigated the calculated effective permittivity as a function of the metal volume fraction (δ_1/δ). The period (δ) was kept constant at 60 nm in this case, and the results are shown in Figure 10.16. As before, we again see that the NLE and SHA results match perfectly. For EMT_1 and EMT_2, the results show that, in the x direction, the error increases as the metal volume fraction decreases. In the y-direction, however, we see the opposite trend. We observe that, depending on the application, EMT_1 can produce substantial errors in both the x- and y-directions of the effective dielectric function. The EMT_2 calculation gives more accurate results for major applications. The NLE and SHA methods are the most accurate for all the cases.

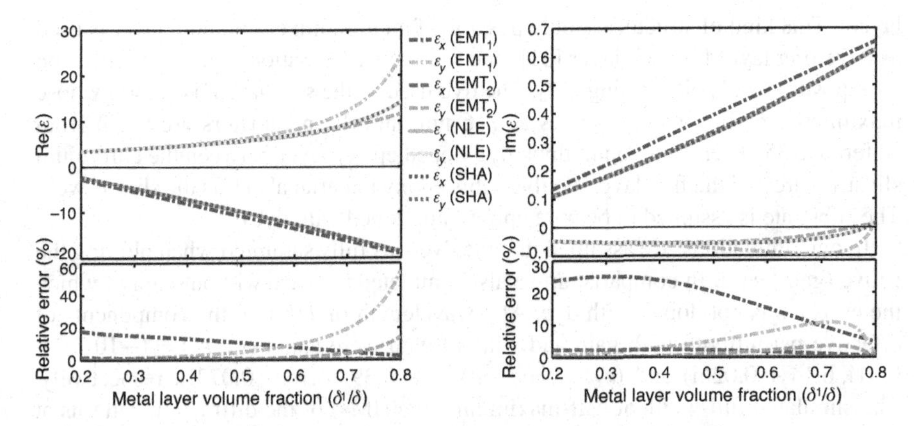

FIGURE 10.16 Effective anisotropic permittivity for the silver–gain HMM with different metal volume fractions at a wavelength of 720 nm at normal incidence ($\theta = 0$) calculated from EMT_1 (dashed-dotted line), EMT_2 (dashed line), NLE (solid line), and SHA (dotted line). The period (δ) is kept constant at 60 nm. Top left panel: real part of the effective permittivity. Top right: imaginary part of the effective permittivity. Bottom panels: relative errors in percent with respect to the SHA results. Note that the results from NLE and SHA are coincident, and their curves overlap.

10.4.2 Subwavelength Focusing with Active Lamellar Structures

In this section, we numerically explore the effect of loss compensation with an active HMM [6]. The structure contains a Cr mask layer that is 50 nm thick with two 40 nm wide slits separated by 250 nm (see Fig. 10.17a). The HMM structure consists of 10 layers of alternating 5 nm thick silver layers and 16 nm thick dye-doped epoxy

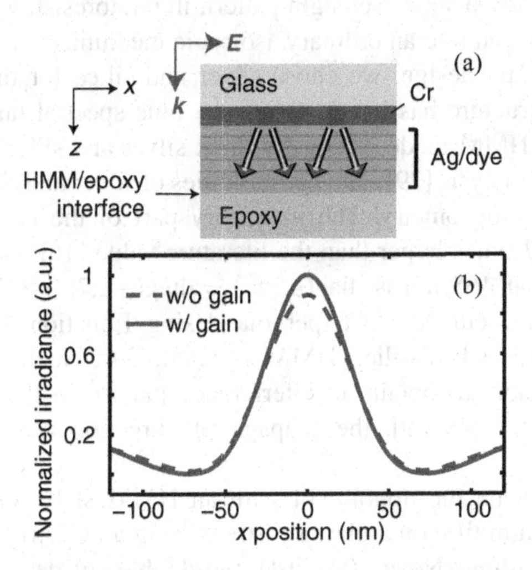

FIGURE 10.17 (a) Schematic diagram for a double-slit design with silver–dye multilayer structure. (b) Normalized irradiance at the interface between the silver–dye HMM and the epoxy substrate calculated with gain (solid line) and without gain (dashed line). The arrow in (a) shows the position of the interface.

layers. This kind of structure is only possible if an ultrathin germanium layer is used as a wetting layer for each layer [31], and multilayer formation requires a repetition of deposition and spin-coating [47]. The dye used in the simulation is Rh800 whose maximum emission wavelength is at 716 nm, and the parameters are taken from Reference 35. There is a 16 nm thick dye-doped epoxy layer between the chromium slit structure and the first layer of silver; this epoxy material also fills the slits as well. The substrate is assumed to be pure epoxy (not doped with dye).

In our simulations, we assumed that the dye was fully saturated when plotting the active (gain) case. In comparison, we also simulated the case without gain in which the epoxy was not doped with dye. At a wavelength of 716 nm, the components of effective permittivity with gain and without gain were about $(\varepsilon_x, \varepsilon_z) = (-10.71 + 0.41i, 6.14 + 0.026i)$ and $(\varepsilon_x, \varepsilon_z) = (-10.71 + 0.39i, 6.14 - 0.073i)$, respectively. The simulated full-width at half-maximums (FWHMs) of the diffraction patterns at the interface of the multilayer structure and the substrate are shown in Figure 10.17b. With gain, the FWHM of the irradiance at the interface was about 47 nm, which is about $\lambda/15$. If the structure contained no gain, the irradiance at the interface was reduced, and the FWHM of the field intensity increased to about 54 nm as indicated by the dashed line in the figure.

10.4.3 Experimental Demonstration of Subwavelength Diffraction

In this section, we present and discuss our experimental results on subwavelength diffraction with a nanoslit device. Our device consists of two metallic slits of subwavelength width placed on top of a planar, lamellar structure of alternating metal and dielectric layers. A photolithography approach is employed to provide subwavelength detection. Our experiments show that light diffracted from a double slit into an HMM indeed can form a subwavelength pattern in photoresist, which is in contrast to double-slit diffraction into an ordinary, isotropic medium.

In our experimental design, we chose silver and silica for the binary lamellar structure. Such a structure has low losses in the blue spectral range. The effective permittivity of our HMM made of equally thick silver and silica layers was calculated from an EMT analysis [29]. The permittivities of silver and silica were obtained with spectroscopic ellipsometry. The imaginary part of the permittivity of silver was found to be 1.7 times larger than the literature value [16]. Hence, the effective permittivity tensor at 465 nm is $\mathrm{diag}(\varepsilon_x, \varepsilon_z) = \mathrm{diag}(-2.78 + 0.22i, 6.31 + 0.15i)$, and the isofrequency curve is a hyperbola. Using Equation 10.3, the propagation direction in our silver–silica HMM is about 34°. Such an oblique beaming angle is preferable to obtain an interference pattern in the HMM. Note that for hyperlens designs [48, 49], the propagation direction was almost normal to the interface.

In order to fabricate the double slit with an HMM slab, we first deposited a 50 nm thick chromium film on a glass substrate using an electron-beam evaporator. Then, using FIB milling (Nova 200, FEI), the double-slit design was milled into the chromium film. A schematic diagram and an SEM image of the double slit are shown in Figures 10.18a and 10.18b, respectively. The separation of the two slits

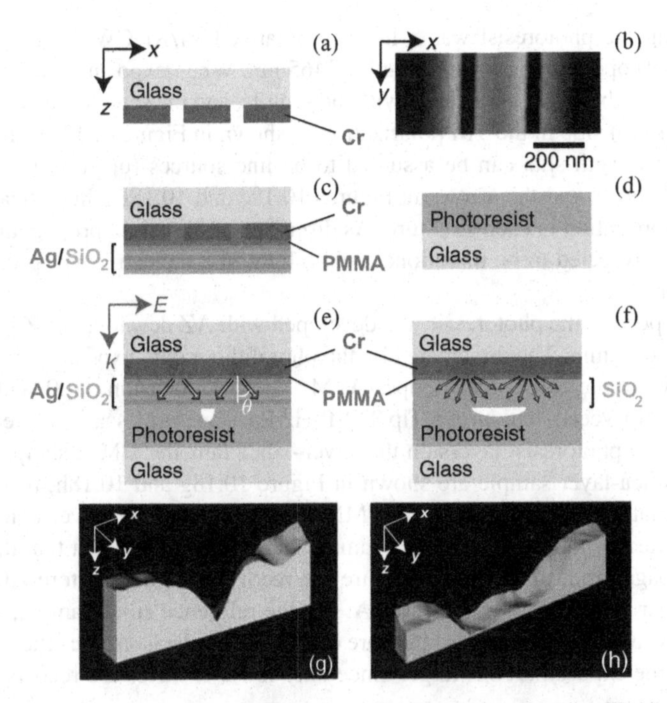

FIGURE 10.18 Schematic diagram (a) and a SEM image (b) of the double slit milled in a chromium film. Schematic diagram of the double slit with a silver–silica lamellar HMM slab (c) and a photoresist layer spin-coated on a glass substrate (d). Schematic diagram of the double slit with a silver–silica lamellar HMM slab in contact with a photoresist layer (e) and the double slit with a reference silica slab in contact with a photoresist layer (f) illuminated by TM-polarized light. Incident and diffracted fields are shown in red, and the dips created after exposure are illustrated. AFM images after development for the silver–silica lamellar HMM sample (g) and the silica-layer sample (h). The dimensions of the images in the x-axis are 600 and 1000 nm for (g) and (h), respectively.

was 270 nm, and the width of each slit was 50 nm. In order to fill the slits, we spin-coated diluted poly(methyl methacrylate) (PMMA) on the chromium film. The PMMA solution was prepared by further diluting 950PMMA A4 (MicroChem Corp.) with anisole in a 1:4 volume ratio. This gives a final film thickness of 18 nm after spin-coating at 3000 rpm for 40 s and a soft bake at 185°C for 5 min on a hotplate. On top of the PMMA layer, we deposited three pairs of alternating silver and silica layers for a total of six 15 nm thick layers. To have smooth and continuous silver films, 1 nm thick germanium layers were deposited before each of the silver layers [31]. Hence, the final structure looks like the images shown in Figure 10.18c, and the total thickness of the multilayer structure including the PMMA layer was 111 nm. We also prepared a control sample consisting of a single 110 nm thick layer of silica instead of the alternating metal–dielectric layers (see Fig. 10.18f).

We used the double-slit HMM samples to expose positive photoresist films (AZ 1518, AZ Electronic Materials) in order to detect the diffraction pattern. The photoresist was spin-coated on a glass substrate (see Fig. 10.18d). The light source used for exposing the photoresist was a linearly polarized Ar/Kr CW laser (43 Series, Melles Griot) operating at a wavelength of 465 nm. When exposing the sample, the photoresist was brought into contact with the sample, and the sample was illuminated from the Cr film side in the TM polarization as shown in Figures 10.18e and 10.18f. The subwavelength slits can be assumed to be line sources (or point sources in a 2D cross section), and the arrows in Figures 10.18e and 10.18f schematically show the light propagation in each structure. As discussed earlier, light propagation inside an HMM is expected to be directional, while light in a dielectric will propagate in all directions.

After exposure, the photoresist was developed with AZ developer (AZ Electronic Materials) for 1 min. The dip created by the photolithography exposure process was scanned by atomic force microscope (AFM, Veeco Dimension 3100 with a high aspect ratio tip Veeco Nanoprobe Tip TESP-HAR). The AFM scans of the exposed and developed photoresist layers on the silver–silica lamellar HMM sample and the reference silica-layer sample are shown in Figure 10.18g and 10.18h, respectively. It is clear that the dip created by the HMM sample is much narrower compared to the dip created by the reference silica sample. This is due to the fact that the high-k waves propagating in the HMM structure are recovered at the photoresist, leading to a narrower exposed dip in the PMMA. For the reference silica sample, however, the high-k waves are lost because they are evanescent in the isotropic silica material, and hence the exposed area is larger since only lower-k waves are recovered at the PMMA interface.

To verify our experimental results on the diffraction of high-k waves inside the HMM and to optimize the performance of the double-slit/HMM device, we have conducted full-wave FEM simulations using COMSOL Multiphysics with all dimensions taken from the fabricated samples. The permittivities of the materials were extracted from spectroscopic ellipsometry measurements. The computational domain was truncated by scattering boundary conditions, and a 465 nm TM-polarized plane wave (electric field polarized across the slits, see Fig. 10.18e) was incident from the top boundary. To directly compare with our experimental results, the control structure with a single silica layer was simulated as well. The irradiance of the transmitted light for the silver–silica lamellar HMM and the silica reference slabs is shown in Figures 10.19a and 10.19b, respectively. While the silica slab creates a diffraction-limited pattern, the HMM directs the diffracted light into a subwavelength-scale spot at the HMM–substrate interface, due to the directional propagation of high-k waves from the slits. Note that the HMM pattern is still "diffraction limited" but this limit is much lower for the hyperbolic media.

For our HMM structure with a relatively large structural period, the limitations of the EMT approximation become significant [6, 50]. One of the indications of that issue is the beam walk-off parallel to the layer interfaces that we observe in the simulations (Fig. 10.19a). We would expect some changes in the results when we decrease the period of the HMM structure while leaving the effective parameters formally the same. To investigate this scenario, we simulated a hypothetical HMM structure consisting of 28 pairs of silver and silica layer with 2 nm layer thicknesses. The total thickness of the HMM was 112 nm, which is approximately the same as the experimental sample

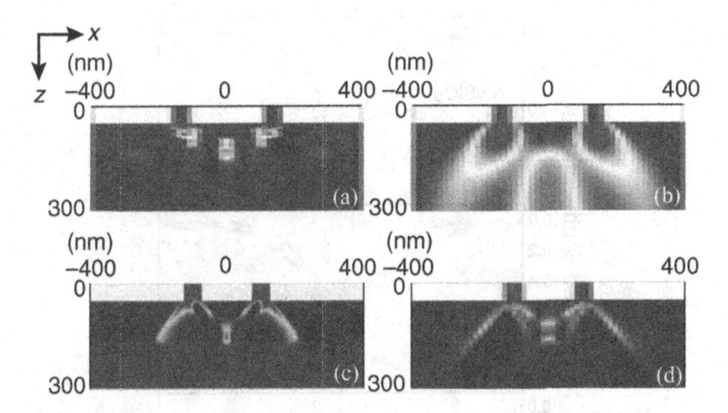

FIGURE 10.19 Simulated irradiance through a double-slit structure at a wavelength of 465 nm into a silver–silica lamellar HMM slab with 15 nm thick layers of silver and silica (a) and in a single layer of silica (b). The structures in (a) and (b) correspond to the experimentally fabricated samples. (c) Same as (a) but with 2 nm thick layers of silver and silica and 28 pairs of layers. (d) A semi-infinite silver–silica HMM slab with 2 nm layer thicknesses. The color scales in (a), (c), and (d) are normalized to each panel's peak diffraction irradiance, and the color scale in (b) is same as in (a).

thickness. In this case the distance between the two slits was set to 200 nm to maximize the irradiance at the center. The calculated irradiance is shown in Figure 10.19c. Since this structure has a smaller period, its performance is closer to the homogeneous effective medium case. Compared to Figure 10.19a, we see more distinct beaming in Figure 10.19c. In terms of the beam propagation angle, the angles in Figures 10.19a and 10.19c are 51° and 40°, respectively. The structure in Figure 10.19c gives a propagation angle that is closer to the angle obtained from Equation 10.3. What is more important is that the pattern size is significantly decreased, as shown below. We have also plotted a semi-infinite slab of alternating 2 nm thick silver/silica layers. These results, plotted in Figure 10.19d, depict the interference of two ideally symmetric diffraction patterns obtained from each slit. Figure 10.19c, in contrast, illustrates additional interference with the backscattered field, which destroys the symmetry of the individual diffraction patterns. These modeling and simulation efforts were actually performed before the experiments, and the initial EMT-based estimates have been followed by full-wave modeling to optimize the distance between the slits that would maximize the center peak intensity.

In Figure 10.20a, we show the AFM scans averaged along the slits for the silver–silica lamellar HMM sample exposed for 8–10 min. When the pattern depth in photoresist is 15 nm, the FWHM of the dip was 83 nm. Longer exposure times resulted in a deeper profile and a wider FWHM. Thus for exposure depths of 35 and 42 nm, the FWHM values are 105 and 135 nm, respectively. The depth of the exposure does not increase, but the FWHM continues to increase above a certain exposure time that depends on the resist extinction and the irradiance. Our results clearly indicate that planar metal–dielectric lamellar structures have the capability to transform the diffracted light from the slits into a subwavelength interference spot. In Figure 10.20b, we show the normalized AFM scans presented in Figure 10.20a

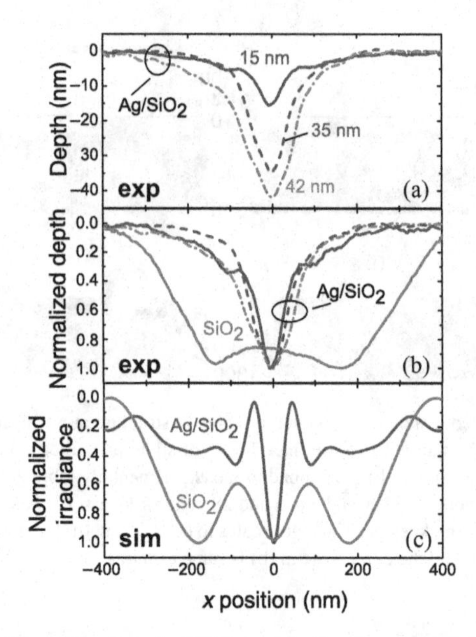

FIGURE 10.20 (a) Depth profiles from the AFM scans for the silver–silica lamellar HMM sample with three different depths depending on the exposure conditions. (b) Normalized depth profiles of the HMM sample shown in (a) and the silica-layer reference sample. (c) Simulated irradiances at the slab–substrate interface for the lamellar HMM slab (Ag/SiO_2) and the uniform slab of silica.

as well as those of the silica-layer sample. As the penetration depth (exposure time) increases, the contrast improves and the FWHM increases. If we were to extrapolate the experimental profile to even smaller depths, the side lobe peaks would grow and the central peak would be narrower. The overall depth profile would approach that of the simulated irradiance just near the resist surface.

Figure 10.20c depicts the irradiances at the interfaces of the layer(s) and the substrate obtained from the simulations shown in Figure 10.19. The FWHMs of the central peaks of the HMM structure and the single silica layer are 45 and 514 nm, respectively, which agree well with our experimental results. When comparing the experimental results and the simulations, we need to keep in mind that the tightest spot occurs at the interface photoresist and the final silica layer for the HMM sample. As the distance from the interface increases, the peak-to-background ratio becomes worse. This happens because the photoresist is not a hyperbolic medium, and therefore the waves entering the photoresist from the multilayers diverge within a few tens of nanometers. Since the depth of the dip is 15 nm even for the shallowest result, it is natural that the FWHM of the experimentally measured dip is wider than that from the simulation results of Figure 10.20c. The overall picture approaches the simulated results if the depth of the dip is shallower. The simulations for the 28-layer structure show potential for improving the resolution. Indeed, the FWHM of the interference peak for this short-period structure is only about 22 nm. However, to reach this resolution, the photoresist quality should be improved. The best available resist at our wavelength has grains of about 60 nm, which limits the attainable resolution [51].

10.5 SUMMARY

We have shown that arrays of nanoslits could exhibit unique features with the aid of plasmonics and metamaterials. By taking advantage of the width-dependent phase of light exiting the slits, we have designed lenses to focus either TM-polarized or TE-polarized light. These same lenses act as concave lenses for the orthogonal polarization of incident light. The fabricated lens prototypes were studied experimentally, and the results were successfully corroborated by numerical simulations. Similar types of lenses could exhibit tuneability in their focal intensities as well as their focal lengths when the slits are filled with index-changing materials.

We have also shown that nanoslits with HMMs work as subwavelength focusing devices. The dispersions of active HMMs were investigated by four different methods, and our analysis showed loss compensation in one of the permittivity components. We have considered practical designs for the devices and have numerically shown that the active HMM can generate a more intense and smaller focus, as small as $\lambda/15$. We observed a $\lambda/6$ diffraction peak from our fabricated sample with a passive HMM. Devices such as these could work as subwavelength probes, and the planar and compact design is beneficial for various applications including lithography and sensing.

ACKNOWLEDGMENTS

This work was supported in part by the United States Army Research Office (ARO) Multidisciplinary University Research Initiative (MURI) program award 50342-PH-MUR; ARO grants W911NF-04-1-4070350, W911NF-09-1-0075, W911NF-09-1-0231, and W911NF-09-1-0539; and National Science Foundation (NSF) Partnerships for Research and Education in Materials grant DMR 0611430. Support was also provided by Samsung Advanced Institute of Technology, United States Office of Naval Research (ONR) MURI grant N00014-10-1-0942, US Air Force Office of Scientific Research (AFOSR) grant FA9550-10-1-0264, Intelligence Community Postdoctoral Research Fellowship grant 104796, and NSF-DMR 1120923. AVK and VPD would also like to cite partial support by US Air Force Research Labs (AFRL) Materials and Manufacturing Directorate – Applied Metamaterials Program.

REFERENCES

1. Cai W, Shalaev V (2009) *Optical Metamaterials: Fundamentals and Applications.* Springer-Verlag.
2. Thongrattanasiri S, Podolskiy VA (2009) Hypergratings: nanophotonics in planar anisotropic metamaterials. *Opt. Lett.* 34: 890–892.
3. Ishii S, Kildishev AV, Shalaev VM, Chen KP, Drachev VP (2011) Metal nanoslit lenses with polarization-selective design. *Opt. Lett.* 36: 451–453.

4. Ishii S, Kildishev AV, Shalaev VM, Drachev VP (2011) Controlling the wave focal structure of metallic nanoslit lenses with liquid crystals, *Laser Phys. Lett.* 8: 828–832.
5. Kildishev AV, Litchinitser NM (2010) Efficient simulation of non-linear effects in 2D optical nanostructures to TM waves. *Opt. Commun.* 283: 1628–1632.
6. Ni X, Ishii S, Thoreson MD, Shalaev VM, Han S, Lee S, Kildishev AV (2011) Loss-compensated and active hyperbolic metamaterials. *Opt. Express* 19: 25255–25262.
7. Ishii S, Kildishev AV, Drachev VP (2012) Diffractive nanoslit lenses for subwavelength focusing. *Opt. Commun.*, 285: 3368–3372.
8. Ishii S, Kildishev AV, Narimanov E, Shalaev VM, Drachev VP (2013) Subwavelength diffraction pattern from a double-slit in hyperbolic media, *Laser Photonics Rev.* 7: 265–271.
9. Sun Z, Kim HK (2004) Refractive transmission of light and beam shaping with metallic nano-optic lenses. *Appl. Phys. Lett.* 85: 642–644.
10. Shi H, Wang C, Du C, Luo X, Dong X, Gao H (2005) Beam manipulating by metallic nano-slits with variant widths. *Opt. Express* 13: 6815–6820.
11. Verslegers L, Catrysse PB, Yu Z, White JS, Barnard ES, Brongersma ML, Fan S (2009) Planar lenses based on nanoscale slit arrays in a metallic film. *Nano Lett.* 9: 235–238.
12. Lin L, Goh XM, McGuinness LP, Roberts A (2010) Plasmonic lenses formed by two-dimensional nanometric cross-shaped aperture arrays for Fresnel-region focusing. *Nano Lett.* 10: 1936–1940.
13. Huang FM, Zheludev N, Chen Y, de Abajo FJG (2007) Focusing of light by a nanohole array. *Appl. Phys. Lett.* 90: 091119.
14. Huang FM, Kao TS, Fedotov VA, Chen Y, Zheludev NI (2008) Nanohole array as a lens. *Nano Lett.* 8: 2469–2472.
15. Kim KY, Cho YK, Tae HS, Lee JH (2006) Light transmission along dispersive plasmonic gap and its subwavelength guidance characteristics. *Opt. Express* 14: 320–330.
16. Johnson PB, Christy RW (1972) Optical constants of the noble metals. *Phys. Rev. B* 6: 4370–4379.
17. Khoo IC, Wu ST (1993) *Optics and Nonlinear Optics of Liquid Crystals.* Vol. 1. World Scientific.
18. Kossyrev PA, Yin A, Cloutier SG, Cardimona DA, Huang D, Alsing PM, Xu JM (2005) Electric field tuning of plasmonic response of nanodot array in liquid crystal matrix. *Nano Lett.* 5: 1978–1981.
19. Chu K, Chao C, Chen Y, Wu Y, Chen C (2006) Electrically controlled surface plasmon resonance frequency of gold nanorods. *Appl. Phys. Lett.* 89: 103107.
20. Evans P, Wurtz G, Hendren W, Atkinson R, Dickson W, Zayats A, Pollard R (2007) Electrically switchable nonreciprocal transmission of plasmonic nanorods with liquid crystal. *Appl. Phys. Lett.* 91: 043101.
21. Dickson W, Wurtz GA, Evans PR, Pollard RJ, Zayats AV (2008) Electronically controlled surface plasmon dispersion and optical transmission through metallic hole arrays using liquid crystal. *Nano Lett.* 8: 281–286.
22. Berthelot J, Bouhelier A, Huang C, Margueritat J, Colas-des-Francs G, Finot E, Weeber J-C, Dereux A, Kostcheev S, Ahrach HIE, Baudrion A-L, Plain J, Bachelot R, Royer P,

Wiederrecht GP (2009) Tuning of an optical dimer nanoantenna by electrically controlling its load impedance. *Nano Lett.* 9: 3914–3921.

23. Xiao S, Chettiar UK, Kildishev AV, Drachev V, Khoo I, Shalaev VM (2009) Tuneable magnetic response of metamaterials. *Appl. Phys. Lett.* 95: 033115.
24. Sanda PN, Dove DB, Ong HL, Jansen SA, Hoffmann R (1989) Role of surface bonding on liquid-crystal alignment at metal surfaces. *Phys. Rev. A* 39: 2653–2658.
25. Boyd RW (2003) *Nonlinear Optics*. Academic.
26. Min C, Wang P, Jiao X, Deng Y, Ming H (2007) Beam manipulating by metallic nano-optic lens containing nonlinear media. *Opt. Express* 15: 9541–9546.
27. Srituravanich W, Fang N, Sun C, Luo Q, Zhang X (2004) Plasmonic nanolithography. *Nano Lett.* 4: 1085–1088.
28. Merlin R (2007) Radiationless electromagnetic interference: evanescent-field lenses and perfect focusing. *Science* 317: 927.
29. Rytov SM (1956) Electromagnetic properties of a finely stratified medium. *Sov. Phys. JETP* 2: 466–475.
30. Naik GV, Schroeder JL, Sands TD, Boltasseva A (2011) Titanium nitride as a plasmonic material for visible wavelengths. Arxiv preprint. arXiv: 1011.4896.
31. Chen W, Thoreson MD, Ishii S, Kildishev AV, Shalaev VM (2010) Ultra-thin ultra-smooth and low-loss silver films on a germanium wetting layer. *Opt. Express* 18: 5124–5134.
32. West P, Ishii S, Naik G, Emani N, Shalaev VM, Boltasseva A (2009) Searching for better plasmonic materials. *Laser Photonics Rev.* 4: 765–808.
33. Bloemer MJ, D'Aguanno G, Scalora M, Mattiucci N, de Ceglia D (2008) Energy considerations for a superlens based on metal/dielectric multilayers. *Opt. Express* 16: 19342–19353.
34. Zhang J, Jiang H, Gralak B, Enoch S, Tayeb G, Lequime M (2009) Compensation of loss to approach-1 effective index by gain in metal-dielectric stacks. *Eur. Phys. J. Appl. Phys.* 46: 32603.
35. Xiao S, Drachev VP, Kildishev AV, Ni X, Chettiar UK, Yuan HK, Shalaev VM (2010) Loss-free and active optical negative-index metamaterials. *Nature* 466: 735–738.
36. Andresen MPH, Skaldebo AV, Haakestad MW, Krogstad HE, Skaar J (2010) Effect of gain saturation in a gain compensated perfect lens. *J. Opt. Soc. Am. B* 27: 1610–1616.
37. Seidel J, Grafström S, Eng L (2005) Stimulated emission of surface plasmons at the interface between a silver film and an optically pumped dye solution. *Phys. Rev. Lett.* 94: 177401.
38. Noginov MA, Zhu G, Bahoura M, Adegoke J, Small CE, Ritzo BA, Drachev VP, Shalaev VM (2006) Enhancement of surface plasmons in an Ag aggregate by optical gain in a dielectric medium. *Opt. Lett.* 31: 3022–3024.
39. De Leon I, Berini P (2010) Amplification of long-range surface plasmons by a dipolar gain medium. *Nat. Photonics* 4: 382–387.
40. Tumkur T, Zhu G, Black P, Barnakov YA, Bonner CE, Noginov MA (2011) Control of spontaneous emission with functionalized multilayered hyperbolic metamaterials. *Proc. SPIE* 8093: 80930O.

41. Yeh P (2005) *Optical Waves in Layered Media*. 2nd ed. Wiley Series in Pure and Applied Optics. Wiley-Interscience. p 416.
42. Elser J, Podolskiy VA, Salakhutdinov I, Avrutsky I (2007) Nonlocal effects in effective-medium response of nanolayered metamaterials. *Appl. Phys. Lett.* 90: 191109.
43. Ni X, Liu Z, Boltasseva A, Kildishev AV (2010) The validation of the parallel three-dimensional solver for analysis of optical plasmonic bi-periodic multilayer nanostructures. *Appl. Phys. A* 100: 365–374.
44. Liu Z, Chen KP, Ni X, Drachev VP, Shalaev VM, Kildishev AV (2010) Experimental verification of two-dimensional spatial harmonic analysis at oblique light incidence. *J. Opt. Soc. Am. B* 27: 2465–2470.
45. Ni X, Liu Z, Kildishev AV (2010), "PhotonicsDB: Optical Constants" https://nanohub.org/resources/PhotonicsDB. (DOI 10.4231/D3FT8DJ4J).
46. Sivan Y, Xiao SM, Chettiar UK, Kildishev AV, Shalaev VM (2009) Frequency-domain simulations of a negative-index material with embedded gain. *Opt. Express* 17: 24060–24074.
47. Tumkur T, Zhu G, Black P, Barnakov YA, Bonner CE, Noginov MA (2011) Control of spontaneous emission in a volume of functionalized hyperbolic metamaterial. *Appl. Phys. Lett.* 99: 151115.
48. Jacob Z, Alekseyev LV, Narimanov E (2006) Optical hyperlens: far-field imaging beyond the diffraction limit. *Opt. Express* 14: 8247–8256.
49. Salandrino A, Engheta N (2006) Far-field subdiffraction optical microscopy using metamaterial crystals: theory and simulations. *Phys. Rev. B* 74: 075103.
50. Wood B, Pendry J, Tsai D (2006) Directed subwavelength imaging using a layered metal-dielectric system. *Phys Rev. B* 74: 115116.
51. Melville DOS, Blaikie RJ (2005) Super-resolution imaging through a planar silver layer. *Opt. Express* 13: 2127–2134.

Printed in the United States
By Bookmasters